Statistical Models of Shape

Rhodri Davies • Carole Twining • Chris Taylor

Statistical Models of Shape

Optimisation and Evaluation

Springer

Rhodri Davies
Division of Imaging Science
and Biomedical Engineering (ISBE)
University of Manchester
UK
rhodri.davies@manchester.ac.uk

Carole Twining
Division of Imaging Science
and Biomedical Engineering (ISBE)
University of Manchester
UK
carole.twining@manchester.ac.uk

Chris Taylor
Division of Imaging Science
and Biomedical Engineering (ISBE)
University of Manchester
UK
chris.taylor@manchester.ac.uk

ISBN: 978-1-84800-137-4 e-ISBN: 978-1-84800-138-1
DOI: 10.1007/978-1-84800-138-1

British Library Cataloguing in Publication Data
A catalogue record for this book is available from the British Library

Library of Congress Control Number: 2008933222

Printed on acid-free paper

9 8 7 6 5 4 3 2 1

Springer Science+Business Media
springer.com

I'r fy rhieni, diolch am eich holl gefnogaeth.
Rhodri

To my wife Jo, and to the memory of my father.
Carole

To Jill, Andrew, and Alison, for keeping me sane.
Chris

Acknowledgements

Whilst undertaking the research that formed the basis for this book, Carole Twining and Rhodri Davies were funded by the MIAS[1] Inter-disciplinary Research Consortium (IRC) project, EPSRC grant No. GR/N14248/01, UK Medical Research Council Grant No. D2025/31.

During his doctoral training, which also contributed to that research, Rhodri Davies was funded on a BBSRC Industrial CASE studentship with AstraZeneca.[2] During the writing of this book, Rhodri Davies was fortunate to receive financial support from the Foulkes Foundation.[3]

A special thank you to Tomos Williams, Anna Mills, Danny Allen, and Tim Cootes (ISBE, Manchester) for their assistance, humour, and forebearance. Thanks also to John Waterton (AstraZeneca) and Alan Brett (Optasia Medical[4]).

A special mathematical thank you to Stephen Marsland (Massey University, New Zealand) for his many contributions to the development of this work, and for many fruitful arguments.

The hippocampus dataset was kindly provided by Martin Styner, Guido Gerig, and co-workers from the University of North Carolina, Chapel Hill, USA; the schizophrenia study from which this data was drawn was supported by The Stanley Foundation. The surfaces of the distal femur were provided by Tomos Williams, Chris Wolstenholme, and Graham Vincent of Imorphics.[5] The authors also wish to extend their thanks to their colleagues within the MIAS IRC, and to their ISBE colleagues at Manchester, for helping to provide the sort of intellectual environment in which excellent research can flourish.

---

1 "From Medical Images and Signals to Clinical Information."

2 AstraZeneca Pharmaceuticals, Alderley Park, Macclesfield, Cheshire, SK10 4TF.

3 Foulkes Foundation Fellowship, 37 Ringwood Avenue, London, N2 9NT.

4 Optasia Medical Ltd, Haw Bank House, High Street, Cheadle, SK8 1AL.

5 Imorphics Ltd, Kilburn House, Lloyd Street North, Manchester Science Park, Manchester, M15 6SE.

Contents

Chapter 1
Introduction

The goal of image interpretation is to convert raw image data into meaningful information. Images are often interpreted manually. In medicine, for example, a radiologist looks at a medical image, interprets it, and translates the data into a clinically useful form. Manual image interpretation is, however, a time-consuming, error-prone, and subjective process that often requires specialist knowledge. Automated methods that promise fast and objective image interpretation have therefore stirred up much interest and have become a significant area of research activity.

Early work on automated interpretation used low-level operations such as edge detection and region growing to label objects in images. These can produce reasonable results on simple images, but the presence of noise, occlusion, and structural complexity often leads to erroneous labelling. Furthermore, labelling an object is often only the first step of the interpretation process. In order to perform higher-level analysis, a priori information must be incorporated into the interpretation process. A convenient way of achieving this is to use a flexible model to encode information such as the expected size, shape, appearance, and position of objects in an image.

The use of flexible models was popularized by the active contour model, or 'snake' [98]. A snake deforms so as to match image evidence (e.g., edges) whilst ensuring that it satisfies structural constraints. However, a snake lacks specificity as it has little knowledge of the domain, limiting its value in image interpretation.

More sophisticated models based on the physical properties of an object have also been proposed (e.g., [134]). However, the expected patterns of variation of the model are usually estimated from only a single prototype, which requires many assumptions to be made. A more promising approach – and that followed in this book – is to use statistical models that attempt to *learn* the actual patterns of variability found in a class of objects, rather than making arbitrary assumptions. The idea is to estimate the population statistics from a *set* of examples instead of using a single prototype. The pattern of variation for a given class of object is established from a training set and

R. Davies et al., *Statistical Models of Shape*,
DOI: 10.1007/978-1-84800-138-1_1, © Springer-Verlag London Limited 2008

statistical analysis is used to give an efficient parameterisation of this variability, providing a compact and efficient representation.

The starting point in the construction of a statistical model is usually a training set of segmented images. In order to calculate statistics across the training set, a correspondence must be established between each member. It is important to choose the correct correspondence, otherwise a poor representation of the modelled object will result.

Correspondence is often established manually, but this is a time-consuming process that presents a major bottleneck in model construction. The manual definition of correspondence is also restricted to two-dimensional objects, which limits their use in interpreting medical images, since many of these are three dimensional. Other approaches to model-building have also been proposed, but these do not produce correspondences that are correct in any obvious way and the models that they produce are of limited utility.

This book presents a generic solution to this correspondence problem by treating it as part of the learning process. We will see that the key is to treat model construction as an optimisation problem, thus automating the process and guaranteeing the effectiveness of the resulting models. The other subject covered in this book is the evaluation of statistical models. This is also an important aspect of modelling since it allows us to quantify the likely utility of the model in practical applications. Model evaluation methods are established for cases with ground truth or in its absence.

In the remainder of this first chapter, we will take a look at some practical problems where statistical models have been applied before an overview of the rest of the book is presented.

1.1 Example Applications of Statistical Models

Statistical models have been used to successfully solve a wide range of practical problems, from Chinese character recognition [163] to cardiac modelling [72]. The number of applications is vast, but here we will focus on a few interesting examples and concentrate on the properties of statistical models that have allowed them to be successfully applied.

1.1.1 Detecting Osteoporosis Using Dental Radiographs

Statistical shape models were originally conceived as a basis for automatic image segmentation [31] – the process of labelling an image so that the labels correspond to real-world objects. A big advantage of using a shape model for this task is that it can produce extremely accurate and objective segmentations. They can therefore be used to detect changes that might otherwise be

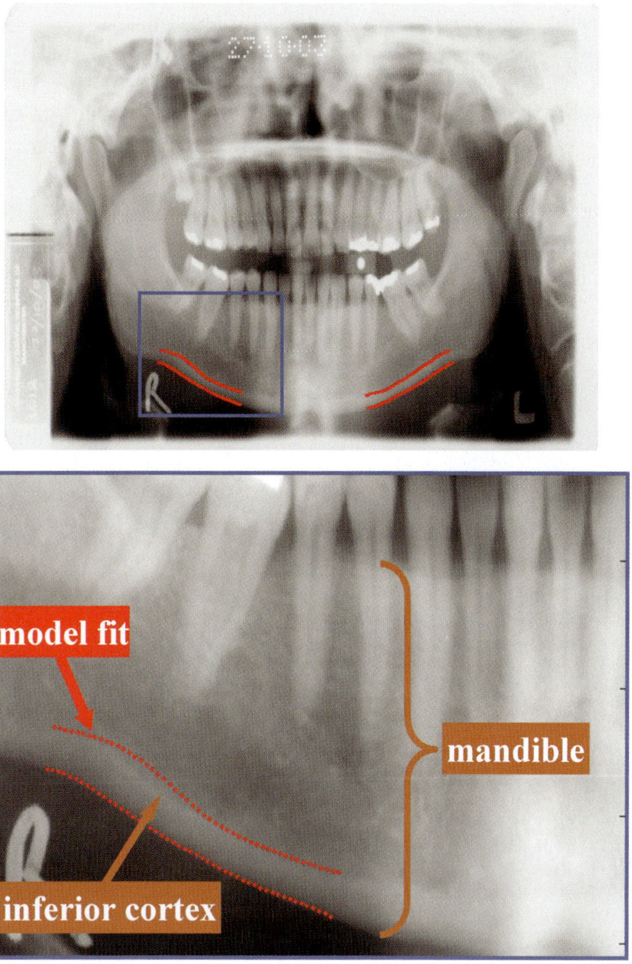

Fig. 1.1 An example of a segmentation of the inferior mandibular cortex on a panoramic dental radiograph using the method described in [1]. **Top**: a panoramic radiograph, where the red lines represent the segmentation of the left and right inferior mandibular cortex using a shape model. **Bottom**: Detail of the segmentation on the right side of the patient's mandible. Figure courtesy of P.D. (Danny) Allen, University of Manchester.

missed by human annotation. An example of where this additional accuracy has proved to be critical is in detecting osteoporosis in dental radiographs [1, 59].

Osteoporosis is a common disorder that causes a reduction in bone tissue density, leading to brittle bones that are prone to fracture. The standard method of diagnosis involves a dual energy x-ray absorbiometry (DXA) scanner, but these are dedicated machines with limited accessibility. Although access is improving, only patients with a high index of clinical suspicion are referred for DXA scans, resulting in missed diagnoses.

It has been reported that osteoporosis can also be detected by careful measurement of the width of the inferior mandibular cortex on panoramic dental radiographs – a common investigation in dental practice. However, if the cortical width is measured manually, the time taken impedes the dental practitioner from performing the test during a routine consultation. Manual measurement also introduces considerable inter- and intra-operator variability, resulting in reduced sensitivity and specificity of detection. However, it has been shown [1, 59] that this variability can be reduced by using a statistical shape model for segmentation – an example of a segmentation of the inferior mandibular cortex using a statistical shape model is shown in Fig. 1.1. The accuracy of the resulting segmentations was shown to be sufficient to diagnose skeletal osteoporosis with good diagnostic ability and reliability [59]. Furthermore, measurement was performed in real time with minimal human interaction. This application thus promises another means of detecting osteoporosis, or at least of flagging high-risk cases for referral to DXA scanning.

1.1.2 Detecting Vertebral Fractures

Statistical models provide a compact and efficient basis for describing the variation of a class of object. Model parameter values can therefore be used as effective features in classifier systems. We already know a bit about osteoporosis from the previous section, so we will now look at an example of where complications of the disease can be detected using a model-based classifier.

A common complication of osteoporosis is a fractured vertebra. Although many of these fractures are asymptomatic, they are an important indicator of more harmful fractures in the future. Diagnosis of a fracture is usually made by a radiologist, but, as with any human operator, they are liable to report subjective results. Also, vertebral fracture assessment by DXA scanners is becoming common in places other than radiology units (e.g., general practice), and may not be carried out by an expert radiologist. Therefore, it is desirable to establish quantitative criteria that capture some of the subtle information used by an expert, since current (vertebral) height-based quantitative methods are insufficiently specific, especially in diagnosing the mild fractures that occur in the early stages of the disease. These height-based measures do not capture subtle shape information, nor other features present in the image. A recent body of work [167, 145, 146] has shown that using statistical models of shape and appearance can offer substantial improvements in diagnostic accuracy over conventional quantitative methods.

The first step in the system is to segment the vertebrae of interest using a statistical model – an example of a typical segmentation result is given in Fig. 1.2. As with the example of the inferior mandibular cortex given

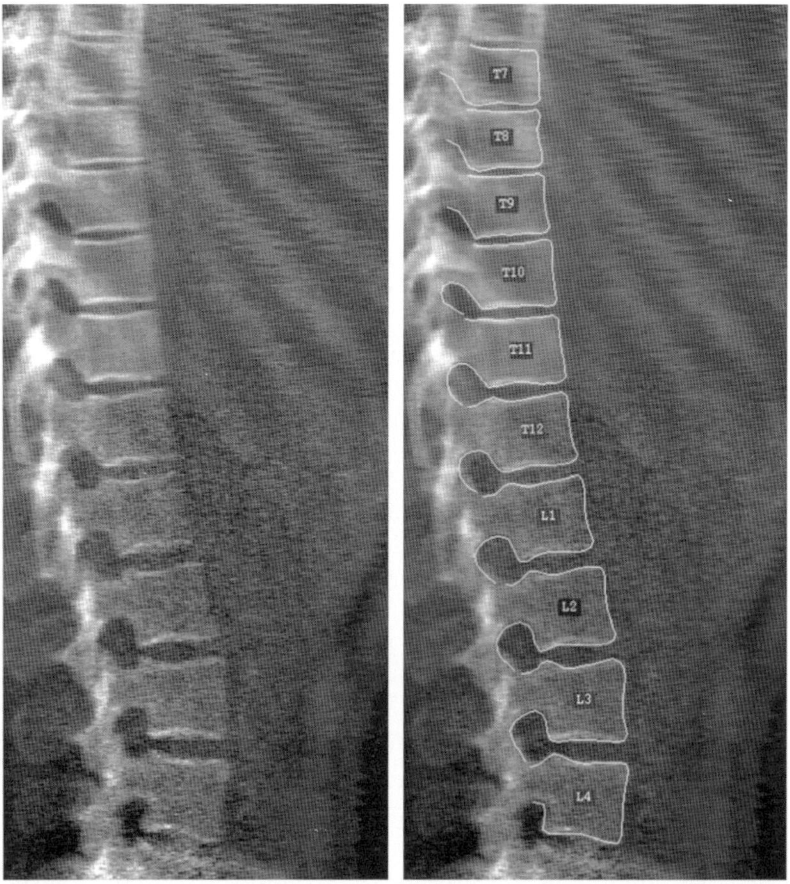

Fig. 1.2 An example of a segmentation of human vertebrae using the method described in [146]. **Left**: a DXA image showing the spine from the seventh thoracic vertebra (T7) down to the fourth lumbar vertebra (L4). **Right**: the segmentation achieved by the model, overlaid on the image. Figure courtesy of Martin Roberts, ISBE, University of Manchester.

above, this segmentation is quick, requires minimal user interaction, and, most importantly, produces more objective and reproducible results than manual segmentation.

The model parameters found in segmentation can then be fed into a trained classifier, which will return a positive or negative diagnosis of vertebral fracture. The advantage of using the model parameters, rather than other established measures (such as vertebral height), is that they capture much more information about the state of the vertebrae such as their shape, appearance, and pose. This information forms a much stronger basis for the classifier system.

1.1.3 Face Identification, Tracking, and Simulation of Ageing

We have said in the introduction that many non-statistical modelling approaches infer object properties from a single prototype. This approach has several disadvantages over the statistical approach – one of these is that it lacks the flexibility to model objects that exhibit significant variability. An example of where this is particularly evident is in human face analysis.

Face analysis, which includes tasks such as finding, identifying, and tracking someone's face in an image, has application in many fields – from aiding human-computer interaction to security surveillance monitoring. However, face analysis is a difficult task, not least because of the huge inherent variability of a face – not just between individuals, but variation in the appearance of the same person's face due to changes in expression, pose, lighting, etc.

A statistical model has the ability to learn this variability from a set of examples – as long as enough examples are presented the model will be generic enough to deal with these variations. Such a model of facial appearance was presented in [64] and was shown to perform well in finding, identifying, and tracking faces in images.

Statistical models can also be used in generative mode, where the model is used to synthesize new examples of the class of object. This opens the possibility of using the models in different types of applications to those that we have seen so far. A good example is described in [104], which shows how facial ageing can be simulated artificially. So, given a photograph of a

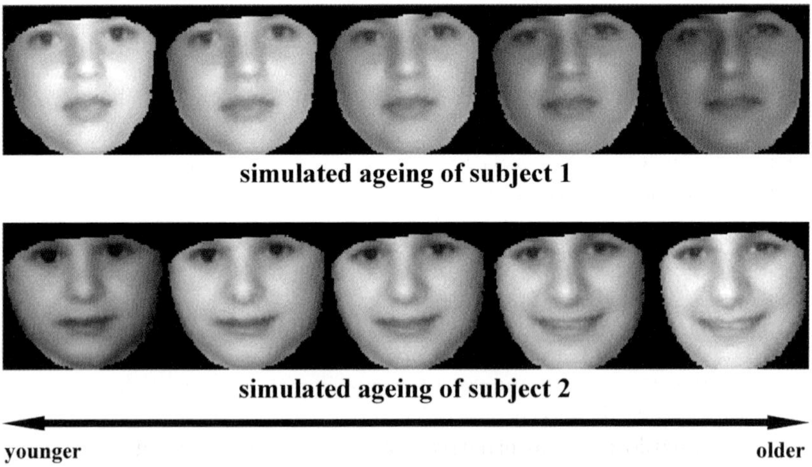

Fig. 1.3 Simulated ageing using the method described in [104]. The figure shows the effect of simulating ageing in photographs of two individuals. Figure reproduced from Fig. 4 in [104] with permission (©2002 IEEE).

person's face, the model can generate impressions of that person at different ages – some examples are shown in Fig. 1.3. This has application in ageing photographs of missing or wanted persons as well as improving robustness of face analysis to ageing.

1.2 Overview

The organization of the remainder of this book is as follows.

We start with a theoretical treatment of statistical shape models in Chap. 2. This contains a comprehensive description of model-building, and considers both discrete and continuous representations of shape. Several other aspects of model-building, such as shape alignment and practical applications, are also covered.

Chapter 3 covers a fundamental problem in shape modelling in greater detail: that of establishing a correspondence between a set of shapes. The chapter begins by illustrating the importance of establishing a suitable correspondence, before looking at various ways in which this can be achieved. The last part of the chapter introduces our approach of model-building within an optimisation framework. This allows correspondence to be established by viewing it as an integral part of the learning process.

One essential component of this optimisation approach to model-building is an objective function that quantifies what is meant by the quality of a model. The subject of objective functions is covered in Chap. 4; we look at various objective functions that can be used to establish correspondence, including a full derivation of the Minimum Description Length objective function (and various approximations to it).

In order to minimise our chosen objective function, we must be able to manipulate correspondence across the set of shapes. At the same time, we must ensure that only valid correspondences are generated. In Chap. 5, we show how this can be achieved by re-parameterising each shape. Several representations of curve re-parameterisation are described in Chap. 5. The extension to surface re-parameterisation is complicated by the need for explicit surface parameterisation, but a generic method of achieving this along with several representations of re-parameterisation for surfaces is given in Chap. 6.

The final component of the optimisation approach is a method of finding the configuration of re-parameterisation functions that lead to the optimal value of the objective function. This problem is explored in Chap. 7, which presents a generic optimisation approach, as well as looking at how it can be tailored for certain situations.

Chapter 8 explores an alternative approach to representing and manipulating correspondence, using a non-parametric approach – the goal being to produce more robust results in less time. A fluid-based regularizer is described as well as an efficient method of optimisation.

Finally, Chap. 9 considers the question of how we should evaluate models of shape or models of appearance. In particular, we address the question of whether ground truth data should be used for evaluation, and how models can be evaluated in the absence of such ground truth data.

Chapter 2
Statistical Models of Shape and Appearance

As was explained in the Introduction, our aim is to start with a set of example shapes (the *training set*), and learn from this the patterns of variability of the shape of the class of objects for which the training set can be considered a representative sample. We will first consider this problem in a rather abstract sense, and illustrate how the question of correspondence *between* different shapes is inextricably linked to the question of representing a set of shapes.

Mathematically, the usual approach is to construct a mapping from an example shape to a point in some shape space. This is the process of constructing a *representation* of shape. The idea is that every physical shape corresponds to a point in shape space, and conversely, each point in shape space corresponds to some physical shape. There are many ways of constructing such representations, but whichever method is chosen, what is then obtained is a mapping from our training set of shapes to a set of points in shape space.

Modelling can then be considered as the process of modelling the distribution of our training points in shape space. However, before we can begin to talk about the distribution of such points, we first need to define a notion of distance on shape space.

A definition of a distance on shape space then leads directly to the notion of correspondence between the physical shapes themselves. Consider two distinct physical shapes, and the two points in shape space which represent those two shapes. We can then imagine a continuous path between the two points in shape space, and, given that we have a definition of distance, the shortest such path between the two shapes. When we map this construction back to the space of physical shapes, what we obtain is a continuous sequence of physical shapes that interpolates between our two original shapes. If we now consider a single point on one shape, we can then follow it through this continuous sequence of shapes, and hence locate the physical point on the other shape to which this point corresponds. This is what is meant by a dense correspondence between shapes.

Let us now return to our physical shapes, and imagine translating and rotating a physical shape (that is, altering the *pose* of the shape). In many

R. Davies et al., *Statistical Models of Shape*,
DOI: 10.1007/978-1-84800-138-1_2, © Springer-Verlag London Limited 2008

cases, the pose of the physical shape is unimportant, and what we mean by *change* of shape is not such a transformation.

But there is another transformation we could consider. Suppose we imagine two *distinct* points in shape space, which give the same *physical* shape, but different correspondence when compared to some other (reference) shape (using the construction defined above). If such a construction is possible, we see that it is possible (at least in theory) to manipulate the correspondence between shapes (moving the point in shape space), whilst leaving the physical shape unaltered. Which then means that we have to answer the question as to what correspondence we should use for our analysis of shape variability (the *correspondence problem* – see also Sect. 3.1).

We hence see that the issue of shape correspondence naturally arises as soon as we consider the steps necessary to represent a set of shapes, and analyse their distribution. Some methods of shape representation do not allow correspondence to be manipulated independently of shape, and in these cases, the correspondence they generate can be considered as *implicit* (for example, the SPHARM method [76], or early M-Rep methods [137]). However, there are other methods of shape representation for which the correspondence is *explicit*, which allow correspondence to be manipulated independently of physical shape.

In the remainder of this book, we will restrict ourselves to such a shape representation, the shape representation which leads to the class of deformable models known as *Statistical Shape Models*[1] (SSMs). We will now describe this shape representation in detail, beginning with the finite-dimensional case.

2.1 Finite-Dimensional Representations of Shape

Let us consider first building a finite-dimensional representation of a single shape S. The most intuitive and simplest way to represent such a shape is a join-the-dots approach.

We take a set of n_P points which lie on the shape S, with positions:

$$\mathbf{x}^{(i)} \in S, \; i = 1, \dots n_P. \tag{2.1}$$

The coordinates of each point position can be concatenated to give a single *shape vector* $\mathbf{x} = \{\mathbf{x}^{(i)}\}$. For example:

$$\mathbf{x} \doteq (x^{(1)}, y^{(1)}, z^{(1)}, x^{(2)}, y^{(2)}, z^{(2)}, \dots, x^{(n_P)}, y^{(n_P)}, z^{(n_P)}), \; S \subset \mathbb{R}^3, \tag{2.2}$$

[1] Note that these were initially called *Point Distribution Models (PDMs)*. However, due to a clash with nomenclature in the statistics literature, they were later re-christened *Statistical Shape Models (SSMs)*. Both terms can be found in the literature.

where $(x^{(i)}, y^{(i)}, z^{(i)})$ are the Cartesian coordinates of the ith point on the shape. For a shape S in $\mathbb{R}^d$, this gives a $d \times n_P$-dimensional representation. In most cases, Cartesian coordinates are sufficient, but in cases where parts of shapes can rotate, it may be useful to instead use angular coordinates [87].

The final representation of the shape is then generated from the shape vector by interpolation. For shapes in $\mathbb{R}^2$ (curves), this is a simple join-the-dots approach, using either straight-line segments (polygonal representation), or by spline interpolants if a smoother shape is preferred. For shapes in $\mathbb{R}^3$ (surfaces), interpolants can similarly also be linear (planes), or a higher-order spline interpolant.

What we have not considered so far is the connectivity of the points, and the topology of the shape. For the case of shapes in $\mathbb{R}^2$, the simplest case is where the shape has only one connected component, with the topology of either an open or closed line. The points are usually numbered so that they are connected consecutively – for the closed shapes, we must also form a loop by connecting the last point to the first. For more complicated multi-part shapes, the points which are members of each part, and the connectivity within each part have to be specified separately.

Similar considerations holds for shapes in $\mathbb{R}^3$. The simplest case is then single-part shapes in $\mathbb{R}^3$, with the topology of either open surfaces or spheres, with the points being part of a triangulated mesh.

Once we have a finite-dimensional representation of a single shape S, we can easily see how this can be extended to form a *common* representation of a *set* of shapes. To be specific, let us take a set of n_S shapes:

$$S_i : i = 1, \dots n_S. \tag{2.3}$$

We suppose that each shape is then represented by a set of n_P points, such that the individual points are placed in *corresponding* positions across the set of shapes. This then gives us a set of initial shape vectors $\{\mathbf{x}_i : i = 1, \dots n_S\}$ which form a representation of the whole set of shapes in a common shape space $\mathbb{R}^{dn_P}$.

2.1.1 Shape Alignment

In many cases, the size, placement, and orientation of an object is arbitrary, and has nothing to do with the actual variation of shape that we are interested in. In mathematical terms, there are degrees of freedom (scaling, translation, and rotation) associated with each shape example, which we wish to factor out of our shape analysis.

Consider a fixed shape $\mathbf{y}$, and a second moving shape $\mathbf{x}$, which we wish to align with the first by means of a similarity transformation. A general similarity transformation acting on $\mathbf{x}$ can be written as:

$$\mathbf{x} \mapsto s\mathbf{R}(\mathbf{x} - \mathbf{t}), \tag{2.4}$$

where $\mathbf{t}$ represents a translation in $\mathbb{R}^d$, $\mathbf{R}$ is a $dn_P \times dn_P$ representation of a rotation in $\mathbb{R}^d$, and $s \in \mathbb{R}^+$ is a scaling. Note that these elements of the representation of a similarity transformation are such that they act on the concatenated set of shape points in the shape vector. They are constructed from a representation that acts on single points in the obvious way, although the exact details depend on the way in which the coordinates of the shape points have been concatenated.

We wish to find the similarity transformation which brings the moving shape $\mathbf{x}$ as close as possible to the fixed shape $\mathbf{y}$. The simplest way to define proximity is just the magnitude of the Euclidean norm of the difference between the two shape vectors in $\mathbb{R}^{dn_P}$:

$$\mathcal{L} \doteq \|\mathbf{y} - s\mathbf{R}(\mathbf{x} - \mathbf{t})\|^2, \tag{2.5}$$

which is the square of the Procrustes distance between the shapes [78]. In terms of the positions of individual points, this expression can be rewritten as:

$$\mathcal{L} = \sum_{i=1}^{n_P} \|\mathbf{y}^{(i)} - s\mathbf{R}(\mathbf{x}^{(i)} - \mathbf{t})\|^2, \tag{2.6}$$

where $\mathbf{t}$ is now just a vector in $\mathbb{R}^d$, and $\mathbf{R}$ is a $d \times d$ rotation matrix.

If we define our origin so that it lies at the centre of mass of the fixed shape:

$$\frac{1}{n_P}\sum_{i=1}^{n_P} \mathbf{y}^{(i)} = \mathbf{0}, \tag{2.7}$$

with rotation defined about this origin, the optimal translation can then be calculated as:

$$\left.\frac{\partial \mathcal{L}}{\partial \mathbf{t}}\right|_{s,\mathbf{R}} = 0 \implies \left.\frac{\partial}{\partial \mathbf{t}}\right|_{s,\mathbf{R}} \sum_{i=1}^{n_P} \|s\mathbf{R}(\mathbf{x}^{(i)} - \mathbf{t})\|^2 = \mathbf{0}, \tag{2.8}$$

$$\implies \left.\frac{\partial}{\partial \mathbf{t}}\right|_{s,\mathbf{R}} \sum_{i=1}^{n_P} \|(\mathbf{x}^{(i)} - \mathbf{t})\|^2 = \mathbf{0}, \tag{2.9}$$

$$\implies \boxed{\mathbf{t} = \frac{1}{n_P}\sum_{i=1}^{n_P} \mathbf{x}^{(i)}} = \frac{1}{n_P}\sum_{i=1}^{n_P} \left(\mathbf{x}^{(i)} - \mathbf{y}^{(i)}\right). \tag{2.10}$$

That is, the centroid/centre of mass of the original moving shape is translated so that it coincides with the centre of mass of the fixed shape.

Once the shapes have been centred, we can then calculate the combined scaling and rotation:

$$\frac{\partial \mathcal{L}}{\partial s\mathbf{R}} = 0 \implies \boxed{\sum_{i=1}^{n_P} y_\mu^{(i)} x_\beta^{(i)} = sR_{\mu\alpha} \sum_{j=1}^{n_P} x_\alpha^{(j)} x_\beta^{(j)},} \tag{2.11}$$

where $\mathbf{x}^{(i)} = \{x_\alpha^{(i)} : \alpha = 1, \dots d\}$ and $\mathbf{y}^{(i)} = \{y_\alpha^{(i)} : \alpha = 1, \dots d\}$ are the Cartesian components of the point positions. This can then be solved for the matrix $s\mathbf{R}$ (for example, see [102] for further details).

Rather than aligning just a pair of shapes, we wish to mutually align an entire set of shapes $\{S_i : i = 1, \dots n_S\}$, $S_i = \{\mathbf{x}_j^{(i)} : j = 1, \dots n_P\}$. We use a similar criterion to that considered above, either by considering the squared Procrustes distances between all pairs of shapes, or between all shapes and the mean shape. This is known as generalized Procrustes analysis. The translations are obtained as before, centering each shape on the origin. However, the general problem of finding the optimal rotations and scalings is not well-posed unless further constraints are placed on the mean [174], as will be explained below.

For statistical shape analysis and statistical shape models, a simple iterative approach is usually sufficient. After first centering all the shapes, a typical algorithm then proceeds [38] as Algorithm 2.1.

Algorithm 2.1 : Mutually Aligning a Set of Shapes.

Initialize:

- Choose one shape as the reference frame, call it $\mathbf{x_{ref}}$, and retain this.
- Normalize the scale so that $\|\mathbf{x_{ref}}\| = \mathbf{1}$.
- Set the initial estimate of the mean shape to be $\mathbf{x_{ref}}$.

Repeat:

- Perform pairwise alignment of all shapes to the current estimate of the mean shape.
- Recompute the mean of the set of shapes:

  $\bar{\mathbf{x}} \doteq \{\bar{\mathbf{x}}^{(i)} : i = 1, \dots n_P\}, \;\; \bar{\mathbf{x}} \doteq \frac{1}{n_S} \sum_{j=1}^{n_S} \mathbf{x_j}.$
- Align $\bar{\mathbf{x}}$ to the initial reference frame $\mathbf{x_{ref}}$.
- Normalize the mean so that $\|\bar{\mathbf{x}}\| = 1$.

Until convergence.

Note that it is necessary to retain the initial reference frame to remove the global degree of freedom corresponding to rotating all the shapes by the same amount. Setting $\|\bar{\mathbf{x}}\| = 1$ similarly removes the degree of freedom associated with scaling all the shapes by the same factor. The degrees of freedom associated with a uniform translation have already been removed by centering all the shapes before we began the rest of the alignment.

There remains the question of what transformations to allow during the iterative refinement. A common approach is to scale all shapes so that $\|\mathbf{x}_i\| = 1$, and allow only rotations during the alignment stage. This means that from the original shape space $\mathbb{R}^{dn_P}$, all shapes have been projected onto the surface of a hypersphere $\|\mathbf{x}\| = 1$. This means that the submanifold of $\mathbb{R}^{dn_P}$ on which the aligned shapes lie is curved, and if large shape changes occur, significant non-linearities can appear. This may be problematic when we come to the next stage of building a statistical model of the distribution of shapes. An alternative is to allow both scaling and rotation during alignment, but this can also introduce significant non-linearities. If this is a problem, the non-linearity can be removed by projecting the aligned shapes onto the tangent hyperplane to the hypersphere at the mean shape. That is:

$$\mathbf{x}_i \mapsto s_i \mathbf{x}_i, \quad s_i \in \mathbb{R}^+ \text{ such that } (\bar{\mathbf{x}} - s_i \mathbf{x}_i) \cdot \bar{\mathbf{x}} = 0. \tag{2.12}$$

See [38] for further details and explicit examples.

2.1.2 Statistics of Shapes

To summarize our progress so far, we have mapped our initial shape vectors (2.2) in $\mathbb{R}^{dn_P}$ to a new set of mutually aligned shape vectors, by factoring out uninteresting degrees of freedom corresponding to pose (scale, orientation, and position). We now wish to analyse the statistics of this distribution of shape vectors. To do this, we first need to find a set of axes specific to the particular set of shapes. We have in some sense already started to perform this, since we have a mean shape $\bar{\mathbf{x}}$ that can be used as an origin.

To see that this is a necessary procedure, consider the extreme case where there is a shape point, $\mathbf{x}^{(i)}$ say, which does not change its position across the set of examples. Since this point does not vary, there is no value in retaining the axes corresponding to the coordinates of this point $\{x_\alpha^{(i)} : \alpha = 1, \ldots d\}$. We wish instead to find a new set of axes in $\mathbb{R}^{dn_P}$ that span the subspace which contains the (aligned) shapes. One simple procedure for performing this task is Principal Component Analysis (PCA).

2.1.3 Principal Component Analysis

We start from our set of shape vectors $\{\mathbf{x}_i : i = 1, \ldots n_S\}$ (we will assume from now on that we are only considering sets of shape vectors which have been aligned), with components relative to our original axes:

$$\mathbf{x}_i = \{x_{i\mu} : \mu = 1, \ldots d \times n_P\}. \tag{2.13}$$

These are the components and axes defined by those in $\mathbb{R}^d$, the original space in which the input shapes reside.

We wish to find a new set of orthogonal axes in $\mathbb{R}^{dn_P}$ that better reflects the actual distribution of the set. The origin of this new set of axes will be set to the mean shape $\bar{\mathbf{x}}$. Let these new axes be described by a set of orthonormal vectors:

$$\{\mathbf{n}^{(a)}\} \text{ such that } \mathbf{n}^{(a)} \cdot \mathbf{n}^{(b)} = \delta_{ab}, \tag{2.14}$$

where δ_{ab} is the Kronecker delta.

We then have the following theorem:

Theorem 2.1. PCA.
The set of orthonormal directions $\{\mathbf{n}^{(a)}\}$ that maximises the quantity:

$$\mathcal{L} \doteq \sum_{a} \sum_{i=1}^{n_S} \left((\mathbf{x}_i - \bar{\mathbf{x}}) \cdot \mathbf{n}^{(a)} \right)^2, \tag{2.15}$$

are given by the eigenvectors of the data covariance matrix $\mathbf{D}$ for the shapes, where we define $\mathbf{D}$ of size $dn_P \times dn_P$ with components:

$$\boxed{D_{\mu\nu} \doteq \sum_{i=1}^{n_S} (\mathbf{x}_i - \bar{\mathbf{x}})_\mu (\mathbf{x}_i - \bar{\mathbf{x}})_\nu.} \tag{2.16}$$

Then the eigenvectors are defined by:

$$\mathbf{D}\mathbf{n}^{(a)} = \lambda_a \mathbf{n}^{(a)}, \; a = 1, \ldots n_S - 1. \tag{2.17}$$

Proof. *Suppose we are extracting these vectors in some sequential manner, so that having found an acceptable subset $\{\mathbf{n}^{(a)} : a = 1, \ldots b-1\}$, we now wish to make the optimum choice of the next vector $\mathbf{n}^{(b)}$. Optimality is then given by maximising:*

$$\mathcal{L} \doteq \sum_{i=1}^{n_S} \left((\mathbf{x}_i - \bar{\mathbf{x}}) \cdot \mathbf{n}^{(b)} \right)^2, \tag{2.18}$$

with respect to $\mathbf{n}^{(b)}$, subject to the orthonormality constraints:

$$\mathbf{n}^{(a)} \cdot \mathbf{n}^{(b)} = \delta_{ab}, \; a = 1, \ldots b. \tag{2.19}$$

Using Lagrange multipliers $\{c_{ba} : a = 1, \ldots b\}$, the solution to this constrained optimisation problem corresponds to the stationary point of the function:

$$\mathcal{L} = \sum_{i=1}^{n_S} \left((\mathbf{x}_i - \bar{\mathbf{x}}) \cdot \mathbf{n}^{(b)} \right)^2 + \sum_{a=1}^{b-1} c_{ba} \mathbf{n}^{(a)} \cdot \mathbf{n}^{(b)} + c_{bb} \left(\mathbf{n}^{(b)} \cdot \mathbf{n}^{(b)} - 1 \right). \quad (2.20)$$

$$\frac{\partial \mathcal{L}}{\partial c_{ba}} = 0 \implies \mathbf{n}^{(a)} \cdot \mathbf{n}^{(b)} = \delta_{ab}, \text{ which are the required constraints.} \quad (2.21)$$

$$\frac{\partial \mathcal{L}}{\partial \mathbf{n}^{(b)}} = 0 \implies 2 \sum_{i=1}^{n_S} (\mathbf{x}_i - \bar{\mathbf{x}})_\nu (\mathbf{x}_i - \bar{\mathbf{x}})_\mu n_\mu^{(b)} + \sum_{a=1}^{b-1} c_{ba} n_\nu^{(a)} + 2 c_{bb} n_\nu^{(b)} = 0, \quad (2.22)$$

where we use the Einstein summation convention[2] *that the repeated index* μ *is summed from* $\mu = 1$ *to* dn_P*. Using the definition of the covariance matrix* $\mathbf{D}$ *(2.16), we can rewrite the condition as:*

$$2\mathbf{D}\mathbf{n}^{(b)} + \sum_{a=1}^{b-1} c_{ba} \mathbf{n}^{(a)} + 2 c_{bb} \mathbf{n}^{(b)} = \mathbf{0}. \quad (2.23)$$

For the case $b = 1$ *(the first direction we choose), this reduces to:*

$$\mathbf{D}\mathbf{n}^{(1)} + c_{11} \mathbf{n}^{(1)} = 0 \quad (2.24)$$

$$\implies \boxed{\mathbf{D}\mathbf{n}^{(1)} = \lambda_1 \mathbf{n}^{(1)}} \;\&\; \mathbf{n}^{(1)}\mathbf{D} = \lambda_1 \mathbf{n}^{(1)}, \quad c_{11} \doteq \lambda_1. \quad (2.25)$$

That is, the vector $\mathbf{n}^{(1)}$ *is a left and right eigenvector of the (symmetric) shape covariance matrix* $\mathbf{D}$*, with eigenvalue* λ_1*. The condition for the second axis can then be written as:*

$$2\mathbf{D}\mathbf{n}^{(2)} + c_{21} \mathbf{n}^{(1)} + 2 c_{22} \mathbf{n}^{(2)} = \mathbf{0}. \quad (2.26)$$

Taking the dot product of this expression with $\mathbf{n}^{(1)}$*, we obtain:*

$$2\mathbf{n}^{(1)}\mathbf{D}\mathbf{n}^{(2)} + c_{21} = 0 \quad (2.27)$$

$$\implies 2\lambda_1 \mathbf{n}^{(1)} \cdot \mathbf{n}^{(2)} + c_{21} = 0 \implies c_{21} = 0. \quad (2.28)$$

$$\therefore \; \mathbf{D}\mathbf{n}^{(2)} + c_{22} \mathbf{n}^{(2)} = \mathbf{0} \implies \boxed{\mathbf{D}\mathbf{n}^{(2)} = -c_{22} \mathbf{n}^{(2)} \doteq \lambda_2 \mathbf{n}^{(2)}.} \quad (2.29)$$

It then follows by induction that the required set of axes $\{\mathbf{n}^{(a)}\}$ *are the orthonormal set of eigenvectors of the shape covariance matrix* $\mathbf{D}$*.* □

The sum of the squares of the components of the shape vectors along each of the PCA directions $\mathbf{n}^{(a)}$ is then given by:

---

[2] Note that, in general, indices that appear in brackets $\cdot^{(a)}$ will *not* be summed over unless explicitly stated. See Glossary.

$$\sum_{i=1}^{n_S}\left((\mathbf{x}_i-\bar{\mathbf{x}})\cdot\mathbf{n}^{(a)}\right)^2 = n_\mu^{(a)}D_{\mu\nu}n_\nu^{(a)} = \lambda_a \geq 0. \tag{2.30}$$

This means that the set of axes can be ordered in terms of relative importance by sorting the eigenvalues in terms of decreasing size. Since there are n_S shapes, there are at most $n_S - 1$ non-zero eigenvalues. This means that for the case $n_S - 1 < dn_P$, we have performed dimensionality reduction by locating the directions with zero eigenvalue that are orthogonal to the subspace spanned by the data.

In practice, we retain not just the directions corresponding to non-zero eigenvalues, but instead that ordered set which encompasses a certain amount of the total variance of the data.

$$\text{Ordered set of eigenvalues:} \quad \lambda_1 \geq \lambda_2, \ldots \geq \lambda_{dn_P}, \tag{2.31}$$

$$\text{Total variance:} \quad \sum_{a=1}^{n_S-1}\lambda_a, \tag{2.32}$$

$$\text{Variance up to } n_m: \quad \sum_{a=1}^{n_m}\lambda_a. \tag{2.33}$$

The number of modes n_m retained is then chosen to be the lowest value such that the variance up to n_m is some specified fraction of the total variance.

We can also transform coordinates to the system defined by the directions $\{\mathbf{n}^{(a)}\}$, with origin $\bar{\mathbf{x}}$. For each shape $\mathbf{x}_i$ this then defines a new vector of shape parameters $\mathbf{b}^{(i)} \in \mathbb{R}^{n_m}$ thus:

$$\mathbf{b}^{(i)} = \{b_a^{(i)} : a = 1, \ldots n_m\}, \; b_a^{(i)} \doteq \left(\mathbf{n}^{(a)}\cdot(\mathbf{x}_i - \bar{\mathbf{x}})\right), \tag{2.34}$$

where the covariance in this frame is now given by the diagonal matrix:

$$\boxed{D_{ab} \doteq \sum_{i=1}^{n_S}(\mathbf{n}^{(a)}\cdot\mathbf{b}^{(i)})(\mathbf{n}^{(b)}\cdot\mathbf{b}^{(i)}) = \lambda_a\delta_{ab}.} \tag{2.35}$$

We define the matrix of eigenvectors:

$$\boxed{\mathbf{N}, \; N_{\mu a} \doteq n_\mu^{(a)},} \tag{2.36}$$

which is then of size $dn_P \times n_m$. We can then form an approximate reconstruction of the shape vector $\mathbf{x}_i$ from the corresponding parameter vector $\mathbf{b}^{(i)}$ thus:

$$\mathbf{x}_i \approx \bar{\mathbf{x}} + \mathbf{N}\mathbf{b}^{(i)}. \tag{2.37}$$

The reconstruction is only approximate, since we have only retained the first n_m eigenvectors, rather than all eigenvectors with non-zero eigenvalue.

The matrix $\mathbf{N}$ performs a mapping from the coordinate axes defined in (shape) parameter space to the original shape space. The mean shape simply performs a translation of the origin, since the origin of parameter space is taken to correspond to the mean shape. The corresponding backwards mapping, from shape space to parameter space, is performed by the matrix $\mathbf{N}^T$. For a general parameter vector $\mathbf{b} \in \mathbb{R}^{n_m}$ and shape vector $\mathbf{x} \in \mathbb{R}^{dn_P}$:

$$\boxed{\mathbf{b} \mapsto \bar{\mathbf{x}} + \mathbf{N}\mathbf{b}, \;\; \mathbf{x} \mapsto \mathbf{N}^T (\mathbf{x} - \bar{\mathbf{x}})\,.} \tag{2.38}$$

Note however that the mappings are not the inverse of each other, even if all the variance is retained, since the dimensionality of parameter space is less than the dimensionality of shape space. For a shape vector $\mathbf{x}$ which is not part of the original training set, the action of $\mathbf{N}^T$ first projects the shape vector into the subspace spanned by the training set, then forms an (approximate) representation of this using the n_m available modes.

If we suppose that the parameter vectors for our original set of shapes are drawn from some probability distribution $p(\mathbf{b})$, then we can sample parameter vectors $\mathbf{b}$ from this distribution. We can then construct the corresponding shapes for each parameter vector $\mathbf{b}$ as above (2.38). This gives us an arbitrarily large set of generated shapes, sharing the same distribution as the original set. This is usually referred to as applying the SSM in a generative mode.

The remaining task is to learn this distribution $p(\mathbf{b})$, given our original set of shapes – in this context, we refer to this set as a *training set.*

2.2 Modelling Distributions of Sets of Shapes

For a simple unimodal distribution of shapes in shape space, PCA generates a coordinate system centred on the distribution, whose axes are aligned with the significant directions of the distribution, and represent modes of variation of that data. If the distribution is not simple, PCA will still enable us to discard dimensions which are orthogonal to the data, that is, perform dimensional reduction. The individual directions $\mathbf{n}^{(a)}$ will not however necessarily correspond to modes of variation of the data.

In the following sections, we consider various methods for studying and representing the distribution of the training data in shape space. We start with the simplest case of a single multivariate Gaussian, where the data is unimodal and the PCA axes do correspond to real modes of variation of the input data. For the case of multimodal or non-linear data distributions, we discuss two types of kernel methods, the classical method of kernel density estimation, and the more recent technique of kernel principal component analysis.

2.2.1 Gaussian Models

We will consider modelling the distribution of the data by a multivariate Gaussian. Having already applied PCA, we now model the parameter space containing the vectors $\{\mathbf{b}^{(i)}\}$ defined above (2.34).

We consider a multivariate Gaussian distribution centred on the origin in parameter space, that is, centred on the mean of the data in shape space.

Theorem 2.2. Maximum Likelihood Method.
Consider a centred Gaussian probability density function (pdf) of the form:

$$p(\mathbf{b}) \propto \left(\prod_{c=1}^{n_m} \frac{1}{\sigma_c}\right) \exp\left(-\frac{1}{2}\sum_{a=1}^{n_m}\left(\frac{\mathbf{b}\cdot\mathbf{m}^{(a)}}{\sigma_a}\right)^2\right), \tag{2.39}$$

where $\{\mathbf{m}^{(a)} : a = 1, \dots n_m\}$ *are some orthonormal set of directions:*

$$\mathbf{m}^{(a)} \cdot \mathbf{m}^{(b)} = \delta_{ab}, \tag{2.40}$$

and $\{\sigma_a\}$ *are the set of width parameters. The fitted Gaussian which maximises the quantity:*

$$\prod_{i=1}^{n_S} p(\mathbf{b}^{(i)}), \tag{2.41}$$

is then given by $\{\mathbf{m}^{(a)}\}$ *equal to the eigenvectors of the covariance matrix of* $\{\mathbf{b}^{(i)}\}$*. If these eigenvectors have corresponding eigenvalues* $\{\lambda_a\}$*, then the optimum width parameters are:*

$$\sigma_a^2 = \frac{1}{n_S}\lambda_a. \tag{2.42}$$

Proof. *We are required to maximise:*

$$\prod_{i=1}^{n_S} p(\mathbf{b}^{(i)}). \tag{2.43}$$

Equivalently, we can maximise instead the logarithm of this:

$$\mathcal{L} = -n_S \sum_{c=1}^{n_m} \ln \sigma_c - \frac{1}{2}\sum_{i=1}^{n_S}\sum_{a=1}^{n_m}\left(\frac{\mathbf{b}^{(i)}\cdot\mathbf{m}^{(a)}}{\sigma_a}\right)^2 + (\textit{constant terms}), \tag{2.44}$$

with the orthonormality constraints as above. For the case of the directions $\{\mathbf{m}^{(a)}\}$*, if we compare this to (2.20), we see that it is essentially the same optimisation problem as the one we encountered previously. Hence we can deduce that the directions* $\{\mathbf{m}_a\}$ *are just the eigenvectors of the covariance* *matrix of the* $\{\mathbf{b}^{(i)}\}$. *And since this covariance matrix is diagonal in the PCA coordinate frame (2.35), we finally have that* $\mathbf{m}^{(a)} = \mathbf{n}^{(a)}\ \forall\ a = 1, \ldots n_m$.

For the parameters $\{\sigma_a\}$, *we then have to optimise:*

$$\mathcal{L} = -n_S \sum_{c=1}^{n_m} \ln \sigma_c - \frac{1}{2} \sum_{a=1}^{n_m} \frac{\lambda_a}{\sigma_a^2} + \textit{(constant terms)}, \tag{2.45}$$

$$\implies \frac{\partial \mathcal{L}}{\partial \sigma_a} = -\frac{n_S}{\sigma_a} + \frac{\lambda_a}{\sigma_a^3}, \tag{2.46}$$

$$\therefore \frac{\partial \mathcal{L}}{\partial \sigma_a} = 0 \implies \boxed{\sigma_a^2 = \frac{1}{n_S}\lambda_a = \frac{1}{n_S} \sum_{i=1}^{n_S} \left((\mathbf{x}_i - \bar{\mathbf{x}}) \cdot \mathbf{n}^{(a)} \right)^2,} \tag{2.47}$$

which is just the mean variance across the set of shapes in the direction $\mathbf{n}^{(a)}$.

□

In many cases, where the shape variation is linear, a multivariate Gaussian density model is sufficient. A single Gaussian cannot however adequately represent cases where there is significant non-linear shape variation, such as that generated when parts of an object rotate, or where there are changes to the viewing angle in a two-dimensional representation of a three-dimensional object. The case of rotating parts of an object can be dealt with by using polar coordinates for these parts, rather than the Cartesian coordinates considered previously [87]. However, such techniques do not deal with the case where the probability distribution is actually multimodal, and in these cases, more general probability distribution modelling techniques must be used. In what follows, we consider kernel-based techniques, the first being classical kernel density estimation, and the second based on the technique of kernel principal component analysis.

2.2.2 Kernel Density Estimation

As before, we start from the set of n_S centred points $\{\mathbf{b}^{(i)}\}$ in shape space $\mathbb{R}^{n_m}$. Kernel density estimation [165] estimates a pdf from data points by essentially smearing out the effect of each data point, by means of a kernel K:

$$p(\mathbf{b}) = \frac{1}{n_S h^{n_m}} \sum_{i=1}^{n_S} K\left(\frac{\mathbf{b} - \mathbf{b}^{(i)}}{h} \right), \tag{2.48}$$

where h is a scaling parameter. In the trivial case where the kernel K is a Dirac δ-function, we obtain the empirical distribution of the data, a pdf $p(\mathbf{b})$ which is zero everywhere except at a data point. A non-trivial choice of kernel would be a multivariate Gaussian:

$$K(\mathbf{b}) \doteq \mathcal{N}(\mathbf{b}; \mathbf{0}, \mathbf{D}), \tag{2.49}$$

where the covariance $\mathbf{D}$ of the kernel can be chosen to match the covariance of the data $\{\mathbf{b}^{(i)}\}$.

A slightly more sophisticated approach is the sample smoothing estimator [15, 175]. Rather than a single global scale parameter h, there is now a local scale parameter, which reflects the local density about each data point, allowing wider kernels in areas where data points are sparse, and narrower kernels in more densely populated areas. Similarly, the kernel covariance can also vary locally [152].

Such kernel methods can give good estimates of the shape distribution. However, the large number of kernels can make them too computationally expensive in an application such as the Active Shape Model (ASM) (Sect. 2.4.1). Cootes et al. [35, 36] developed a method of approximating the full kernel density estimate using a smaller number of Gaussians within a Gaussian mixture model:

$$p_{mix}(\mathbf{b}) \doteq \sum_{i=1}^{n_{mix}} w_i \mathcal{N}(\mathbf{b}; \boldsymbol{\mu}_i, \mathbf{D}_i), \tag{2.50}$$

where n_{mix} is the number of Gaussians within the mixture model, w_i is the weight of the ith Gaussian, with center $\boldsymbol{\mu}_i$ and covariance $\mathbf{D}_i$. The fitting of the parameters can be achieved using a modification [36] to the standard Expectation Maximisation (EM) algorithm method [117].

2.2.3 Kernel Principal Component Analysis

The previous method aims to fit a non-linear or multimodal shape distribution by constructing a parametric non-linear and multimodal distribution within the original shape space.

The Kernel Principal Component Analysis (KPCA) method takes a different approach. KPCA [156, 157] is a technique for non-linear feature extraction, closely related to methods applied in Support Vector Machines [194, 188] (SVMs). Rather than working within the original data space with non-linear and multimodal distributions, KPCA seeks to construct a non-linear mapping of input space $\mathcal{I}$ to a new feature space.

Let $\mathbf{b}$ represent a point in our input data space[3] $\mathcal{I} = \mathbb{R}^{n_m}$, which is mapped to a feature space $\mathcal{F}$:

$$\boldsymbol{\Phi} : \mathbb{R}^{n_m} \mapsto \mathcal{F}, \; \mathbb{R}^{n_m} \ni \mathbf{b} \mapsto \boldsymbol{\Phi}(\mathbf{b}) \in \mathcal{F}, \tag{2.51}$$

[3] Here, we start from the dimensionally reduced space $\mathbb{R}^{n_m}$ rather than the original shape space $\mathbb{R}^{dn_P}$ in order to also include the infinite-dimensional case $n_P \mapsto \infty$ that is considered in Sect. 2.3.

where $\mathcal{F}$ is typically of very high (even infinite) dimensionality. Rather than constructing the mapping $\mathbf{\Phi}$ explicitly, we instead employ the kernel trick, that dot products of mapped points, and hence the implicit mapping $\mathbf{\Phi}$, can be specified by giving the Mercer kernel function [122] $\mathcal{K}$, where:

$$\mathcal{K} : \mathcal{I} \times \mathcal{I} \mapsto \mathbb{R}, \tag{2.52}$$

$$\mathbf{\Phi}(\mathbf{b}) \cdot \mathbf{\Phi}(\mathbf{c}) \doteq \mathcal{K}(\mathbf{b}, \mathbf{c}) \equiv \mathcal{K}(\mathbf{c}, \mathbf{b}) \;\forall\; \mathbf{b}, \mathbf{c} \in \mathcal{I}. \tag{2.53}$$

If we recall the definition of PCA given earlier (Theorem 2.1), we see that computation of the PCA axes depends on our being able to compute dot products between data points and the PCA axis vectors. We hence deduce that since we can compute dot products in the feature space (2.53) by use of the kernel trick, we can then perform PCA in the feature space $\mathcal{F}$ without having to explicitly construct the kernel mapping $\mathbf{\Phi}$. And since the kernel mapping is a non-linear mapping, PCA in the feature space $\mathcal{F}$ then corresponds to a method of non-linear components analysis in the original data space $\mathcal{I}$.

Suppose we have n_S data points $\{\mathbf{b}^{(i)}\}$ in our data space $\mathcal{I}$. The non-linear KPCA components are then given by the following theorem.

Theorem 2.3. KPCA: Centred Components.
Suppose we have data points $\{\mathbf{b}^{(i)} : i = 1, \ldots n_S\}$ in a data space $\mathcal{I}$, and that there exists a mapping $\mathbf{\Phi}$ to a feature space $\mathcal{F}$, the mapping being defined by a Mercer kernel $\mathcal{K}$ as follows:

$$\mathbf{\Phi} : \mathbf{b} \mapsto \mathbf{\Phi}(\mathbf{b}) \;\forall\; \mathbf{b} \in \mathcal{I}, \;\; \mathbf{\Phi}(\mathbf{b}) \cdot \mathbf{\Phi}(\mathbf{c}) \doteq \mathcal{K}(\mathbf{b}, \mathbf{c}) \;\forall\; \mathbf{b}, \mathbf{c} \in \mathcal{I}. \tag{2.54}$$

We define the following:

$$\mathbf{\Phi}^{(i)} \doteq \mathbf{\Phi}(\mathbf{b}^{(i)}), \;\; i = 1, \ldots n_S \tag{2.55}$$

$$\widetilde{\mathbf{\Phi}}^{(i)} \doteq \mathbf{\Phi}^{(i)} - \frac{1}{n_S} \sum_{j=1}^{n_S} \mathbf{\Phi}^{(j)}. \tag{2.56}$$

$$\mathbf{K}_{ij} \doteq \mathbf{\Phi}^{(i)} \cdot \mathbf{\Phi}^{(j)} \equiv \mathcal{K}(\mathbf{b}^{(i)}, \mathbf{b}^{(j)}), \tag{2.57}$$

$$\widetilde{\mathbf{K}}_{ij} \doteq \widetilde{\mathbf{\Phi}}^{(i)} \cdot \widetilde{\mathbf{\Phi}}^{(j)}. \tag{2.58}$$

The centred KPCA components $\{\widetilde{\Phi}^{(i)}_{\alpha} : \alpha = 1, \ldots n_K, n_K \leq n_S - 1\}$ of a data point $\mathbf{b}^{(i)}$ are then extracted from the set of solutions of the eigenproblem:

$$\boxed{\lambda_\alpha n_i^{(\alpha)} = \widetilde{\mathbf{K}}_{ij} n_j^{(\alpha)}, \;\; \widetilde{\Phi}^{(i)}_{\alpha} = n_i^{(\alpha)}.} \tag{2.59}$$

Proof. *As stated above, KPCA applied to the data points $\{\mathbf{b}^{(i)}\}$ is just ordinary linear PCA applied to the mapped data points $\{\mathbf{\Phi}^{(i)}\}$. Centering the mapped data points then gives the $\{\widetilde{\mathbf{\Phi}}^{(i)}\}$ as defined above.*

In Theorem 2.1, linear PCA was defined by maximising the quantity given in (2.15). For KPCA, we define a set of orthogonal (but not necessarily orthonormal) direction vectors $\{\mathbf{n}^{(\alpha)} : \alpha = 1, \dots n_K\}$ *which lie in the subspace of* $\mathcal{F}$ *spanned by the* n_S *vectors* $\{\widetilde{\mathbf{\Phi}}^{(i)}\}$*, and maximise the analogous quantity:*

$$\mathcal{L} = \frac{1}{2n_S}\sum_{i=1}^{n_S}\left(\mathbf{n}^{(\alpha)}\cdot\widetilde{\mathbf{\Phi}}^{(i)}\right)^2 - \frac{c_\alpha}{2}\left(\mathbf{n}^{(\alpha)}\cdot\mathbf{n}^{(\alpha)} - (a^{(\alpha)})^2\right), \tag{2.60}$$

where $c_\alpha > 0$ *is a Lagrange multiplier for maximisation under the normalization constraint:*

$$\mathbf{n}^{(\alpha)}\cdot\mathbf{n}^{(\alpha)} \equiv \|\mathbf{n}^{(\alpha)}\|^2 = (a^{(\alpha)})^2. \tag{2.61}$$

We solve this problem by setting the first derivative of $\mathcal{L}$ *with respect to* $\mathbf{n}^{(\alpha)}$ *to zero as follows:*

$$\frac{\partial\mathcal{L}}{\partial\mathbf{n}^{(\alpha)}} = \frac{1}{n_S}\sum_{i=1}^{n_S}\left(\mathbf{n}^{(\alpha)}\cdot\widetilde{\mathbf{\Phi}}^{(i)}\right)\widetilde{\mathbf{\Phi}}^{(i)} - c_\alpha\mathbf{n}^{(\alpha)}. \tag{2.62}$$

$$\therefore \quad \frac{\partial\mathcal{L}}{\partial\mathbf{n}^{(\alpha)}} = 0 \implies \frac{1}{n_S}\sum_{i=1}^{n_S}\widetilde{\mathbf{\Phi}}^{(i)}\left(\mathbf{n}^{(\alpha)}\cdot\widetilde{\mathbf{\Phi}}^{(i)}\right) = c_\alpha\mathbf{n}^{(\alpha)}. \tag{2.63}$$

Taking the dot product with $\widetilde{\mathbf{\Phi}}^{(j)}$*:*

$$\implies \frac{1}{n_S}\sum_{i=1}^{n_S}\left(\widetilde{\mathbf{\Phi}}^{(j)}\cdot\widetilde{\mathbf{\Phi}}^{(i)}\right)\left(\mathbf{n}^{(\alpha)}\cdot\widetilde{\mathbf{\Phi}}^{(i)}\right) = c_\alpha\left(\mathbf{n}^{(\alpha)}\cdot\widetilde{\mathbf{\Phi}}^{(j)}\right), \tag{2.64}$$

$$\implies \sum_{i=1}^{n_S}\widetilde{\mathbf{K}}_{ji}\left(\mathbf{n}^{(\alpha)}\cdot\widetilde{\mathbf{\Phi}}^{(i)}\right) = (n_S c_\alpha)\left(\mathbf{n}^{(\alpha)}\cdot\widetilde{\mathbf{\Phi}}^{(j)}\right). \tag{2.65}$$

The interpretation of $\left(\mathbf{n}^{(\alpha)}\cdot\widetilde{\mathbf{\Phi}}^{(j)}\right)$ *is that it is the PCA component of* $\widetilde{\mathbf{\Phi}}^{(j)}$ *along the direction* $\mathbf{n}^{(\alpha)}$*, hence the* α^{th} *centred KPCA component of* $\mathbf{b}^{(j)}$*. If we define:*

$$n_j^{(\alpha)} \doteq \left(\mathbf{n}^{(\alpha)}\cdot\widetilde{\mathbf{\Phi}}^{(j)}\right), \tag{2.66}$$

then PCA in feature space reduces to the eigenproblem:

$$\boxed{\widetilde{\mathbf{K}}_{ji}n_i^{(\alpha)} = (n_S c_\alpha)n_j^{(\alpha)} = \lambda_\alpha n_j^{(\alpha)}.} \tag{2.67}$$

□

Note that as in linear PCA, we choose the define the index α so that the eigenvalues are ordered in decreasing order.

If we recall the definitions of the kernel matrices $\mathbf{K}$ (2.57) and $\widetilde{\mathbf{K}}$ (2.58), and rewrite $\widetilde{\mathbf{K}}$ in terms of $\mathbf{K}$, we have that:

$$\mathbf{K}_{ij} \doteq \mathcal{K}(\mathbf{b}^{(i)}, \mathbf{b}^{(j)}),$$
$$\widetilde{\mathbf{K}}_{ij} = \mathbf{K}_{ij} - \frac{1}{n_S}\sum_p \mathbf{K}_{pj} - \frac{1}{n_S}\sum_q \mathbf{K}_{iq} + \frac{1}{n_S^2}\sum_{p,q} \mathbf{K}_{pq}. \tag{2.68}$$

Looked at in this way, in terms of kernels involving the input data points, the significance of $\widetilde{\mathbf{K}}$ and its eigenvectors is obscured. It is only the identification between Mercer kernels and mappings that enables $\widetilde{\mathbf{K}}$ to be seen as just the covariance matrix for the mapped data points.[4]

We can also define non-centred KPCA components $\{\Phi_\alpha^{(i)}\}$, where:

$$\boxed{\Phi_\alpha^{(i)} \doteq \mathbf{n}^{(\alpha)} \cdot \mathbf{\Phi}^{(i)}.} \tag{2.69}$$

Theorem 2.4. KPCA: Non-centred Components.
With definitions as above, the non-centred KPCA components of a data point $\mathbf{b}^{(j)}$ *are given by:*

$$\boxed{\Phi_\alpha^{(j)} \doteq \mathbf{n}^{(\alpha)} \cdot \mathbf{\Phi}^{(j)} = \frac{1}{\lambda_\alpha} n_i^{(\alpha)} \mathbf{K}_{ij}.} \tag{2.70}$$

Proof. Remember that the $\{\widetilde{\mathbf{\Phi}}^{(i)}\}$ *are centred points, hence:*

$$\sum_{i=1}^{n_s} \widetilde{\mathbf{\Phi}}^{(i)} \equiv \mathbf{0} \;\Rightarrow\; \sum_{i=1}^{n_s} \mathbf{n}^{(\alpha)} \cdot \widetilde{\mathbf{\Phi}}^{(i)} = 0 \;\Rightarrow\; \sum_{i=1}^{n_s} n_i^{(\alpha)} = 0. \tag{2.71}$$

For all $\lambda_\alpha \neq 0$ *(2.59), the corresponding eigenvector* $\mathbf{n}^{(\alpha)}$ *lies in the space spanned by the set* $\{\widetilde{\mathbf{\Phi}}^{(i)}\}$*, hence* $\mathbf{n}^{(\alpha)}$ *can be expanded in this basis:*

$$\mathbf{n}^{(\alpha)} = \sum_{i=1}^{n_S} w_i^{(\alpha)} \widetilde{\mathbf{\Phi}}^{(i)}. \tag{2.72}$$
$$\therefore\; \mathbf{n}^{(\alpha)} \cdot \widetilde{\mathbf{\Phi}}^{(j)} \doteq n_j^{(\alpha)} = w_i^{(\alpha)} \widetilde{\mathbf{K}}_{ij} \tag{2.73}$$
$$\Rightarrow\; \lambda_\alpha n_j^{(\alpha)} = \lambda_\alpha w_i^{(\alpha)} \widetilde{\mathbf{K}}_{ij}. \tag{2.74}$$

Comparison with the eigenvector equation gives that:

$$w_i^{(\alpha)} = \frac{1}{\lambda_\alpha} n_i^{(\alpha)} \;\Rightarrow\; \mathbf{n}^{(\alpha)} = \frac{1}{\lambda_\alpha} \sum_{i=1}^{n_S} n_i^{(\alpha)} \widetilde{\mathbf{\Phi}}^{(i)}. \tag{2.75}$$

Substituting $\mathbf{\Phi}^{(i)} - \frac{1}{n_S}\sum_{k=1}^{n_S} \mathbf{\Phi}^{(k)}$ *for* $\widetilde{\mathbf{\Phi}}^{(i)}$ *and using (2.71) then gives:*

[4] Note that this result, and the definition of a finite-dimensional covariance matrix $\widetilde{\mathbf{K}}_{ij}$ in a space $\mathcal{F}$ that is possibly infinite-dimensional is actually an application of a result for covariance matrices that is presented later in this chapter (Theorems 2.6 and 2.7).

$$\mathbf{n}^{(\alpha)} = \frac{1}{\lambda_\alpha}\sum_{i=1}^{n_S} n_i^{(\alpha)}\widetilde{\boldsymbol{\Phi}}^{(i)} = \frac{1}{\lambda_\alpha}\sum_{i=1}^{n_S} n_i^{(\alpha)}\boldsymbol{\Phi}^{(i)}. \tag{2.76}$$

Hence:

$$\Phi_\alpha^{(j)} \doteq \mathbf{n}^{(\alpha)} \cdot \boldsymbol{\Phi}^{(j)} = \frac{1}{\lambda_\alpha} n_i^{(\alpha)} \mathbf{K}_{ij}. \tag{2.77}$$

□

As well as the centred and non-centred components for data points, we can now also compute the centred and non-centred KPCA components for an arbitrary point $\mathbf{b}$ in the input space.

Theorem 2.5. KPCA: Components of a Test Point.
A general point $\mathbf{b}$ *in the input space maps to a point* $\boldsymbol{\Phi}(\mathbf{b})$ *in feature space, with non-centred KPCA components:*

$$\Phi_\alpha(\mathbf{b}) \doteq \mathbf{n}^{(\alpha)} \cdot \boldsymbol{\Phi}(\mathbf{b}) = \frac{1}{\lambda_\alpha}\sum_{i=1}^{n_S} n_i^{(\alpha)} \mathcal{K}(\mathbf{b}^{(i)}, \mathbf{b}), \tag{2.78}$$

and centred KPCA components:

$$\widetilde{\boldsymbol{\Phi}}(\mathbf{b}) \doteq \boldsymbol{\Phi}(\mathbf{b}) - \frac{1}{n_S}\sum_{j=1}^{n_S} \boldsymbol{\Phi}^{(j)},$$

$$\widetilde{\Phi}_\alpha(\mathbf{b}) \doteq \mathbf{n}^{(\alpha)} \cdot \widetilde{\boldsymbol{\Phi}}(\mathbf{b}) = \Phi_\alpha(\mathbf{b}) - \frac{1}{\lambda_\alpha n_S}\sum_{i,j=1}^{n_S} n_i^{(\alpha)} \mathbf{K}_{ij}. \tag{2.79}$$

Proof. This follows straightforwardly from the previous results, and is left as an exercise for the reader.

2.2.4 Using Principal Components to Constrain Shape

We now need to consider the ways that PCA and KPCA components are used in shape applications, and consider in detail the significant differences between them.

For shape spaces $\mathbb{R}^{n_m}$, it is obvious that PCA components can increase without limit, since the mapping from shape parameters to shapes (2.38) is defined for any point in the parameter space. Hence we can always exclude test shapes $\mathbf{b}$ far from the training data by setting an *upper* bound on each of the PCA components, creating a bounding parallelepiped. This does not necessarily exclude all shapes which are unlike the training data for cases where the training set shape variation is non-linear or multimodal.

For the case of Gaussian distributions of shapes (2.39), when the PCA axes are appropriately scaled, the pdf becomes spherically symmetric:

$$p(\mathbf{b}) \propto \exp\left(-\frac{1}{2}\sum_{a=1}^{n_m}\left(\frac{\mathbf{b}\cdot\mathbf{m}^{(a)}}{\sigma_a}\right)^2\right), \quad \mathbf{m}^{(a)}\cdot\mathbf{m}^{(b)} = \delta_{ab}, \tag{2.80}$$

$$b_a \doteq \mathbf{m}^{(a)}\cdot\mathbf{b}, \; a = 1, \ldots n_m. \tag{2.81}$$

$$\tilde{b}_a \doteq \frac{b_a}{\sigma_a}, \tag{2.82}$$

$$\therefore \; p(\tilde{\mathbf{b}}) \propto \exp\left(-\frac{1}{2}\sum_{a=1}^{n_m}\left(\tilde{\mathbf{b}}\cdot\mathbf{m}^{(a)}\right)^2\right) = \exp\left(-\frac{1}{2}\|\tilde{\mathbf{b}}\|^2\right), \tag{2.83}$$

where:

$$\|\tilde{\mathbf{b}}\|^2 \doteq \sum_{a=1}^{n_m}\left(\frac{b_a}{\sigma_a}\right)^2, \tag{2.84}$$

is the squared Mahalanobis distance [110] between the point $\mathbf{b}$ and the origin (mean shape). Hence surfaces of constant Mahalanobis distance correspond to probability isosurfaces, with a simple monotonic relationship between Mahalanobis distance and probability density, and we can create bounding ellipsoids by placing an upper bound on the Mahalanobis distance.

Our initial trivially obvious observation that PCA components can increase without limit is not however generally true for KPCA components.

Consider Theorem 2.5, (2.78):

$$\Phi_\alpha(\mathbf{b}) = \frac{1}{\lambda_\alpha}\sum_{i=1}^{n_S} n_i^{(\alpha)}\mathcal{K}(\mathbf{b}^{(i)}, \mathbf{b}). \tag{2.85}$$

We see that the way the non-centred or centred components of a test point behave as the test point moves away from the data depends on the way the kernel function $\mathcal{K}(\mathbf{b}^{(i)}, \mathbf{b})$ behaves. As was noted by Schölkopf et al. [156]:

$$\Phi_\alpha(\mathbf{b}) \doteq \mathbf{n}^{(\alpha)}\cdot\boldsymbol{\Phi}(\mathbf{b}) \leq \|\mathbf{n}^{(\alpha)}\| \; \|\boldsymbol{\Phi}(\mathbf{b})\| = a^{(\alpha)}\left(\mathcal{K}(\boldsymbol{\Phi}(\mathbf{b}), \boldsymbol{\Phi}(\mathbf{b}))\right)^{\frac{1}{2}}, \tag{2.86}$$

where the normalization of the vector $\mathbf{n}^{(\alpha)}$ is as defined in (2.61).

In Table 2.1, we give examples of some commonly used kernels, and the range of allowed values of $\|\boldsymbol{\Phi}(\mathbf{b})\|$.

We see that for the polynomial kernels (which of course contain linear PCA as a limiting case when $m = 1$), the values of the KPCA components

Table 2.1 Examples of Mercer kernels.

Kernel Type	$\mathcal{K}(\mathbf{b},\mathbf{c})$	$\|\boldsymbol{\Phi}(\mathbf{b})\|$
Polynomial	$(\mathbf{b}\cdot\mathbf{c})^m$,	$0 \leq \|\boldsymbol{\Phi}(\mathbf{b})\|^2 < \infty$.
	$(\mathbf{b}\cdot\mathbf{c}+r)^m$	$r^m \leq \|\boldsymbol{\Phi}(\mathbf{b})\|^2 < \infty$.
Radial Basis Function (RBF)	$\exp\left(-\frac{1}{2\sigma^2}\|\mathbf{b}-\mathbf{c}\|^2\right)$	$\|\boldsymbol{\Phi}(\mathbf{b})\|^2 \equiv 1$.
Sigmoid	$\tanh(\mathbf{b}\cdot\mathbf{c}+r)$	$0 \leq \|\boldsymbol{\Phi}(\mathbf{b})\|^2 \leq 1$.

are unlimited. However, for both the sigmoid and RBF kernels, $\|\boldsymbol{\Phi}(\mathbf{b})\|$ and hence the values of the components are strictly bounded.

The RBF kernel is particularly interesting. Note that the modulus of the mapped vector $\boldsymbol{\Phi}(\mathbf{b})$ in feature space is *identically* one. This means that the mapped input space $\varPhi(\mathcal{I})$ is an embedded submanifold of feature space $\mathcal{F}$. If we consider the modulus of the difference vector between two mapped points, we have:

$$\|\boldsymbol{\Phi}(\mathbf{b}) - \boldsymbol{\Phi}(\mathbf{c})\|^2 \equiv 2\left(1 - \exp\left(-\frac{1}{2\sigma^2}\|\mathbf{b}-\mathbf{c}\|^2\right)\right). \tag{2.87}$$

We hence see that as $\mathbf{b}$ moves away from $\mathbf{c}$ in input space:

$$\|\boldsymbol{\Phi}(\mathbf{b}) - \boldsymbol{\Phi}(\mathbf{c})\|^2 \to 2 \text{ as } \boldsymbol{\Phi}(\mathbf{b})\cdot\boldsymbol{\Phi}(\mathbf{c}) \to 0,$$

which is what we would expect for orthogonal points on a unit hypersphere.

Consider now the projection from this embedded submanifold to KPCA space, which is the space of KPCA components.[5] The explicit expression for the non-centred KPCA components (2.78) for the case of an RBF kernel is:

$$\varPhi_\alpha(\mathbf{b}) = \frac{1}{\lambda_\alpha}\sum_{i=1}^{n_S} n_i^{(\alpha)} \exp\left(-\frac{1}{2\sigma^2}\|\mathbf{b}^{(i)} - \mathbf{b}\|^2\right), \quad \sum_{i=1}^{n_s} n_i^{(\alpha)} = 0. \tag{2.88}$$

The first trivially obvious point to note is that all the KPCA components tend to zero for any test point far from all the data. Let us now focus on the case where the kernel width σ is sufficiently small, and for a fixed value of α. Because of the summation constraint on $\{n_i^{(\alpha)}\}$, at least one of the elements $\{n_i^{(\alpha)}\}$ must be of opposite sign to the others. The set of $\{\varPhi_\alpha^{(i)} = \varPhi_\alpha(\mathbf{b}^{(i)})\}$ (the non-centred KPCA components of the data points), will hence take both negative and positive values across the data. We can hence conclude that for sufficiently small values of σ, the extrema of any KPCA component

[5] It should be noted that whilst we can always move from input space (the point $\mathbf{b}$) to KPCA space (the space of KPCA components) using (2.78), we cannot necessarily do the inverse. So, it is not necessarily the case that an arbitrary point in KPCA space (defined by a set of KPCA components) possesses a pre-image in input space (although various approximation methods do exist, see [156] Sect. 4.5 for further details).

will tend to lie in the vicinity of the data points, taking both negative and positive values, and that these values *bracket* the values obtained for points far from the data (that is, zero). Since this is an ordering property, it persists if we switch to centred KPCA components, since this just corresponds to a translation in KPCA space.

Consider a path in input space which starts at infinity, then approaches some part of the data, then moves out to infinity again. At first, *all* KPCA components will have vanishingly small modulus. As the point approaches the data, some component(s) acquire a value of larger modulus, which then shrinks again as we move away from all the data. In linear PCA, proximity to data was described by placing an *upper* bound to the modulus of each component, but such a procedure will not be generally valid for KPCA components.[6] This argument rests on the kernel width σ being in some sense small, but this behaviour persists for some finite range of σ (indeed, if it did not, RBF KPCA would be of no use as a *non-linear* feature extractor).

It hence suggests that an appropriate proximity-to-data measure for KPCA components would involve a sum over the moduli of non-centred KPCA components. This was the approach taken by Twining and Taylor [185] as follows.

We first define the normalization of the eigenvectors. In contrast to Mika et al. [124] and Schölkopf et al. [156], the eigenvectors are normalized (2.61) with respect to the data:

$$\sum_{i=1}^{n_S}\left(\tilde{\Phi}_i^{(\alpha)}\right)^2 \equiv \sum_{i=1}^{n_S}\left(n_i^{(\alpha)}\right)^2 \doteq 1 \;\Rightarrow\; \|\mathbf{n}^{(\alpha)}\|^2 = \left(a^{(\alpha)}\right)^2 = \frac{1}{\lambda_\alpha} \;\forall\; \alpha, \tag{2.89}$$

which hence gives an orthogonal but not orthonormal basis $\{\mathbf{n}^{(\alpha)}\}$. We then introduce scaled non-centred components:

$$\Psi_\alpha(\mathbf{b}) \doteq \lambda_\alpha \Phi_\alpha(\mathbf{b}) = \sum_{i=1}^{n_S} n_i^{(\alpha)} \exp\left(-\frac{1}{2\sigma^2}\|\mathbf{b}^{(i)} - \mathbf{b}\|^2\right). \tag{2.90}$$

As to why scaled components are used rather than the original components, consider the following. For a general test point, we have the usual definition of non-centred components (2.78):

$$\Phi_\alpha \doteq \mathbf{n}^{(\alpha)} \cdot \mathbf{\Phi}(\mathbf{b}). \tag{2.91}$$

We can hence expand the vector $\mathbf{\Phi}(\mathbf{b})$ in terms of the eigenvectors thus:

---

[6] This point was not sufficiently appreciated by Romdhani et al. [148] when they considered shape spaces for faces. They defined their valid shape region by placing an *upper bound* on their KPCA components, by analogy with the case of linear PCA. In the limit of large σ, RBF KPCA does indeed approach linear PCA (see [185] Appendix A for details), but this behaviour and hence the analogy is not generally valid for RBF KPCA (nor for KPCA in general, as noted in the text).

$$\mathbf{\Phi}(\mathbf{b}) = \sum_{\alpha=1}^{n_K} \lambda_\alpha \Phi_\alpha(\mathbf{b})\mathbf{n}^{(\alpha)} + \mathbf{V}(\mathbf{b}), \tag{2.92}$$

where n_K denotes that we have chosen only the n_K largest eigenvalues to include, and $\mathbf{V}(\mathbf{b})$ is some vector perpendicular to the set of eigenvectors so chosen. We can then deduce that since $\|\mathbf{\Phi}(\mathbf{b})\| \equiv 1$ for all test points, we have that:

$$\|\mathbf{\Phi}(\mathbf{b})\|^2 = \sum_{\alpha=1}^{n_K} \lambda_\alpha |\Phi_\alpha(\mathbf{b})|^2 + \|\mathbf{V}(\mathbf{b})\|^2 \equiv 1 \Rightarrow \sum_{\alpha=1}^{n_K} \lambda_\alpha |\Phi_\alpha(\mathbf{b})|^2 \leq 1. \tag{2.93}$$

This hence places a strict upper bound on the modulus of the components, with the larger the eigenvalue the *smaller* the maximum allowed value of $|\Phi_\alpha(\mathbf{b})|$. For the purposes of proximity-to-data estimation using sums of squares of components, this is not an ideal situation, since we feel intuitively that the modes with larger eigenvalues should make the largest contribution. Hence the use of the scaled components $\Psi_\alpha(\mathbf{b}) \doteq \lambda_\alpha \Phi_\alpha(\mathbf{b})$, since $|\Psi_\alpha(\mathbf{b})|^2 \leq \lambda_\alpha$.

A general proximity-to-data measure is then of the form:

$$f^{(m)}(\mathbf{b}) \propto \left(\sum_{\alpha=1}^{n_K} \Psi_\alpha(\mathbf{b})\Psi_\alpha(\mathbf{b}) \right)^{\frac{m}{2}}, \quad m = 1, 2, \ldots \infty. \tag{2.94}$$

Of particular interest is the simplest case where $m = 2$, where the proximity-to-data measure can be exactly normalized, giving the pseudo-density:

$$\boxed{\widehat{p}(\mathbf{b}) = \frac{1}{A} \sum_{i,j=1}^{n_S} \sum_{\alpha=1}^{n_K} n_i^{(\alpha)} n_j^{(\alpha)} \mathcal{K}(\mathbf{b}, \mathbf{b}^{(i)}) \mathcal{K}(\mathbf{b}, \mathbf{b}^{(j)})} \tag{2.95}$$

$$\text{Normalization factor: } A = \sigma^{n_m} \pi^{\frac{n_m}{2}} \operatorname{Tr}(\mathbf{K}^{\frac{1}{2}}\mathbf{B})$$

$$\text{where } \mathbf{K}_{ij}^{\frac{1}{2}} \doteq \sqrt{\mathbf{K}_{ij}} \text{ and } \mathbf{B}_{ij} \doteq \sum_{\alpha=1}^{n_K} n_i^{(\alpha)} n_j^{(\alpha)}.$$

It has been shown, with the aid of exactly-solvable data distributions ([185] Appendix B), that this does indeed provide a smoothed estimate of the data density, with the kernel width σ acting as a smoothing parameter. On artificial noisy data, it was also shown to give quantitatively better estimates of the density than either naïve or adaptive kernel density methods such as those described in Sect. 2.2.2.

In [183], Twining and Taylor used a lower bound on the pseudo-density to define the class of allowed shapes for statistical shape models, and showed that this gave improved performance when compared to standard linear models.

From a theoretical point of view, the above density estimate is interesting since it differs from those described earlier (Sect. 2.2.2) in that it is a quadratic rather than a linear combination of kernel functions. The kernel width plays the same sort of rôle in both types of kernel methods. The KPCA pseudo-density possesses a further advantage, in that the number of KPCA components n_K included in the summation can be varied. This means that for noisy data, the higher modes (which tend to contain the noise) can be neglected in the density estimate by truncating the summation. As noted above, this has been shown to produce superior results on noisy input data to the standard kernel methods which do not possess such a feature.

However, the basic algorithm as summarized here is still computationally intensive, in that we have to calculate the kernel function between a test point and all data points. We note that several authors (e.g., [128, 166]) have already addressed the problem of optimising KPCA as applied to large data sets ($n_S > 3000$). The question of constructing reduced-set approximations to the exact KPCA results [155] has also been addressed, so we anticipate that considerable improvement on this basic algorithm is possible.

So far, we have considered training sets of shapes represented by a finite number of points, and shown how various methods of principal component analysis and density estimation can be used to describe the distribution of such shapes in shape space. However, in the real world, natural shapes we encounter are continuous. We hence proceed to discuss how the analysis can be extended to the case of continuous shapes.

2.3 Infinite-Dimensional Representations of Shape

In Sect. 2.1, we considered finite-dimensional representations of shape, and saw that both PCA and Gaussian modelling rests on the properties of the covariance matrix $\mathbf{D}$.

However, our input shapes are actually continuous, so we are interested in the limit where $n_P \to \infty$. The problem with our original covariance matrix $\mathbf{D}$ (2.16) is that it is of size $dn_P \times dn_P$, hence becomes infinitely large in this limit. However, the number of non-zero eigenvalues is at most $n_S - 1$. These eigenvalues can be calculated even in the limit by considering the following known result for covariance matrices [51].

Theorem 2.6. Equivalence of Eigenvectors: Finite-Dimensional Case.

Consider a set of shapes $\{\mathbf{x}_i : i = 1, \ldots n_S\}$ *with shape covariance matrix* $\mathbf{D}$ *(2.16) with elements:*

$$\boxed{D_{\mu\nu} \doteq \sum_{i=1}^{n_S} (\mathbf{x}_i - \bar{\mathbf{x}})_\mu (\mathbf{x}_i - \bar{\mathbf{x}})_\nu,} \tag{2.96}$$

and the corresponding eigenvectors and eigenvalues:

$$\mathbf{D}\mathbf{n}^{(a)} \doteq \lambda_a \mathbf{n}^{(a)}, \quad \mathbf{n}^{(a)} \cdot \mathbf{n}^{(b)} = \delta_{ab}, \tag{2.97}$$

where there are at most $n_S - 1$ non-zero eigenvalues.

Also consider the matrix $\widetilde{\mathbf{D}}$ with components:

$$\boxed{\widetilde{D}_{ij} \doteq (\mathbf{x}_i - \bar{\mathbf{x}}) \cdot (\mathbf{x}_j - \bar{\mathbf{x}}).} \tag{2.98}$$

There then exists a one-to-one mapping between the eigenvectors of $\mathbf{D}$ with non-zero eigenvalue, and the eigenvectors of $\widetilde{\mathbf{D}}$ with non-zero eigenvalues, and the eigenvalues for corresponding eigenvectors are equal.

Proof. *We start from the eigenvector equation for $\mathbf{D}$, and the definition of $\mathbf{D}$ (2.16)*[7]*:*

$$D_{\mu\nu} n_\nu^{(a)} = \lambda_a n_\mu^{(a)} \;\Rightarrow\; (\mathbf{x}_j - \bar{\mathbf{x}})_\mu (\mathbf{x}_j - \bar{\mathbf{x}})_\nu n_\nu^{(a)} = \lambda_a n_\mu^{(a)}, \;\; \lambda_a \neq 0. \tag{2.99}$$

Multiplying both sides by $(\mathbf{x}_i - \bar{\mathbf{x}})_\mu$, and summing over μ gives:

$$(\mathbf{x}_i - \bar{\mathbf{x}})_\mu (\mathbf{x}_j - \bar{\mathbf{x}})_\mu (\mathbf{x}_j - \bar{\mathbf{x}})_\nu n_\nu^{(a)} = \lambda_a (\mathbf{x}_i - \bar{\mathbf{x}})_\mu n_\mu^{(a)}, \tag{2.100}$$

$$\Rightarrow\; (\mathbf{x}_i - \bar{\mathbf{x}}) \cdot (\mathbf{x}_j - \bar{\mathbf{x}}) \left[(\mathbf{x}_j - \bar{\mathbf{x}}) \cdot \mathbf{n}^{(a)} \right] = \lambda_a \left[(\mathbf{x}_i - \bar{\mathbf{x}}) \cdot \mathbf{n}^{(a)} \right]. \tag{2.101}$$

$$\textit{Define: } \widetilde{\mathbf{n}}^{(a)} \doteq \{ n_i^{(a)} = (\mathbf{x}_i - \bar{\mathbf{x}}) \cdot \mathbf{n}^{(a)} : i = 1, \dots n_S \}. \tag{2.102}$$

$$\therefore\; \widetilde{D}_{ij} \widetilde{n}_j^{(a)} = \lambda_a \widetilde{n}_i^{(a)}, \;\; \widetilde{\mathbf{D}} \widetilde{\mathbf{n}}^{(a)} = \lambda_a \widetilde{\mathbf{n}}^{(a)}. \tag{2.103}$$

So either $\widetilde{\mathbf{n}}^{(a)} = \mathbf{0}$, or $\widetilde{\mathbf{n}}^{(a)}$ is an eigenvector of $\widetilde{\mathbf{D}}$ with eigenvalue λ_a. If $\widetilde{\mathbf{n}}^{(a)} = \mathbf{0}$, it then follows from the definition (2.102) that the corresponding $\mathbf{n}^{(a)}$ would be orthogonal to the data, with $\lambda_a = 0$, which contradicts our original condition that $\lambda_a \neq 0$.

We can hence conclude that all eigenvectors of $\mathbf{D}$ with non-zero eigenvalue have a corresponding eigenvector of $\widetilde{\mathbf{D}}$ with the same eigenvalue. What remains is the converse claim, that all eigenvectors of $\widetilde{\mathbf{D}}$ correspond to an eigenvector of $\mathbf{D}$.

The proof follows in an exactly similar fashion. We start from the eigenvector equation:

$$\widetilde{\mathbf{D}} \widetilde{\mathbf{n}}^{(a)} = \lambda_a \widetilde{\mathbf{n}}^{(a)}, \tag{2.104}$$

where $\widetilde{\mathbf{n}}^{(a)}$ is any eigenvector of $\widetilde{\mathbf{D}}$ with a non-zero eigenvalue, $\lambda_a \neq 0$.

Inserting the definition of $\widetilde{\mathbf{D}}$ (2.98):

$$\widetilde{D}_{ij} \widetilde{n}_j^{(a)} = \lambda_a \widetilde{n}_i^{(a)} \;\Rightarrow\; (\mathbf{x}_i - \bar{\mathbf{x}})_\mu (\mathbf{x}_j - \bar{\mathbf{x}})_\mu \widetilde{n}_j^{(a)} = \lambda_a \widetilde{n}_i^{(a)}. \tag{2.105}$$

[7] Note our summation convention (see Glossary) that repeated indices are summed over, except where indices appear in brackets $\cdot^{(a)}$, which are only summed over if explicitly stated.

Multiplying by $(\mathbf{x}_i - \bar{\mathbf{x}})_\nu$ *and summing over* i*:*

$$(\mathbf{x}_i - \bar{\mathbf{x}})_\nu (\mathbf{x}_i - \bar{\mathbf{x}})_\mu (\mathbf{x}_j - \bar{\mathbf{x}})_\mu \widetilde{n}_j^{(a)} = \lambda_a \widetilde{n}_i^{(a)} (\mathbf{x}_i - \bar{\mathbf{x}})_\nu \tag{2.106}$$

$$\Rightarrow (\mathbf{x}_i - \bar{\mathbf{x}})_\nu (\mathbf{x}_i - \bar{\mathbf{x}})_\mu \left[(\mathbf{x}_j - \bar{\mathbf{x}})_\mu \widetilde{n}_j^{(a)} \right] = \lambda_a \left[\widetilde{n}_i^{(a)} (\mathbf{x}_i - \bar{\mathbf{x}})_\nu \right]. \tag{2.107}$$

$$\textit{Define:} \quad \mathbf{n}^{(a)} = \{ n_\mu^{(a)} = \widetilde{n}_i^{(a)} (\mathbf{x}_i - \bar{\mathbf{x}})_\mu \}, \tag{2.108}$$

$$\therefore \quad D_{\nu\mu} n_\mu^{(a)} = \lambda_a n_\nu^{(a)}, \ \mathbf{D}\mathbf{n}^{(a)} = \lambda_a \mathbf{n}^{(a)}. \tag{2.109}$$

The solution $\mathbf{n}^{(a)} = 0$ *can be excluded: if we consider (2.105), this can be rewritten as:*

$$(\mathbf{x}_i - \bar{\mathbf{x}})_\mu (\mathbf{x}_j - \bar{\mathbf{x}})_\mu \widetilde{n}_j^{(a)} = \lambda_a \widetilde{n}_i^{(a)} \ \Rightarrow \ (\mathbf{x}_i - \bar{\mathbf{x}})_\mu n_\mu^{(a)} = \lambda_a \widetilde{n}_i^{(a)}. \tag{2.110}$$

So if $\widetilde{\mathbf{n}}^{(a)} \neq \mathbf{0}$ *and* $\lambda_a \neq 0$*, it then follows that* $\mathbf{n}^{(a)} \neq 0$*.*

We hence see that all eigenvectors of $\widetilde{\mathbf{D}}$ *with non-zero eigenvalue correspond to eigenvectors of* $\mathbf{D}$ *with the same eigenvalue. And since the correspondence has now been established in both directions, it then follows that all the eigenvectors of* $\widetilde{\mathbf{D}}$ *and* $\mathbf{D}$ *with non-zero eigenvalues can be placed into a one-to-one correspondence, and the eigenvalues are the same.* □

Earlier (2.34), we defined shape parameter vectors $\mathbf{b}$, where:

$$\mathbf{b}^{(i)} = \{ b_a^{(i)} : a = 1, \ldots n_m \}, \ b_a^{(i)} \doteq \left(\mathbf{n}^{(a)} \cdot (\mathbf{x}_i - \bar{\mathbf{x}}) \right). \tag{2.111}$$

If we consider the eigenvectors $\{\widetilde{\mathbf{n}}^{(a)}\}$ defined above (2.102):

$$\widetilde{n}_i^{(a)} = \mathbf{n}^{(a)} \cdot (\mathbf{x}_i - \bar{\mathbf{x}}), \tag{2.112}$$

we see that:

$$\boxed{b_a^{(i)} \equiv \widetilde{n}_i^{(a)}.} \tag{2.113}$$

That is, it seems that we can obtain the parameter vectors for the training set of shapes directly from the eigenvectors of $\widetilde{\mathbf{D}}$. There is a final point that needs to be noted. The eigenvectors $\{\widetilde{\mathbf{n}}^{(a)}\}$ defined in (2.102) inherited their normalization from the *orthonormal* set of eigenvectors $\{\mathbf{n}^{(a)}\}$. It hence follows that:

$$\boxed{\|\widetilde{\mathbf{n}}^{(a)}\|^2 = \lambda_a.} \tag{2.114}$$

This means that we can indeed extract the needed eigenvalues and the shape parameter vectors directly from $\widetilde{\mathbf{D}}$.

As regards the shape generated from an arbitrary parameter vector $\mathbf{b}$, we have:

$$\mathbf{x} \doteq \bar{\mathbf{x}} + \mathbf{N}\mathbf{b}, \ \mathbf{N} = \{ N_{\mu a} = n_\mu^{(a)} \}. \tag{2.115}$$

This can then be re-written as:

$$\boxed{\mathbf{x} \doteq \bar{\mathbf{x}} + \sum_{i=1}^{n_S} (\mathbf{b}^{(i)} \cdot \mathbf{b})(\mathbf{x}_i - \bar{\mathbf{x}}).} \tag{2.116}$$

This means that a generated shape is to be considered as a linear combination of the training shapes, where the weight for each shape is related to the parameter vectors for the training shape and the generated shape.

2.3.1 Parameterised Representations of Shape

Let us consider for the moment the simple case of shapes in $\mathbb{R}^2$. The finite set of n_P points that describe such a shape can be indexed by an integer $\{\mathbf{x}^{(j)} : j = 1, \ldots n_P, \ \mathbf{x}^{(j)} \in S.\}$. For (single-part) shapes with the simplest topology (open or closed lines), we can include the connectivity information in with the indexing, so that $\mathbf{x}^{(j-1)}$ is joined to $\mathbf{x}^{(j)}$ and so on. The same indexing is applied across a set of shapes, so that the jth point on any one shape corresponds with the jth point on any other shape.

The subset of the integers $\{j : j = 1, \ldots n_P\}$ can then be considered as the parameter space for our shape representation, a parameter space consisting of just a set of discrete points. To construct an *infinite-dimensional* shape representation, we just need to consider a continuous parameter space.[8]

For our simple one-dimensional shapes, we have a continuous shape S_i which is then sampled at n_P points. S_i is then represented by a finite-dimensional shape vector:

$$\mathbf{x}_i \doteq \{\mathbf{x}_i^{(j)} : j = 1, \ldots n_P\}.$$

The associated discrete parameter space is:

$$\{1, 2, 3, \ldots n_P\} \equiv \left\{\frac{1}{n_P}, \frac{2}{n_P}, \ldots, 1\right\},$$

with the associated mapping:

$$\{1, 2, 3, \ldots n_P\} \ni j \mapsto \mathbf{x}_i^{(j)} \in S_i.$$

The mapping respects the topology of the shape, so that the ordering of the integers respects the ordering of points along the shape.

[8] Note that the reader should not confuse this usage of *parameter space* in terms of parameterised shape with the space of *shape* parameters $\mathbf{b}$ (2.34). *Parameter space*, when used in this latter context, should be understood as a convenient shorthand for *shape*-parameter space.

The continuous analog of this parameter space is just the real line between 0 and 1. For an arbitrary point on this line, with parameter value $x \in [0, 1]$, we then have the mapping to the shape:

$$[0,1] \ni x \mapsto \mathbf{S}_i(x) \in S_i.$$

$\mathbf{S}_i(\cdot)$ is then the vector-valued *shape function* associated with shape S_i. The mapping between the real line and the shape has to respect the topology of the shape. This means that if we traverse the shape in one direction, the parameter value then either strictly decreases or strictly increases. The only exception is if the shape has the topology of a circle. This means that the point $x = 0$ in parameter space is connected to $x = 1$, so that the ends of the real line between 0 and 1 have now been joined to form a circle. In general, the mapping from the parameter space to the shape has to be continuous, and one-to-one, so that each parameter value corresponds to a single point on the shape, and each point on the shape has a single associated parameter value. In mathematical terms, such a mapping is a homeomorphism.[9]

In the general case, we have a vector-valued parameter $\mathbf{x} \in X$, where the topology of the parameter space X matches the topology of the shapes. So, for example, simple shapes in $\mathbb{R}^3$ might have the topology of a sphere, where the parameter space is then the appropriate topological primitive – that is, a sphere. For each shape in the training set, there is a continuous, one-to-one mapping $\mathfrak{X}_i$ from the parameter space to the shape S_i.

$$\boxed{X \stackrel{\mathfrak{X}_i}{\longmapsto} S_i, \ \mathbf{x} \stackrel{\mathfrak{X}_i}{\longmapsto} \mathbf{S}_i(\mathbf{x}).} \tag{2.117}$$

The continuity of the mapping means that as the point $\mathbf{x}$ moves on some continuous path around the parameter space, the point $\mathbf{S}_i(\mathbf{x})$ similarly moves in a continuous path on the surface of the shape S_i, with no sudden jumps allowed.

In the finite-dimensional shape representation, shape correspondence is between points with the same index, so that $\mathbf{x}_i^{(j)} \sim \mathbf{x}_k^{(j)}, \ \forall \ i, k$, where $\cdot \sim \cdot$ denotes correspondence. In the continuous case, correspondence is defined analogously:

$$\boxed{\mathbf{S}_j(\mathbf{x}) \sim \mathbf{S}_k(\mathbf{x}),} \tag{2.118}$$

which gives a dense correspondence between any pair of shapes. It is the details of the set of mappings $\{\mathfrak{X}_i\}$ which defines the dense correspondence across the set of shapes.

The covariance matrix $\widetilde{\mathbf{D}}$ for finite-dimensional shape representations is given by (2.98):

$$\widetilde{D}_{jk} \doteq (\mathbf{x}_j - \bar{\mathbf{x}}) \cdot (\mathbf{x}_k - \bar{\mathbf{x}}). \tag{2.119}$$

[9] Technically, a mapping which is one-to-one and continuous in both directions. If the mapping and its inverse is also constrained to be *differentiable* to some order, then this mapping is a *diffeomorphism*.

The continuous analog of the mean shape vector $\bar{\mathbf{x}}$ is the mean shape:

$$\bar{S}(\mathbf{x}) \doteq \frac{1}{n_S}\sum_{i=1}^{n_S}\mathbf{S}_i(\mathbf{x}). \tag{2.120}$$

To take the limit $n_P \to \infty$, we imagine sampling the shapes ever more densely. To make sure this limit is well-defined, we take the n_P points to be equally spaced on the mean shape. This then gives the infinite-dimensional limit of the covariance matrix:

$$\widetilde{D}_{jk} \doteq (\mathbf{x}_j - \bar{\mathbf{x}})\cdot(\mathbf{x}_k - \bar{\mathbf{x}}) \;\to\; \int (\mathbf{S}_j(\mathbf{x}) - \bar{\mathbf{S}}(\mathbf{x}))\cdot(\mathbf{S}_k(\mathbf{x}) - \bar{\mathbf{S}}(\mathbf{x}))dA(\mathbf{x}), \tag{2.121}$$

where $dA(\mathbf{x})$ is the length/area element on the mean shape at the point $\bar{\mathbf{S}}(\mathbf{x})$.

In order to have a smooth transition as regards eigenvalues, it is convenient to introduce the normalized covariance matrices, so that:

$$D_{\mu\nu} \doteq \frac{1}{n_P}\sum_{i=1}^{n_S}(\mathbf{x}_i - \bar{\mathbf{x}})_\mu(\mathbf{x}_i - \bar{\mathbf{x}})_\nu, \tag{2.122}$$

$$\widetilde{D}_{ij} \doteq \frac{1}{n_P}(\mathbf{x}_i - \bar{\mathbf{x}})\cdot(\mathbf{x}_j - \bar{\mathbf{x}}), \tag{2.123}$$

$$\Rightarrow \quad \boxed{\widetilde{D}_{ij} \doteq \frac{1}{A}\int (\mathbf{S}_i(\mathbf{x}) - \bar{\mathbf{S}}(\mathbf{x}))\cdot(\mathbf{S}_j(\mathbf{x}) - \bar{\mathbf{S}}(\mathbf{x}))dA(\mathbf{x}),} \tag{2.124}$$

where A is the total surface area/length of the mean shape. In practice, such integrals can be evaluated by numerical integration techniques. We have chosen a common normalization for $\mathbf{D}$ and $\widetilde{\mathbf{D}}$ so that the equivalence of eigenvalues is maintained (Theorem 2.6). The connection between this common set of eigenvalues for the *normalized* covariance matrices and the variance along the PCA axes is now given by:

$$\boxed{\frac{1}{n_P}\sum_{i=1}^{n_S}\left((\mathbf{x}_i - \bar{\mathbf{x}})\cdot\mathbf{n}^{(a)}\right)^2 = \lambda_a.} \tag{2.125}$$

We would like to maintain the mapping from shape parameter vectors to shapes as just the generalization of (2.116):

$$\boxed{\mathbf{S}(\mathbf{x}) \doteq \bar{\mathbf{S}}(\mathbf{x}) + \sum_{i=1}^{n_S}(\mathbf{b}^{(i)}\cdot\mathbf{b})(\mathbf{S}_i(\mathbf{x}) - \bar{\mathbf{S}}(\mathbf{x})).} \tag{2.126}$$

However, to do this we need to look at the covariance matrix $\mathbf{D}$ and the meaning of the PCA directions $\{\mathbf{n}^{(a)}\}$ and the parameter vectors $\{\mathbf{b}^{(i)}\}$ in the infinite-dimensional limit, and construct the infinite-dimensional analog of Theorem 2.6.

Theorem 2.7. Equivalence of Eigenvectors/Eigenfunctions: Infinite-Dimensional Case.
Consider the covariance matrix (2.124):

$$\widetilde{\mathbf{D}} \doteq \{\widetilde{D}_{ij} : i, j = 1, \ldots n_S\}, \tag{2.127}$$

$$\widetilde{D}_{ij} \doteq \frac{1}{A} \int (\mathbf{S}_i(\mathbf{x}) - \bar{\mathbf{S}}(\mathbf{x})) \cdot (\mathbf{S}_j(\mathbf{x}) - \bar{\mathbf{S}}(\mathbf{x})) dA(\mathbf{x}), \tag{2.128}$$

with eigenvectors and (non-zero) eigenvalues:

$$\widetilde{\mathbf{D}}\widetilde{\mathbf{n}}^{(a)} = \lambda_a \widetilde{\mathbf{n}}^{(a)}, \ \lambda_a \neq 0. \tag{2.129}$$

Consider also the matrix-valued shape covariance function $\mathbf{D}(\mathbf{y}, \mathbf{x})$ *with elements:*

$$\boxed{D_{\nu\mu}(\mathbf{y}, \mathbf{x}) \doteq \frac{1}{A}(S_{i\nu}(\mathbf{y}) - \bar{S}_\nu(\mathbf{y}))(S_{i\mu}(\mathbf{x}) - \bar{S}_\mu(\mathbf{x}))} \tag{2.130}$$

and the general integral eigenproblem:

$$\int D_{\nu\mu}(\mathbf{y}, \mathbf{x}) n_\mu^{(a)}(\mathbf{x}) dA(\mathbf{x}) = \lambda_a n_\nu^{(a)}(\mathbf{y}) \tag{2.131}$$

$$\Rightarrow \boxed{\int \mathbf{D}(\mathbf{y}, \mathbf{x}) \mathbf{n}^{(a)}(\mathbf{x}) dA(\mathbf{x}) = \lambda_a \mathbf{n}^{(a)}(\mathbf{y}), \ \lambda_a \neq 0.} \tag{2.132}$$

There then exists a one-to-one mapping between the eigenvectors of $\widetilde{\mathbf{D}}$ *with non-zero eigenvalue, and the eigenfunctions of* $\mathbf{D}(\mathbf{y}, \mathbf{x})$ *with non-zero eigenvalues, and the eigenvalues for corresponding eigenvectors/functions are equal.*

Proof. *For parameterised shapes* $\{S_i\}$ *in* $\mathbb{R}^d$*, we define:*

$$\widetilde{\mathbf{S}}_i(\mathbf{x}) \doteq \mathbf{S}_i(\mathbf{x}) - \bar{\mathbf{S}}(\mathbf{x}), \ \widetilde{\mathbf{S}}_i(\mathbf{x}) \doteq \{\widetilde{S}_{i\mu}(\mathbf{x}) : \mu = 1, \ldots d\}. \tag{2.133}$$

Then the eigenvector equation for $\widetilde{\mathbf{D}}$ *can be written as:*

$$\frac{1}{A} \int \widetilde{S}_{i\mu}(\mathbf{x}) \widetilde{S}_{j\mu}(\mathbf{x}) \widetilde{n}_j^{(a)} dA(\mathbf{x}) = \lambda_a \widetilde{n}_i^{(a)}. \tag{2.134}$$

If we multiply by $\widetilde{S}_{i\nu}(\mathbf{y})$ *and sum over* i*, we obtain:*

$$\frac{1}{A} \int \widetilde{S}_{i\mu}(\mathbf{x}) \widetilde{S}_{i\nu}(\mathbf{y}) \left(\widetilde{S}_{j\mu}(\mathbf{x}) \widetilde{n}_j^{(a)} \right) = \lambda_a \left(\widetilde{S}_{i\nu}(\mathbf{x}) \widetilde{n}_i^{(a)} \right). \tag{2.135}$$

If we define the vector-valued function:

$$\mathbf{n}^{(a)}(\mathbf{x}) = \{n_\mu^{(a)}(\mathbf{x}) : \mu = 1, \ldots d\}, \; n_\mu^{(a)}(\mathbf{x}) \doteq \left(\widetilde{S}_{i\mu}(\mathbf{x})\widetilde{n}_i^{(a)}\right), \tag{2.136}$$

and using the definition of the matrix-valued covariance function:

$$\mathbf{D}(\mathbf{y}, \mathbf{x}) \doteq \{D_{\nu\mu}(\mathbf{y}, \mathbf{x}) : \nu, \mu = 1, \ldots d\}, \; D_{\nu\mu}(\mathbf{y}, \mathbf{x}) \doteq \frac{1}{A}\widetilde{S}_{i\nu}(\mathbf{y})\widetilde{S}_{i\mu}(\mathbf{x}), \tag{2.137}$$

then the eigenvector equation can be rewritten as:

$$\int D_{\nu\mu}(\mathbf{y}, \mathbf{x}) n_\mu^{(a)}(\mathbf{x}) dA(\mathbf{x}) = \lambda_a n_\nu^{(a)}(\mathbf{y}) \tag{2.138}$$

$$\Rightarrow \int \mathbf{D}(\mathbf{y}, \mathbf{x})\mathbf{n}^{(a)}(\mathbf{x}) dA(\mathbf{x}) = \lambda_a \mathbf{n}^{(a)}(\mathbf{y}). \tag{2.139}$$

Hence $\mathbf{n}^{(a)}(\mathbf{y})$ *is a solution of the required eigenproblem, with matching eigenvalue* λ_a.

Similarly, if we start from a solution to this eigenproblem, and take the dot product of both sides with $\widetilde{S}_i(\mathbf{y})$ *and integrate over* $dA(\mathbf{y})$, *we obtain:*

$$\frac{1}{A}\int \widetilde{S}_{j\nu}(\mathbf{y})\widetilde{S}_{j\mu}(\mathbf{x}) n_\mu^{(a)}(\mathbf{x})\widetilde{S}_{i\nu}(\mathbf{y}) dA(\mathbf{x}) dA(\mathbf{y}) = \lambda_a \int n_\nu^{(a)}(\mathbf{y})\widetilde{S}_{i\nu}(\mathbf{y}) dA(\mathbf{y}). \tag{2.140}$$

If we define:

$$\widetilde{\mathbf{n}}^{(a)} \doteq \{\widetilde{n}_i^{(a)} : i = 1, \ldots n_S\}, \; \widetilde{n}_i^{(a)} \doteq \int n_\nu^{(a)}(\mathbf{y})\widetilde{S}_{i\nu}(\mathbf{y}) dA(\mathbf{y}), \tag{2.141}$$

then we have:

$$\frac{1}{A}\int \widetilde{S}_{j\nu}(\mathbf{y})\widetilde{S}_{i\nu}(\mathbf{y})\widetilde{n}_j^{(a)} dA(\mathbf{y}) = \lambda_a \widetilde{n}_i^{(a)}, \tag{2.142}$$

$$\Rightarrow \widetilde{D}_{ij}\widetilde{n}_j^{(a)} = \lambda_a \widetilde{n}_i^{(a)} \Rightarrow \widetilde{\mathbf{D}}\widetilde{\mathbf{n}}^{(a)} = \lambda_a \widetilde{\mathbf{n}}^{(a)}, \tag{2.143}$$

which gives a solution to the other eigenproblem. We hence have a one-to-one mapping between the solutions of the two eigenproblems with non-zero eigenvalues. □

We hence have shown the equivalence of the two eigenproblems in the infinite-dimensional case, just as Theorem 2.6 showed their equivalence in the finite-dimensional case.

The integral eigenproblem (2.130) is just the infinite-dimensional analog of the finite-dimensional eigenvector problem for the normalized covariance matrix $\mathbf{D}$ (2.122). The eigenfunctions $\{\mathbf{n}^{(a)}(\mathbf{x})\}$ are then the infinite-dimensional analog of the eigenvectors $\{\mathbf{n}^{(a)}\}$ that we introduced when we considered PCA ((2.17) and Theorem 2.1). As in the finite-dimensional case, these eigenfunctions are both left and right eigenfunctions (since $\mathbf{D}(\mathbf{y},\mathbf{x})$ is symmetric), and it then follows that eigenfunctions belonging to different eigenvalues are orthogonal, where we define the equivalent of the dot product between these vector-valued eigenfunctions as follows:

$$\int \mathbf{n}^{(a)}(\mathbf{x}) \cdot \mathbf{n}^{(b)}(\mathbf{x}) dA(\mathbf{x}). \tag{2.144}$$

We can hence define an orthonormal set of eigenfunctions, so that:

$$\boxed{\int \mathbf{n}^{(a)}(\mathbf{x}) \cdot \mathbf{n}^{(b)}(\mathbf{x}) dA(\mathbf{x}) \doteq \delta_{ab}.} \tag{2.145}$$

Following the finite-dimensional case (2.34), we introduce parameter vectors $\{\mathbf{b}^{(i)} : i = 1, \ldots n_m\}$, which are defined as:

$$\boxed{\mathbf{b}^{(i)} \doteq \{b_a^{(i)} : a = 1, \ldots n_S\},\ b_a^{(i)} \doteq \int \mathbf{n}^{(a)}(\mathbf{x}) \cdot \widetilde{\mathbf{S}}_i(\mathbf{x}) dA(\mathbf{x}).} \tag{2.146}$$

From (2.141), we see that as before (2.113):

$$\boxed{\widetilde{n}_i^{(a)} = b_a^{(i)}.} \tag{2.147}$$

The size of the vectors $\widetilde{\mathbf{n}}^{(a)}$ is however slightly different:

$$\boxed{\|\widetilde{\mathbf{n}}^{(a)}\|^2 = A\lambda_a.} \tag{2.148}$$

Putting all this together, it means that rather than trying to solve the integral eigenproblem, we can instead solve for the eigenvalues and eigenvectors of the covariance matrix $\widetilde{\mathbf{D}}$ (2.124). The integral over the mean shape in the covariance matrix can be solved using a numerical approximation. We then apply the above normalization to the vectors $\widetilde{\mathbf{n}}^{(a)}$, and hence obtain the shape parameter vectors $\mathbf{b}^{(i)}$.

In the finite-dimensional case, the shapes vectors could be expanded in terms of the PCA eigenvector basis, the coefficients being the shape parameter vectors. In the infinite-dimensional case, the shape functions can be expanded in terms of the eigenfunctions. If we define:

$$\tilde{\mathbf{S}}_i(\mathbf{x}) \approx \sum_{a=1}^{n_m} c_{ia} \mathbf{n}^{(a)}(\mathbf{x}), \tag{2.149}$$

then by taking the dot product with $\mathbf{n}^{(a)}(\mathbf{x})$ we find that:

$$c_{ia} = \int \widetilde{\mathbf{S}}_i(\mathbf{x}) \cdot \mathbf{n}^{(a)}(\mathbf{x}) dA(\mathbf{x}) = b_a^{(i)}, \tag{2.150}$$

$$\widetilde{\mathbf{S}}_i(\mathbf{x}) \approx \sum_{a=1}^{n_m} b_a^{(i)} \mathbf{n}^{(a)}(\mathbf{x}) \;\Rightarrow\; \mathbf{S}_i(\mathbf{x}) \approx \bar{\mathbf{S}}(\mathbf{x}) + \sum_{a=1}^{n_m} b_a^{(i)} \mathbf{n}^{(a)}(\mathbf{x}). \tag{2.151}$$

This shape representation is only approximate since we are not necessarily using all the eigenvectors, but only the first n_m (note that, as before, we assume the eigenvalues are arranged in decreasing order).

For a general shape generated by a parameter vector:

$$\mathbf{b} \doteq \{b_a : a = 1, \ldots n_m\},$$

we have that:

$$\mathbf{S}(\mathbf{x}) \doteq \bar{\mathbf{S}}(\mathbf{x}) + \sum_{a=1}^{n_m} b_a \mathbf{n}^{(a)}(\mathbf{x}). \tag{2.152}$$

As in (2.116), this can also be rewritten as follows:

$$\text{From (2.136:)} \quad \mathbf{n}^{(a)}(\mathbf{x}) = \{n_\mu^{(a)}(\mathbf{x}) : \mu = 1, \ldots d\}, \tag{2.153}$$

$$n_\mu^{(a)}(\mathbf{x}) \doteq \left(\widetilde{S}_{i\mu}(\mathbf{x}) \widetilde{n}_i^{(a)}\right) = \sum_{i=1}^{n_S} \left(\widetilde{S}_{i\mu}(\mathbf{x}) b_a^{(i)}\right). \tag{2.154}$$

$$\therefore \quad \mathbf{S}(\mathbf{x}) \doteq \bar{\mathbf{S}}(\mathbf{x}) + \sum_{a=1}^{n_m} b_a \mathbf{n}^{(a)}(\mathbf{x}) \tag{2.155}$$

$$= \bar{\mathbf{S}}(\mathbf{x}) + \sum_{a=1}^{n_m} b_a \sum_{i=1}^{n_S} \left(\widetilde{\mathbf{S}}_i(\mathbf{x}) b_a^{(i)}\right) \tag{2.156}$$

$$\Rightarrow \quad \boxed{\mathbf{S}(\mathbf{x}) = \bar{\mathbf{S}}(\mathbf{x}) + \sum_{i=1}^{n_S} \left(\mathbf{b} \cdot \mathbf{b}^{(i)}\right) \widetilde{\mathbf{S}}_i(\mathbf{x}).} \tag{2.157}$$

To summarize, in the infinite-dimensional case, we have the infinite-dimensional parameterised shapes $\{\mathbf{S}_i(\mathbf{x})\}$, and we can perform PCA on these as is given in Algorithm 2.2.

The point to note is that here we have used PCA to perform a radical dimensional reduction, taking us from the space of infinite-dimensional shapes, to the finite-dimensional space of shape parameter vectors. The use of the alternative covariance matrix means that the only infinite-dimensional objects we need to consider are the input shapes themselves, since all further objects are finite-dimensional. The only approximation required is in the initial calculation of the covariance matrix, where we have to use numerical methods to perform the area integral. The previous statements about the link between the shape parameter vectors and variance in the various PCA directions still hold, given the definition we have already used for vector dot products in the

Algorithm 2.2 : PCA for Infinite-Dimensional Shapes.

- Construct the *finite-dimensional* covariance matrix $\widetilde{\mathbf{D}}$ (2.124) by performing numerical integration over the mean shape.
- Solve for the finite-dimensional eigenvectors $\{\widetilde{\mathbf{n}}^{(\mathbf{a})}\}$ and eigenvalues $\{\lambda_a\}$ of $\widetilde{\mathbf{D}}$.
- Normalize the eigenvectors so that $\|\widetilde{\mathbf{n}}^{(\mathbf{a})}\|^{\mathbf{2}} = \mathbf{A}\lambda_{\mathbf{a}}$.
- Construct the shape parameter vectors, where $\widetilde{n}_i^{(a)} = b_a^{(i)}$.
- We then can generate shapes from a shape vector $\mathbf{b}$:

$$\mathbf{S}(\mathbf{x}) = \bar{\mathbf{S}}(\mathbf{x}) + \sum_{\mathbf{i}=1}^{\mathbf{n_S}} \left(\mathbf{b} \cdot \mathbf{b}^{(\mathbf{i})}\right) \widetilde{\mathbf{S}}_{\mathbf{i}}(\mathbf{x}). \tag{2.158}$$

infinite-dimensional space (2.144). Specifically, the mean variance per shape in the direction $\mathbf{n}^{(a)}(\mathbf{x})$ is given by:

$$\widetilde{\mathbf{S}}_i \cdot \mathbf{n}^{(a)} \doteq \int \widetilde{\mathbf{S}}_i(\mathbf{x}) \cdot \mathbf{n}^{(a)}(\mathbf{x}) dA(\mathbf{x}) = b_a^{(i)}. \tag{2.159}$$

$$\therefore \sum_{i=1}^{n_S} \left(\widetilde{\mathbf{S}}_i \cdot \mathbf{n}^{(a)}\right)^2 = \sum_{i=1}^{n_S} \left(b_a^{(i)}\right)^2 = \sum_{i=1}^{n_S} \left(\widetilde{n}_i^{(a)}\right)^2 = \|\widetilde{\mathbf{n}}^{(a)}\|^2 = A\lambda_a.$$

$$\Rightarrow \boxed{\frac{1}{A} \sum_{i=1}^{n_S} \left((\mathbf{S}_i - \bar{\mathbf{S}}) \cdot \mathbf{n}^{(a)}\right)^2 = \lambda_a.} \tag{2.160}$$

Previously (2.125), we had that λ_a represented the summed variance for all shapes about the mean shape in the direction $\mathbf{n}^{(a)}$, normalized by the number of shape points n_P. Here we see that we have the corresponding expression, but normalized to give the variance per unit area of the mean shape.

Since we have now projected our original data from the infinite-dimensional space of shapes to the finite-dimensional space of shape parameter vectors, the modelling of the distribution of parameter vectors proceeds as before.

The notation used for finite and infinite-dimensional shape representations, and the details of the PCA eigenproblems are summarized in Table 2.2.

2.4 Applications of Shape Models

In the previous sections (Sects. 2.1 and 2.3), we have shown how statistical shape models can be built from training sets of shapes, and how principal component and density estimation techniques can be applied to characterize the distribution of shapes. If all we wish to do is analyse the distribution of

Table 2.2 Summary of the notation and conventions used in the text for finite and infinite-dimensional shape representations, covariance matrices, and PCA eigenproblems. Continues on next page.

Finite Dimensional		Infinite Dimensional
SHAPES AND SHAPE REPRESENTATION		
$\{S_i \subset \mathbb{R}^d : i = 1, \ldots n_S\}$ $\{1, 2, \ldots n_P\} \stackrel{\boldsymbol{x}_i}{\longmapsto} S_i,\ j \stackrel{\boldsymbol{x}_i}{\longmapsto} \mathbf{x}_i^{(j)} \in S_i$ $S_i \to \mathbf{x}_i \doteq \{\mathbf{x}_i^{(j)} : j = 1, \ldots n_P\}$ $\bar{\mathbf{x}} \doteq \frac{1}{n_S} \sum_{i=1}^{n_S} \mathbf{x}_i$		$\{S_i \subset \mathbb{R}^d : i = 1, \ldots n_S\}$ $X \stackrel{\boldsymbol{x}_i}{\longmapsto} S_i,\ \mathbf{x} \stackrel{\boldsymbol{x}_i}{\longmapsto} \mathbf{S}_i(\mathbf{x})$ $S_i \to \mathbf{S}_i \doteq \{\mathbf{S}_i(\mathbf{x}) : \mathbf{x} \in X\}$ $\bar{\mathbf{S}}(\mathbf{x}) \doteq \frac{1}{n_S} \sum_{i=1}^{n_S} \mathbf{S}_i(\mathbf{x})$ $\widetilde{\mathbf{S}}_i(\mathbf{x}) \doteq \mathbf{S}_i(\mathbf{x}) - \bar{\mathbf{S}}(\mathbf{x})$
COVARIANCE MATRICES		
$\mathbf{D}$, $D_{\mu\nu} \doteq (\mathbf{x}_i - \bar{\mathbf{x}})_\mu (\mathbf{x}_i - \bar{\mathbf{x}})_\mu$ $\widetilde{\mathbf{D}}$, $\widetilde{D}_{ij} \doteq (\mathbf{x}_i - \bar{\mathbf{x}}) \cdot (\mathbf{x}_j - \bar{\mathbf{x}})$	$\mathbf{D}$, $D_{\mu\nu} \doteq \frac{1}{n_P} (\mathbf{x}_i - \bar{\mathbf{x}})_\mu (\mathbf{x}_i - \bar{\mathbf{x}})_\mu$ $\widetilde{\mathbf{D}}$, $\widetilde{D}_{ij} \doteq \frac{1}{n_P} (\mathbf{x}_i - \bar{\mathbf{x}}) \cdot (\mathbf{x}_j - \bar{\mathbf{x}})$	$\mathbf{D}(\mathbf{y}, \mathbf{x})$, $D_{\mu\nu}(\mathbf{x}, \mathbf{y}) \doteq \frac{1}{A} (S_{i\mu}(\mathbf{x}) - \bar{S}_\mu(\mathbf{x}))(S_{i\nu}(\mathbf{y}) - \bar{S}_\nu(\mathbf{y}))$ $\widetilde{\mathbf{D}}$, $\widetilde{D}_{ij} \doteq \frac{1}{A} \int (\mathbf{S}_i(\mathbf{x}) - \bar{\mathbf{S}}(\mathbf{x})) \cdot (\mathbf{S}_j(\mathbf{x}) - \bar{\mathbf{S}}(\mathbf{x})) dA(\mathbf{x})$
EIGENPROBLEMS		
$\mathbf{D}\mathbf{n}^{(a)} = \lambda_a \mathbf{n}^{(a)}$		$\int \mathbf{D}(\mathbf{y}, \mathbf{x}) \mathbf{n}^{(a)}(\mathbf{x}) dA(\mathbf{x}) = \lambda_a \mathbf{n}^{(a)}(\mathbf{y})$
$\widetilde{\mathbf{D}}\widetilde{\mathbf{n}}^{(a)} = \lambda_a \widetilde{\mathbf{n}}^{(a)}$		

Table 2.2 Summary of the notation and conventions used in the text for finite and infinite-dimensional shape representations, covariance matrices, and PCA eigenproblems. Continued from previous page.

Finite Dimensional	Infinite Dimensional
EIGENVECTORS AND EIGENVALUES	
$\mathbf{n}^{(a)} \cdot \mathbf{n}^{(b)} = \delta_{ab}$	$\int \mathbf{n}^{(a)}(\mathbf{x}) \cdot \mathbf{n}^{(b)}(\mathbf{x}) dA(\mathbf{x}) = \delta_{ab}$
$\lambda_a = \sum_{i=1}^{n_S} \left((\mathbf{x}_i - \bar{\mathbf{x}}) \cdot \mathbf{n}^{(a)} \right)^2$ $\lambda_a = \frac{1}{n_P} \sum_{i=1}^{n_S} \left((\mathbf{x}_i - \bar{\mathbf{x}}) \cdot \mathbf{n}^{(a)} \right)^2$	$\lambda_a = \frac{1}{A} \sum_{i=1}^{n_S} \left((\mathbf{S}_i - \bar{\mathbf{S}}) \cdot \mathbf{n}^{(a)} \right)^2$ $(\mathbf{S}_i - \bar{\mathbf{S}}) \cdot \mathbf{n}^{(a)} \doteq \int (\mathbf{S}_i(\mathbf{x}) - \bar{\mathbf{S}}(\mathbf{x})) \cdot \mathbf{n}^{(a)}(\mathbf{x}) dA(\mathbf{x})$
$\tilde{\mathbf{n}}^{(a)} \doteq \{\tilde{n}_i^{(a)} : i = 1, \ldots n_S\}$	
$\tilde{n}_i^{(a)} \doteq \mathbf{n}^{(a)} \cdot (\mathbf{x}_i - \bar{\mathbf{x}})$	$\tilde{n}_i^{(a)} \doteq \int \mathbf{n}^{(a)} \cdot (\mathbf{S}_i(\mathbf{x}) - \bar{\mathbf{S}}(\mathbf{x})) dA(\mathbf{x})$
$\Vert\tilde{\mathbf{n}}^{(a)}\Vert^2 \doteq \lambda_a$ $\Vert\tilde{\mathbf{n}}^{(a)}\Vert^2 \doteq n_P \lambda_a$	$\Vert\tilde{\mathbf{n}}^{(a)}\Vert^2 \doteq A \lambda_a$
PARAMETER VECTORS	
$\mathbf{b}^{(i)} = \{b_a^{(i)} : a = 1, \ldots n_m\}$	
$b_a^{(i)} \doteq \mathbf{n}^{(a)} \cdot (\mathbf{x}_i - \bar{\mathbf{x}})$ $\mathbf{x}_i \approx \bar{\mathbf{x}} + \sum_{a=1}^{n_S} b_a^{(i)} \mathbf{n}^{(a)}, \ \mathbf{x} \doteq \bar{\mathbf{x}} + \sum_{a=1}^{n_S} b_a \mathbf{n}^{(a)} = \bar{\mathbf{x}} + \sum_{i=1}^{n_S} \left(\mathbf{b}^{(i)} \cdot \mathbf{b} \right) (\mathbf{x}_i - \bar{\mathbf{x}})$	$b_a^{(i)} \doteq \int \mathbf{n}^{(a)}(\mathbf{x}) \cdot (\mathbf{S}_i(\mathbf{x}) - \bar{\mathbf{S}}(\mathbf{x})) dA(\mathbf{x})$ $\mathbf{S}(\mathbf{x}) = \bar{\mathbf{S}}(\mathbf{x}) + \sum_{i=1}^{n_S} \left(\mathbf{b}^{(i)} \cdot \mathbf{b} \right) (\mathbf{S}_i(\mathbf{x}) - \bar{\mathbf{S}}(\mathbf{x}))$

shapes across the training set, this is often sufficient. For example, we can use the shape of the estimated density or information from principal components to classify subsets of shapes within the training set. Principal component analysis can also tell us about the major and minor modes of shape variation seen across the training set, and provide an intuitive picture of the way the shapes vary.

If we are given an unseen shape, one which was not included in our training set, we can use the same techniques to analyse this new shape. For example, we can decide whether it is like or unlike those seen previously, to what category of shape it belongs, or describe in what way it varies from what we have already seen. We can hence describe this new shape within a wider context of learnt information about this type of shape.

This however presumes that we already have our unseen shape. In many computer vision or medical imaging applications that study shape, the shapes themselves are obtained from images. The most intuitive, and the simplest, way of extracting the shape from the image is to use manual annotation. However, when there are a large number of examples to process, this can become extremely time-consuming. For the case of images in three dimensions, such as those encountered in medical imaging, this annotation can become very difficult.

There are many basic methods for automatically segmenting images [169]. These typically use information such as the colour/greyscale values and texture in the image to identify regions of the image, and information about edge structures in the image to try to delineate the boundaries of such regions or sets of regions that constitute the shape of the imaged object. This can work well provided the shapes are relatively simple, or have good texture/colour cues, or where we do not know what shapes we expect to see in an image. However, for cases where we are looking for a particular known object, the most promising approaches are those which adopt a learning framework. Such systems proceed in much the same way that a human annotator would proceed. The trainee human annotator or computer system is first presented with a set of training examples which have been previously annotated by some expert. Based on what has been learnt from these examples as to the shape which is required, and how it varies, the trainee system or human then annotates examples, possibly with some continuing feedback from an expert to correct mis-annotation.

Such a system can be constructed using a statistical shape model to encode the learnt information about shape. Two algorithms which use such a system for automatic image annotation are the Active Shape Model (ASM) [29, 39] and Active Appearance Model (AAM) [26, 25, 27].

2.4.1 Active Shape Models

Suppose we have an image that contains an example of an object we are trying to annotate with a set of shape points. In general terms, there are several components that help us differentiate a successful from an unsuccessful annotation.

First, we have the global constraint that the set of points should describe a shape which is, according to what we have learnt about shape, a valid example of that object. Secondly, we also have the local constraint that the shape points should lie on edges or structures in the image that look like the locations where such points have been placed in the annotated training examples.

Given an initial guess as to the shape, these two constraints can be used in tandem to home in on the correct position of the shape. Essentially, for each current shape point, we search in the neighbourhood of that point to see if there is a candidate position which better fits the expected appearance of the image. Given such a set of candidate positions, we then apply the global constraint, by replacing the candidate shape by a shape which is as close as possible to the candidate shape (hence fits the local constraints), yet is a valid shape as far as the global constraint is concerned. The process is then iterated until it converges on a final answer. This is the basic Active Shape Model search algorithm.

The global constraint is applied by quantifying the training information about shape and shape variation in terms of a statistical shape model. For a candidate shape, the positions of the candidate points are encoded in terms of a shape vector $\mathbf{x}$ as described previously (2.2). This shape is then Procrustes aligned (Sect. 2.1.1) with the mean shape $\bar{\mathbf{x}}$ from the SSM, to remove unimportant details such as the precise scale and orientation of the object. We then project this shape vector into the subspace of shape space described by the SSM, and evaluate its proximity to the training set of shapes. The first stage typically means extracting the PCA components of the candidate shape as in (2.38), which gives us an approximation representation of the candidate shape in terms of a set of shape parameters $\mathbf{b}$.

We then have to evaluate the proximity of the point $\mathbf{b}$ to the training set of shapes. For PCA components, we can constrain each component individually, forcing the shape to lie within a bounding parallelepiped as described in Sect. 2.2.4. Alternatively, for cases where a density estimate is available (e.g., as in Sects. 2.2.1–2.2.3), we can restrict the minimum allowed value of $p(\mathbf{b})$. For points that do not initially satisfy the constraint, we can evolve the point $\mathbf{b}$ through gradient ascent of $p(\mathbf{b})$ until the constraint is satisfied. For Gaussian density models, this process is considerably simplified, given the monotonic relationship of Mahalanobis distance and the probability density as described in Sect. 2.2.4, and it is sufficient to move the point $\mathbf{b}$ inwards along the ray connecting $\mathbf{b}$ to the origin of parameter space until the constraint is satisfied. It should be noted that setting the appropriate limits is important. Setting them too high at the beginning of the search can overly constrain the shape, and not allow enough freedom in moving through the search space to locate the optimum fitted shape, or allow only solutions which are very close to the training shapes. Whereas too loose a constraint can allow the search process to get stuck in local minima, fitting to shapes far from optimum.

The local part of the ASM search is built on learning about the local image appearance in the neighbourhood of the shape points. For each example of a specific shape point on each training example, the normal to the shape at that point is constructed. The image intensity values are then sampled along this normal to form an image profile. This set of image profiles from each example is then combined into a statistical profile model in the same general manner as for shape models. When searching for a new candidate position for a shape point on an unseen image during search, profiles are sampled in the vicinity, and the profile that best fits the profile model for that point is selected as the new candidate position.

There is an extensive and still growing research literature as regards Active Shape Models, with various variations on the basic ASM described above (e.g., see [40, 33, 39, 28, 34], and the reader should consult the appropriate literature for full details (see [32, 37] for reviews).

2.4.2 Active Appearance Models

The ASM search performs extremely well on some types of data. However, the model uses only limited image information to locate the shape. In the Active Appearance Model (AAM) [26, 25, 27], the training process incorporates information about the expected appearance within the entire interior of the annotated shape, rather than just samples of image information in the neighbourhood of the shape points.

For the annotated training images, the shape part of the model is constructed as before. We obviously cannot just combine the raw image information from the interiors of all the training shapes, but have first to convert this information into a common frame of reference. This is done using the shape information, since this tells us the varying positions of corresponding shape points across the whole set of training examples. If we interpolate the interior of each shape, based on the shape points, this then gives us, by interpolation, a correspondence between the interiors of each shape across the whole set. We then map each training shape to the mean shape, and resample the image information from each shape into the frame of the mean shape. This gives us a set of shape-free texture examples, one from each training image. The pixel-value information for each shape-free texture example is then concatenated into a vector, with the entire training set then giving us a set of data points in a shape-free texture space. The distribution of points in shape-free texture space can then be analysed and modelled using the same techniques as those used for modelling shape spaces. We then have both a statistical model of shape, and a statistical model of texture (essentially a type of shape-free eigenface model [179]). Using these statistical models in generative mode, we can then create shape-free variations of texture and modes of variation of texture. Using the reverse of the initial mapping from texture on a shape to texture on the mean shape, we can also vary shape whilst leaving texture unchanged.

For many imaging examples, there is a significant correlation between shape and texture. One obvious example is two-dimensional images of faces. It is obvious that, for a fixed illumination, as the pose of the subject changes, the texture (i.e., the positions of highlights and shadows) changes, and is correlated with the changes in shape. Even without change of pose, the shape of the face changes under changes of expression, and the texture changes in a correlated fashion.

These correlations can be modelled by concatenating the shape vector and the texture vector of each training example into a single vector. A statistical model of appearance is then built in the usual manner in this combined space of shape and texture, and generates modes of variation that capture the correlations noted above.

The search algorithm for the AAM is slightly more complicated than for the ASM (and we refer readers to the literature for the details [26, 25, 27]). However, the basic rationale is the same as for the ASM, where the learnt information about permissable levels of variation is incorporated into and constrains the search process.

The statistical appearance model, like the statistical shape model, can also be applied in a generative mode. By sampling the space of parameters according to the modelled pdf, we can generate an arbitrarily large number of artificial examples of the modelled object.

For the case of faces, these artificially generated examples can be almost photo-realistic. Analysis of the space of the model can separate out the subspaces corresponding to varying lighting, pose, identity, and expression [44, 42]. This means that given an image of an unseen individual, we can generate examples of this same individual, but apply different expressions. If information about the gender of subjects across the training set is available, it is also possible to manipulate the perceived gender, making a face appear more masculine or more feminine according to what has been learnt from the training set about the way faces tend to differ with gender [43]. Similarly, it is also possible to simulate ageing [104] (see Chap. 1).

The power and flexibility of the ASM/AAM approach has led to their usage in a large (and still growing) number of applications in computer vision and medical imaging. Both approaches require an initial statistical shape model for their implementation, and the quality of this initial model is a prime determining factor in ensuring the quality of the final system. Hence establishing a suitable correspondence is a key step in the model-building process, and one that we address in greater detail in the next chapter.

Chapter 3
Establishing Correspondence

In the previous chapter, we described methods for modelling shape, based on statistical learning techniques. Such shape models have been used successfully in many applications, some of which were described in Chap. 1. However, in order to build such a model, a dense correspondence must be established between all shapes in our training set. As we will see in Sect. 3.1, the utility of the model depends critically on an appropriate definition of groupwise correspondence.

There are various ways in which such a correspondence can be established, and some of these will be reviewed in Sect. 3.2. However, these approaches have various limitations, and it is not clear that the correspondence that they produce is necessarily the correct one when it comes to model-building.

The most promising approach, and the one that is followed in this book, is to treat correspondence as an optimisation problem. The objective function for this optimisation is based on the properties of the model built using the groupwise correspondence. A general framework for this approach to groupwise correspondence and model-building is established in Sect. 3.3. This framework can be broken down into three main components:

- An objective function, which assesses the quality of the model built from the groupwise correspondence.
- A method of manipulating the groupwise correspondence.
- An optimisation algorithm, in order to find the optimum groupwise correspondence.

A brief introductory summary of each component will be provided in this chapter, with a detailed treatment provided in later chapters.

We begin with the problem of correspondence.

DOI: 10.1007/978-1-84800-138-1_3,

3.1 The Correspondence Problem

In the previous chapter on building statistical models of shape, it was assumed that a dense correspondence had already been established across the training set of shapes. This is not generally the case for the sets of training shapes themselves, whether they are obtained by automatic or manual segmentation of images, or by some other method, since the primary consideration in this initial data-gathering stage is obtaining a faithful representation of the shape of the physical object being considered.

The groupwise correspondence matters, since the utility of the resulting model depends critically on the appropriateness of the correspondence. For example, an inappropriate choice of correspondence can result in a distribution of shapes that cannot be well-represented by a reasonable number of parameters, or where the modes of variation of the resulting shape model do not correspond well with the actual modes of variation of the objects from which the shapes were derived. This can mean that representing the training shapes themselves to some reasonable degree of accuracy requires an inordinately large number of shape parameters, and that seemingly 'legal' parameter values entered into the model produces 'illegal' shape examples, totally unlike those seen in the training set.

This correspondence problem can be demonstrated using a simple example. We take a training set consisting of 17 examples of the outline of a hand. For the first model (model A), correspondence was established by using natural landmarks, such as the tips of the fingers, with a dense correspondence then being assigned by interpolation. For the second model (model B), only one such landmark was used. Correspondence along each shape outline was assigned in terms of the fractional distance along each shape from the single initial landmark (usually referred to as arc-length parameterisation). The shapes were Procrustes aligned, as in Algorithm 2.1, then a multivariate Gaussian model built for each choice of correspondence.

Let us now compare these two models, A and B. If we first look at the variance captured by each mode of shape variation, the three leading modes of model A have variances of 1.06, 0.58, and 0.30. Whereas for model B, the variances are 2.19, 0.78, and 0.54, respectively. This suggests that model A is more compact than model B. If we now consider example shapes generated by the models (see Fig. 3.1), using shape parameters within the range found across the training set, we see that model A produces examples that look like plausible examples of hand outlines. In contrast, model B generates implausible examples.

In this case, the difference between the two methods of assigning correspondence can be clearly seen, and all that is required is visual inspection of the shapes generated by the model. But in more realistic cases, different correspondence can still produce significantly different models. This difference may not be discernable using such a simple visual inspection, both models may appear equally plausible, but different correspondence can produce sig-

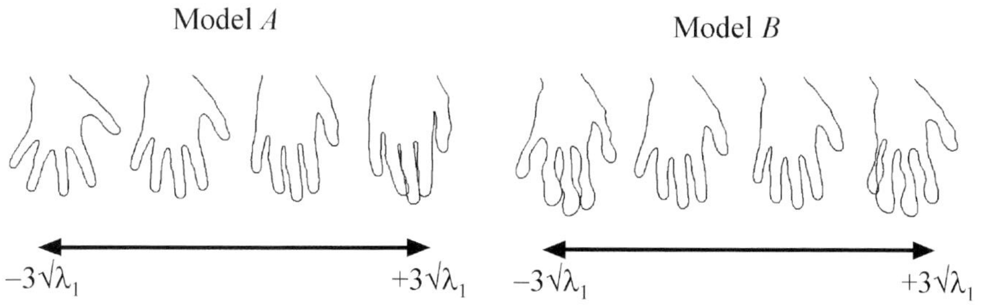

Fig. 3.1 The first mode of variation ($\pm 3\sqrt{\lambda_1}$) of two shape models built from the same training set of hand outlines, but with different groupwise correspondence. Model A interpolates correspondence from a small set of manually placed landmarks, whereas model B uses fractional arc-length distance to assign correspondence. It can clearly be seen that the examples generated by model A are plausible hand outlines, whereas those from model B are not.

nificant differences in terms of detailed model performance, whether this be using the model to classify shape, or when the model is used in an ASM or AAM for extracting shapes from unseen images.

3.2 Approaches to Establishing Correspondence

We have seen already that is important to establish an appropriate correspondence between members of the training set. In this section, we review several common approaches to establishing correspondence. Although object-specific solutions have been proposed (for example, [21]), we will confine our attention to generic approaches.

3.2.1 Manual Landmarking

When statistical shape models were first introduced by Cootes et al. [38], correspondence was defined using manual landmarks, placed at points of anatomical significance. The landmarks used for the hand training set above are an example of this. Such manual landmarking can often give acceptable results. However, manual landmarking can be subjective, with different annotators giving slightly different results. It is also a very time-consuming process, prone to error, and cannot be guaranteed to produce good models. For simple cases, such as the hand example above, or images of faces, it is relatively simple to place reasonable landmarks. However, for many applications (such as medical images), specialist anatomical knowledge is needed to identify consistent and reproducible points of anatomical significance. It could be argued that this need for specialist knowledge is an advantage, since it is information that is then incorporated into the model. But in practice, it can further complicate the process of model-building.

The problems are particularly acute for the case of shape surfaces in three dimensions, since not only can reproducible points be difficult to define on smooth surfaces, but also, just the difficulties with visualization can make such landmarking impractical for large datasets.

For certain classes of shapes obtained from images, where manual landmarking is feasible, the tedium can be somewhat relieved by means of a semi-automatic, bootstrapping procedure. The idea is that an initial shape model is built from the first few annotated examples. This is then incorporated into an ASM or AAM. The ASM or AAM is then used to search on an unseen image, and produces an initial estimate of the segmented shape and its landmarks. This can then be refined by the annotator, and this new example then included in an updated model. It hence assists with the two tasks of manual segmentation and manual landmarking.

3.2.2 Automatic Methods of Establishing Correspondence

Given the above limitations of manual landmarking, even with the inclusion of semi-automatic methods, a fully automatic approach to establishing correspondence is desirable. Many such automated approaches have been proposed. In the following sections, we provide a brief overview of the field by considering some of the most prominent approaches.

3.2.2.1 Correspondence by Parameterisation

The simplest approach to defining correspondence between one-dimensional shape contours is to select a starting point on each example and equally space an equal number of points on each boundary. A similar scheme was presented in [5], where spline control points were equally spaced around the contour of each training shape.

For shape surfaces however, equally spacing a fixed number of points is much more difficult. It can, however, be achieved by finding an explicit parameterisation of the surfaces (using a method such as that described in [14]), then equally spacing points in parameter space.

However, as we saw in the example in Sect. 3.1, equally spacing points on each shape does not necessarily give a reasonable groupwise correspondence, and can lead to very poor models.

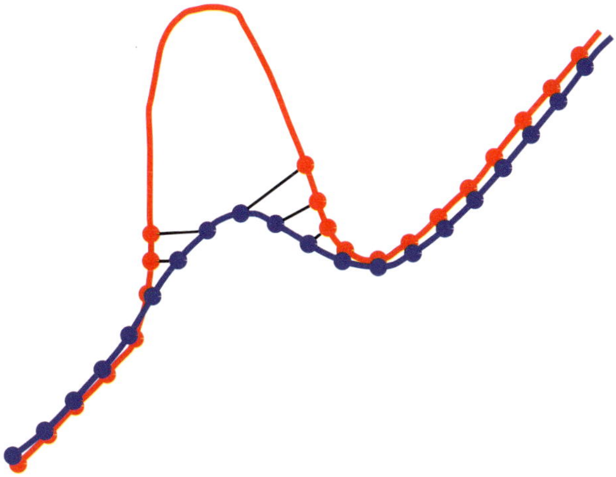

Fig. 3.2 An example of a poor distance-based correspondence between two shapes. For each point on the blue shape, the nearest point on the red shape was chosen as the corresponding point.

3.2.2.2 Distance-Based Correspondence

An intuitive notion of appropriate correspondence is that corresponding points should be in physical proximity when the shapes are mutually aligned. The iterative closest point (ICP) algorithm [8] for alignment of unlabelled point sets is one algorithm that implements such a definition of correspondence. As the name suggests, it is an iterative procedure that assigns an initial correspondence between the point sets, refines the alignment by means of minimising the distances between pairs of points, then re-computes the correspondence and repeats the alignment step until convergence.

More sophisticated algorithms that are also based on minimising point-to-point distances have been suggested by several authors (e.g., [63, 140, 92]). However, a general difficulty with such distance-based approaches is that the relative position of equivalent points may vary considerably over the training set, invalidating proximity as a satisfactory basis for establishing correspondence. An example of where a distance-based method produces an unsatisfactory correspondence is shown in Fig. 3.2.

A slightly different distance-based approach, but one worthy of mention, is the use of distance maps as a basis for establishing correspondence [77, 106]. Consider the shape and the space in which the shape lies. For each point in the space, we can assign a number, which is just the distance to the nearest point on the shape. This then defines a scalar field over the space, which is the *distance map*. If we regularly sample this distance map over some region, it reduces to a distance map image which includes the original shape. Correspondence is then established by performing image registration on the distance maps. However, image registration itself involves various arbitrary choices in terms of the choice of image similarity measure and the representation of the deformation field.

3.2.2.3 Feature-Based Correspondence

We have seen that distance-based correspondences can fail to establish a satisfactory correspondence in some situations. An alternative approach is to extract some features of the object and use these as a basis for establishing correspondence. Correspondence between these features can then be established using a generic numerical optimisation algorithm (e.g., the least-squares method) or by a specialized matching algorithm (e.g., [159, 161]).

A common approach to establishing correspondence is to use shape-based features that capture local properties of the object. Curvature is a popular shape-based feature (e.g., [24, 173, 192]) but, although it ties in with human intuition, equivalent points may not in practice lie on regions of similar curvature. Furthermore, curvature is a measure that tends to be sensitive to noise: errors in segmentation can lead to areas of (artificial) high curvature that complicate the matching process. Curvature has also been used as a means of obtaining an initial correspondence, which is subsequently refined (e.g., [17, 16, 93]).

Other shape-based features have also been proposed. Pitiot et al. [136], for example, describe the observed transport measure and Belongie et al. [7] use shape context. We will be considering these and other feature-based methods in greater detail in Chap. 4.

For the case where shapes are obtained from images, image data supplements the shape data, and can be used to create features. For example, Baumberg and Hogg [6] manipulated the positions of corresponding points so that image data was similar in the vicinity of corresponding points on different images. Several other approaches have also used image data as a basis for establishing correspondence, and these are reviewed below.

Features can also be combined. For example, in [118], a mixture of curvature, point-to-point Euclidian distance, and angle between normal vectors were combined to form an objective function. From a theoretical point of view however, such objective functions have the problem that they are a combination of incommensurate terms, with arbitrary weighting of individual terms.

In most cases, the most appropriate choice of feature depends on the nature of the object being modelled. It is therefore impossible to devise a feature that is generic, suitable for any class of object.

3.2.2.4 Correspondence Based on Physical Properties

We can also assign physical properties to an object, in order to guide the selection of correspondences. For example, Sclaroff and Pentland [158] assigned a dynamics to the points of each shape example, so that the original shape was the equilibrium configuration of the points. They then considered the non-rigid modes of vibration of each shape example, computed by building a finite-element model for each shape. Correspondence between points was then assigned based on similarity of displacement when the lowest modes of vibration were considered.

The thin-plate spline (TPS) has also been used as a basis for establishing correspondence. The thin plate spline is a spline interpolant, based on the physical analogy of bending a thin sheet of metal. It has been widely used in image registration and in shape analysis (see Appendix A and Sect. 4.1.2 for further details of the thin plate spline). The bending energy of the thin-plate spline deformation field can be used as an objective function for correspondence. Suppose we have two corresponding point sets on two shapes. We can then compute the TPS bending energy required to bring the two point sets into alignment. The correspondence can then be adjusted by sliding one set of points around on the shape boundary, so as to minimise this bending energy [12, 142], and hence locate the optimal TPS correspondence. Paulsen and Hilger [132] also used a thin plate spline warping. A model mesh was fitted to each training shape, controlled by the positions of some small set of manual anatomical landmarks. The fitted model mesh was then projected onto each surface, and a Markov random field was used to regularize the resultant correspondence. Lorenz and Krahnstover [109] also used a small number of manual landmark points to align a template to a training set of surfaces. An elastic relaxation was then performed to fit the template to each training shape, thus establishing correspondence.

It should be noted that even in the case where the training set of shapes is derived from the deformation of an actual physical object, the physical properties we assign to our shapes will not in general be the same as the physical properties of the actual object, hence the correspondence we obtain will not be that of the actual deforming object. And in many cases, the observed variations of shape do not correspond to the deformation of an actual object, but to the shape variation that arises as the endpoint of a whole process of biological development.

We hence see that the physical properties we assign to our shapes are either arbitrary (where different dynamics or a different choice of spline interpolant will give a slightly different result), or inappropriate. The resulting correspondence hence has to be considered as essentially arbitrary, despite the intuitive appeal of this approach.

3.2.2.5 Image-Based Correspondence

Most of the methods we have considered so far find correspondences on the boundaries of objects. If we have shapes derived from images of objects, then potentially, we have much greater information available about the object than just the shape of its boundary, we have all the information in the image of the object from which the shape was segmented.

In image registration, the image information is used to define a dense correspondence across the entirety of the image. It hence also defines a correspondence throughout the interior and exterior of the imaged objects. A volumetric shape model can then be built of this entire deformation field. Alternatively, the deformation of the surface of a specific structure can be derived from the volumetric deformation field.

If the original image is not available, a shape boundary itself can be used to generate a related image. This could be by use of the distance map mentioned previously, or more simply, by just using a binary image, which distinguishes between the exterior and interior of a closed shape. These images derived from the shape can then be used in image registration, and a correspondence hence derived [72]. A volumetric shape model can then be built of this entire deformation field (e.g., see [150]).

Where the original image containing the shape is available, the question we are actually posing is whether we should study the set of images themselves, or the shapes segmented from the images. This is not quite the same question as how to determine the correspondence across a set of shapes when no further information is present. The second scenario, of creating images from shapes, is potentially interesting. But again, there is an arbitrary element introduced, inasmuch as we have a choice as to what image we create from our shape. Do we use a simple binary interior/exterior image, do we use a distance map which uses the distance to the closest point on the shape, or some function of this distance? Do we include further information, such as the relation between the direction from the closest point to the local surface normal, and create a non-scalar image from our shape? We hence have an arbitrary choice to make as regards the image created from the shape. We then have additional arbitrary factors, in terms of the objective function we use for image registration, and the representation we choose for the dense deformation field of the image. All of these factors lead to an arbitrariness as regards the found correspondence.

We note here that the case of groupwise correspondence for images (that is, groupwise non-rigid image registration) is actually a slightly more complicated groupwise correspondence problem than the problem for shapes. For a shape, the only variation is spatial deformation, whereas for an image, we can both deform the image spatially by warping, as well as deforming the pixel/voxel values themselves. This type of groupwise correspondence problem can be tackled using the same ideas as those that we will develop for shape correspondence (for example, see [115, 180]), but further discussion is beyond the scope of the current volume.

In what follows, we will restrict ourselves to the case where all the information we have is the set of shape boundaries.

3.2.3 Summary

We have seen that there are various ways in which correspondence can be assigned. Although some of these methods produce acceptable results, there are still significant drawbacks. Many of the methods described here are essentially pairwise techniques, which deal with correspondence across a group by repeated application of pairwise correspondence. Other methods are limited to one-dimensional shapes or curves, which restricts their use in medical imaging applications where three-dimensional images, hence shape surfaces, are available. Some methods require manual intervention, which has the possibility of introducing operator error, even where the intervention is minimal. And other methods require a set of somewhat arbitrary assumptions and choices to be made. It is clear that none of these methods can be considered generic, suitable in principle for any class of object. Finally, and most importantly, the arguments as to why the correspondences generated by the above methods should be considered correct in some sense (rather than just acceptable in practice), are either weak or lacking.

In the next section, we begin to address these issues. We establish a generic framework for fully groupwise shape correspondence, that integrates the choice of correspondence into the model-building process.

3.3 Correspondence by Optimisation

To recap, we require a dense groupwise correspondence across our set of training shapes in order to build a statistical model from those shapes. As we have seen, different choices of correspondence give models of varying quality. The obvious approach is to introduce a feedback loop into the process. Given a correspondence, we construct a model using that correspondence. The properties of the resulting model should then feedback into our algorithm for determining the groupwise correspondence.

Such a process is inherently groupwise. There are tasks where pairwise correspondences between shapes are all that is needed. Pairwise correspondence algorithms can obviously be generalized to include the groupwise case, by either considering the correspondence between each training example and some reference shape, or by considering some large subset of the set of all possible pairwise correspondences between the training shapes. Such approaches are,

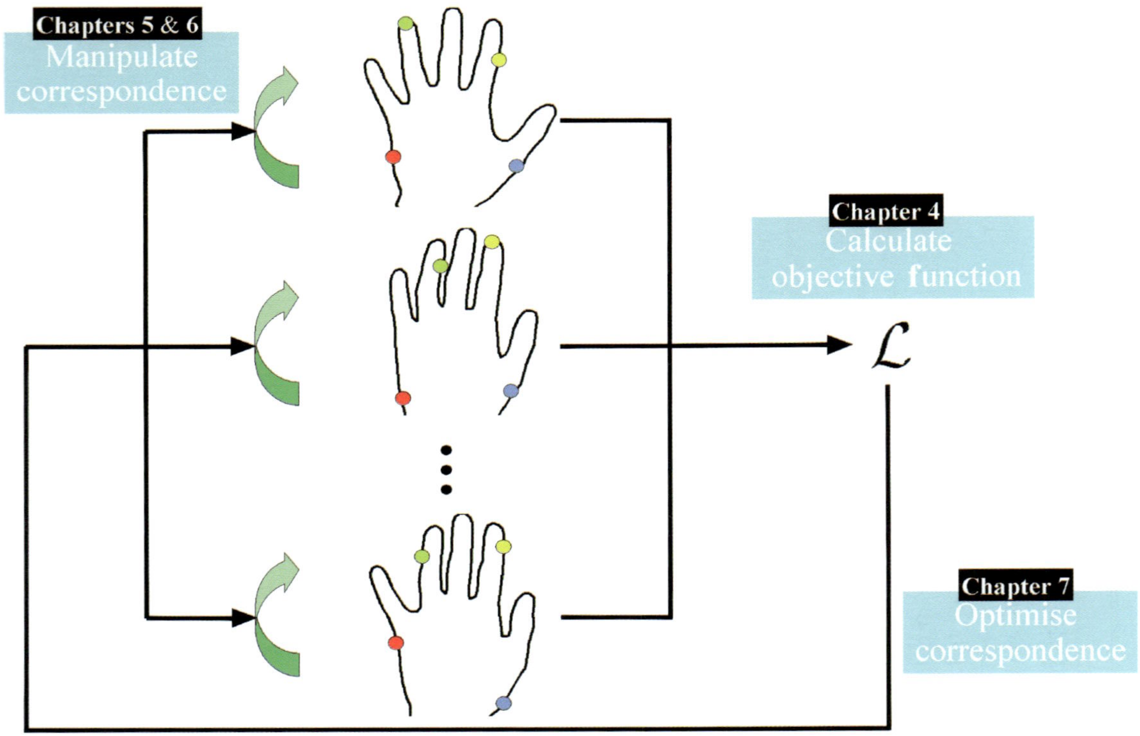

Fig. 3.3 A generic framework for establishing correspondence by optimisation. Correspondence (denoted by the coloured circles) is manipulated so as to minimise the value of an objective function, $\mathcal{L}$. Each component is labeled in the blue boxes, along with the chapter in which they are covered.

however, not ideal. Using a reference shape means that the groupwise correspondence is then consistent, but the choice of reference shape can itself bias the found correspondence. For a more complicated repeated pairwise method, it is difficult to ensure that the found pairwise correspondences are consistent across the entire set. Including the model in the location of correspondence is inherently groupwise, and given the target application, the fact that we cannot reduce this method to deal with the pairwise case is not important.

One advantage of including the model in the location of the correspondence is that the process of modelling is a generic one. Hence an algorithm for groupwise correspondence based on a model should also be generic.

Given the feedback loop proposed, the method is obviously going to be an iterative one. The obvious way to control such a process is by the use of an objective function. This then casts our correspondence problem as an explicit optimisation problem. Viewed in these terms, there are three essential components to our proposed method:

- **An objective function:** This assesses the qualities of the model built from a given set of correspondences.
- **A method for manipulating correspondences:** Given that we are proposing to optimise the correspondence across the entire training set, we obviously require a way to manipulate the groupwise correspondence that is both flexible and efficient.
- **An optimisation algorithm:** We need an optimisation algorithm that is able to locate the optimum correspondence in a reasonable time, even when we have large training sets.

Each of these components will be introduced below. Figure 3.3 illustrates diagrammatically how each of these components fits into the overall scheme. A detailed treatment of each of the components is given in the chapters indicated in the figure.

If we look back at the methods reviewed in the previous section, most of these can be cast in the form of optimisation problems. However, the problem is that most of these methods are inherently pairwise, which is not desirable, as was detailed above. The objective functions used are also rather ad hoc. Finally, these methods also lack an efficient means of manipulating correspondence.

3.3.1 Objective Function

The appropriate choice of an objective function is at the core of our approach. Note that we will assume that a lower value of the objective function is more desirable, hence we will refer to *minimisation* of the value of the objective function. As is shown in Fig. 3.3, objective functions are dealt with in detail in Chap. 4. We begin by looking at shape-based objective functions, some of which were encountered in the previous section. These objective functions measure only properties that can be evaluated directly on the training shapes and usually consider only a pair of shapes at a time.

However, what we are really interested in is building good models, and hence model-based objective functions. Rather than confining ourselves to measuring the utility of a model in a particular application, we instead aim to measure properties that any good model should have. These properties can be summarized as:

- **generalization ability**, so that it can represent any instance of the class of object;
- **specificity**, so that it can *only* represent valid instances of the modelled class of object;
- **compactness**, so that the model can be represented with few parameters.

Various ways of quantifying these properties are described in the second half of Chap. 4. The objective functions that perform best in practice are those based on ideas from information theory. The idea is that a good model should allow a concise description of all members of the training set. As in any branch of science, this is the essential rôle of a model – to account for a possibly large number of observations as manifestations of some underlying pattern which is itself described as simply as possible. Chapter 4 describes in detail how this idea can be formalized using the minimum description length (MDL) principle [144].

3.3.2 Manipulating Correspondence

The approach we have proposed entails a large-scale optimisation problem, which requires manipulation of the correspondence for each shape in the training set. We hence require an efficient method for manipulating correspondence so that we can locate the optimum of our chosen objective function within a reasonable time.

One possible approach is to place our n_P shape points on the surface of each training shape, and manipulate them directly. In effect, every time we move the points we would be trying to generate a diffeomorphism of the shape surface into itself, which is a different problem for each shape in the training set. Even just sliding a single point on the surface is complicated, since, in general, the point could move off the surface and would have to be projected back onto the shape surface.

A much more efficient and flexible approach is to treat the problem of corresponding continuous curves/surfaces as one of re-parameterisation [101]. In order to review the idea behind this approach, we need to revisit the parametric representation of shape introduced in Chap. 2. Recall that we represent each training shape S_i using a parametric shape-function $\mathbf{S}_i(\cdot)$ (2.117) that is defined by an initial one-to-one mapping $\mathfrak{X}_i$ from the parameter space X to the shape S_i:

$$X \overset{\mathfrak{X}_i}{\longmapsto} S_i, \ \mathbf{x} \overset{\mathfrak{X}_i}{\longmapsto} \mathbf{S}_i(\mathbf{x}). \tag{3.1}$$

The mapping $\mathfrak{X}_i$ thus associates a parameter value $\mathbf{x}$ to each point on the ith shape, with the coordinates of that point on the shape being the value of that shape function, $\mathbf{S}_i(\cdot)$. A correspondence between shapes can then be defined at points of the same parameter value $\mathbf{x}$, so that:

$$\mathbf{S}_i(\mathbf{x}) \sim \mathbf{S}_j(\mathbf{x}), \tag{3.2}$$

where $\sim$ denotes correspondence.

Given this parametric representation of shape, we are now in a position to manipulate correspondence by re-parameterisation. If ϕ_i is a re-parameterisation function for the ith shape, then the re-parameterisation is given by:

$$\mathbf{x} \overset{\phi_i}{\longmapsto} \mathbf{x}' \doteq \phi_i(\mathbf{x}), \tag{3.3}$$

where ϕ_i is a diffeomorphism of the *parameter space*. This mapping also acts on the shape-function $\mathbf{S}_i(\cdot)$, so that:

$$\mathbf{S}_i(\cdot) \overset{\phi_i}{\longmapsto} \mathbf{S}'_i(\cdot), \boxed{\mathbf{S}'_i(\mathbf{x}') \equiv \mathbf{S}'_i(\phi_i(\mathbf{x})) \doteq \mathbf{S}_i(\mathbf{x}).} \tag{3.4}$$

To clarify: the point under consideration on the actual shape $\mathbf{S}_i(\mathbf{x})$ does not change, but both the parameter value for that point $(\mathbf{x} \mapsto \mathbf{x}')$, and the shape-function describing the shape $\mathbf{S}_i(\cdot) \mapsto \mathbf{S}'_i(\cdot)$ do change.

This means that, when compared to another shape S_j (that is not being re-parameterised at the moment), the correspondence can be manipulated by varying ϕ_i:

$$\mathbf{S}_j(\mathbf{x}) \sim \mathbf{S}_i(\mathbf{x}) \overset{\phi_i}{\longmapsto} \mathbf{S}_j(\mathbf{x}) \sim \mathbf{S}'_i(\mathbf{x}) \ \Rightarrow \ \mathbf{S}_j(\mathbf{x}) \sim \mathbf{S}_i(\phi_i^{-1}(\mathbf{x})). \tag{3.5}$$

Note that it is sometimes more efficient from an implementational point of view to work with $\phi_i(\mathbf{x})$ rather than $\phi_i^{-1}(\mathbf{x})$, so that:

$$\mathbf{S}_j(\mathbf{x}) \sim \mathbf{S}_i(\phi_i(\mathbf{x})).$$

This is valid, since, by definition, any valid re-parameterisation function $\phi_i(\mathbf{x})$ is invertible, hence we are free to chose to explicitly represent either $\phi_i(\mathbf{x})$ or $\phi_i^{-1}(\mathbf{x})$ as convenient.

The general concept is illustrated in Fig. 3.4, which shows an open curve sampled according to different parameterisations. In practice, each shape in the training set $\{S_i : i = 1, \dots n_S\}$ has its own re-parameterisation function ϕ_i, which are all capable of changing correspondence.

In effect, what we have done is taken the difficult problem of directly generating a diffeomorphic mapping for each individual shape into itself every

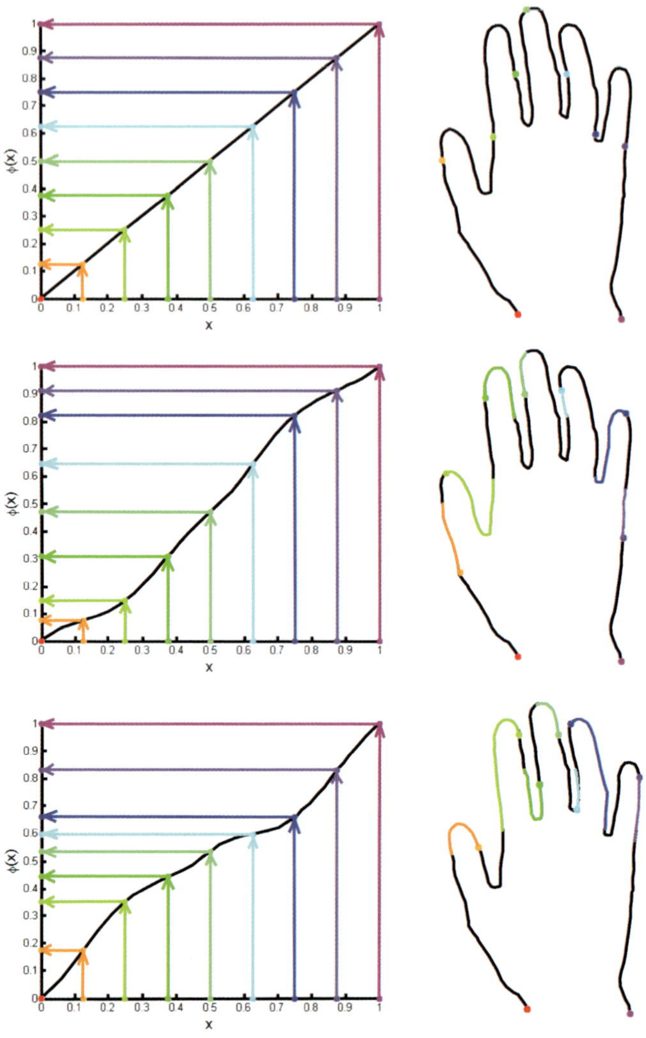

Fig. 3.4 Sampling a shape according to three different parameterisations. The left column shows the parameterisation and the right column shows the sample points on the shape. The top row shows the (original) identity parameterisation and the other two rows show how the points can be redistributed around the shape by changing the shape of the parameterisation function. Each sample point of the parameterisation and the corresponding point on the shape has been colour coded. On the shapes, the coloured lines show the displacement of the point caused by the manipulation of the original parameterisation.

time we adjust the correspondence, and replaced it by the task of only performing it once. This single diffeomorphic mapping for each shape is from the shape to the parameter space, performed once at the start during the initialization phase. Once this initial parameterisation has been constructed, diffeomorphic mappings of each shape surface into itself are generated by considering instead diffeomorphic mappings of the parameter space into itself, and mapping the result back onto the shape surface using the *fixed* initial parameterisation. This is computationally efficient, since the parameter space is much *simpler* than the shapes themselves, and is the *same* for each shape in the training set. For example, for a set of shapes topologically equivalent to spheres, the parameter space is just the unit sphere. Whereas for open or closed lines, the parameter space is either an open line segment, or the unit circle.

We hence need to be able to generate diffeomorphic mappings $\{\phi_i\}$ of the appropriate parameter space. This can be achieved in two ways. The first approach is to develop a representation that is limited to some set of parametric transformations – this can be thought of as hard regularization and is covered in Chaps. 5 and 6 for curves and surfaces, respectively. Soft regularization is an alternative approach where more flexible transformations can be generated, providing it is supported by the data, and this alternative approach is covered in Chap. 8.

3.3.3 Optimisation

Now that we have chosen an objective function and constructed a representation of our re-parameterisation functions, we next need to optimise the objective function with respect to the set of re-parameterisation functions. A simple illustration of this process is given in Fig. 3.5.

For the case of parametric re-parameterisation functions, the number of parameters required to represent the re-parameterisation functions is large. This leads to a difficult high-dimensional optimisation problem. Furthermore, the groupwise objective functions harbour many false minima, which causes standard optimisation algorithms to fail. As a result, a specialized algorithm that exploits properties specific to statistical shape models is required to tackle the problem.

Much work has been done on finding a tractable algorithm and this is described in Chap. 7. The chapter also considers practical issues of the optimisation, and gives detailed implementation examples, including step-by-step details of how a model can be built from typical sets of shapes.

Alternative representations of shape and a non-parametric representation of re-parameterisation are explored in Chap. 8. These representations then require a different approach to optimisation, which is also covered in the chapter.

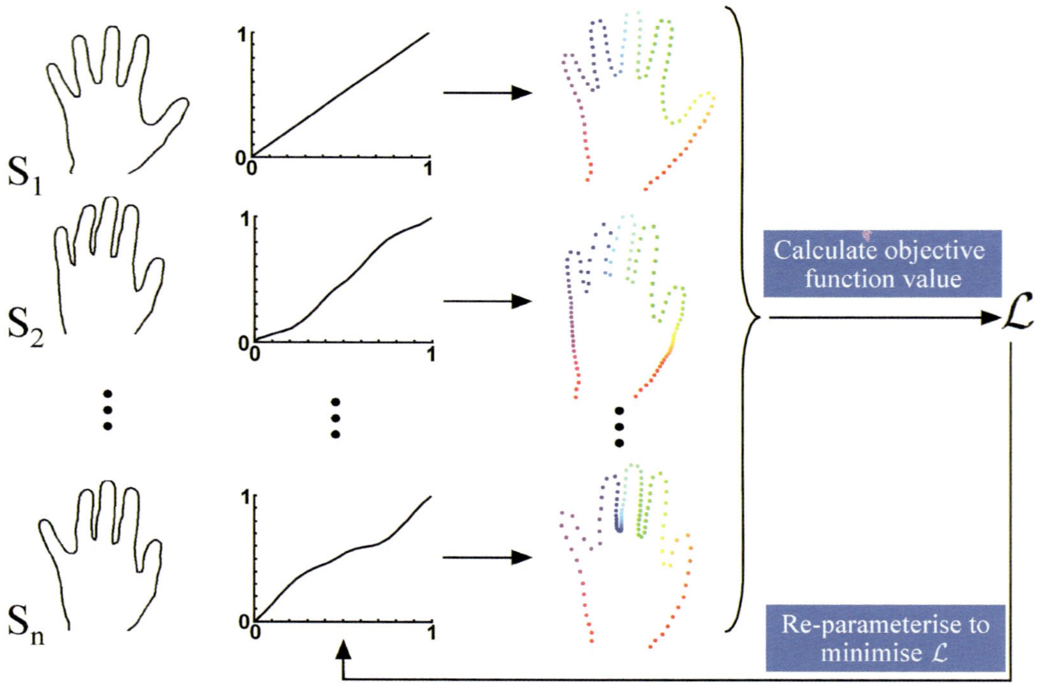

Fig. 3.5 An illustration of the adopted optimisation framework for establishing correspondence. A set of shapes, $\{S_i\}$, are sampled according to their parameterisations. The parameterisation of each shape can be manipulated by a re-parameterisation function, ϕ_i, which also changes the position of the sampled points. Optimisation is performed using an optimisation algorithm to find the set of re-parameterisation functions $\{\phi_i\}$ that minimise the objective function value, $\mathcal{L}$. Colour denotes correspondence.

To summarize, we have provided a brief review of conventional approaches to shape correspondence, and shown why these approaches our not suitable for our groupwise task of model-building. We then described the salient details of our framework for groupwise correspondence by optimisation. The details of each of the components of our framework will be given in the relevant later chapters, as detailed above.

Chapter 4
Objective Functions

An essential component of the optimisation approach to establishing groupwise correspondence is of course the objective function that quantifies what we mean by the quality of the correspondence.

In the literature, correspondence between shapes or images is often meant purely in the pairwise sense, using pairwise objective functions. Some of these pairwise objective functions can then be generalized to the case of groupwise correspondence. There also exist objective functions that are defined purely for the groupwise case. However, this is not the classification that we will adopt here.

We will instead consider the two classes of either *shape-based* or *model-based* objective functions. Shape-based correspondence, as the name implies, tends to assign correspondence by considering localized properties that can be measured directly on the training shapes. Whereas model-based objective functions consider instead the groupwise model that can be built from the correspondence.

Another way to view this distinction is that shape-based objective functions tend to maximise some measure of *similarity* between shapes, whereas the model-based approach includes not just the similarities between shapes, but also the statistics of the *dissimilarities*. In fact, a model can be viewed as a mathematical description of the essential dissimilarity across a group of shapes. This difference means that it is simple to define similarity between even a pair of shapes, whereas the statistics of dissimilarity can only be meaningfully evaluated if we have a group of examples, and hence is limited to the case of groupwise correspondence.

We begin by considering the shape-based objective functions.

R. Davies et al., *Statistical Models of Shape*,
DOI: 10.1007/978-1-84800-138-1_4,

4.1 Shape-Based Objective Functions

The notion of similarity between shapes is a very intuitive one, and a concept that the human visual processing system implements rather well. Objective functions built based on these intuitive notions are usually pairwise, since this is the simplest way to define similarity. But these can of course be trivially extended to the groupwise case, by quantifying groupwise similarity as just the appropriate sum over pairwise similarity measures.

All the shape-based objective functions that will be considered here can be expressed in terms of either discrete (Sect. 2.1) or continuous (Sect. 2.3) representations of shape and shape correspondence, and can be generalized to encompass both curves in two dimensions, or surfaces in three dimensions. For brevity, we will focus on the case where the shapes are finite-dimensional representations of curves in two dimensions, and the extensions to the other cases will only receive explicit mention when they are non-trivial.

4.1.1 Euclidian Distance and the Trace of the Model Covariance

We have already encountered a case of maximising similarity between shapes. In Sect. 2.1.1 a measure of similarity was used to align pairs or groups of shapes, with the correspondence held fixed. The transformation used was composed of translations, scalings and rotations (the appropriately named *similarity* transformations).

As before, we take $\mathbf{x}^{(i)}$ to denote the position of the ith shape point on the shape $\mathbf{x}$. If $\mathbf{y}^{(i)}$ is the corresponding point on the *transformed* version of the other shape, then our previous objective function for alignment can be written in the form:

$$\mathcal{L} = \sum_{i=1}^{n_P} ||\mathbf{x}^{(i)} - \mathbf{y}^{(i)}||^2, \tag{4.1}$$

which is just the square of the Euclidean distance between the shape vectors $\mathbf{x} = \{\mathbf{x}^{(i)} : i = 1, \ldots n_P\}$, and $\mathbf{y} = \{\mathbf{y}^{(i)} : i = 1, \ldots n_P\}$. The meaning of this objective function as quantifying some intuitive concept of similarity is obvious.

The key point about the similarity transformations used for alignment is that they change the positions of the shape points, whilst retaining the shape. But there is another way of manipulating the shape points whilst maintaining the shape, which is just the manipulation of correspondence. Hence this objective function can also be used to optimise correspondence, not just for the case of shape alignment.

We could generalize this pairwise Euclidean distance to the groupwise case by considering the distances between all possible pairs of shapes. However, this quickly becomes prohibitive, since the number of pairs rises quadratically with the number of examples. A simpler method is to consider instead just the distances between each shape and the mean, which is linear in the number of examples. So, just as this Euclidean distance was used in Algorithm 2.1 to align a set of shapes, so it can also be used to quantify similarity, hence determine correspondence, across a group of shapes.

If $\mathbf{x}_j$ now denotes the jth shape in our training set,[1] then summing the squares of the distances between each training shape and the mean, we obtain:

$$\bar{\mathbf{x}} \doteq \frac{1}{n_S}\sum_{j=1}^{n_S}\mathbf{x}_j, \quad \mathcal{L} = \sum_{i=1}^{n_S}\sum_{j=1}^{n_P}||\mathbf{x}_i^{(j)} - \bar{\mathbf{x}}^{(j)}||^2, \tag{4.2}$$

where $\bar{\mathbf{x}}^{(j)}$ is the jth point of the mean shape. This is then just the square of the Euclidean distance in the full shape space $\mathbb{R}^{dn_P}$.

By considering PCA (Theorem 2.1 and in particular (2.30)), it is simple to show that this objective function is just the trace of the covariance matrix $\mathbf{D}$ of the training set. That is:

$$\mathcal{L} = \text{Tr}(\mathbf{D}) = \sum_{a=1}^{n_S-1}\lambda_a, \tag{4.3}$$

where $\{\lambda_a\}$ are the eigenvalues of the covariance matrix. In this form, this objective function has been used extensively to establish correspondence (e.g., [92, 140, 6]). The objective function is minimised by moving points as close as possible to the corresponding points on the mean shape, whilst sliding them along the shape, and it directly minimises the total variance of the model.

However, its use is not without problems. An example of the failure of this objective function was given in [101], an example that we will recreate here.

Consider the set of artificial shapes shown in Fig. 4.1. They consist of a rectangular box, with a semi-circular bump of constant size placed on the upper edge (hence the name *box-bumps*). A set of such shapes is generated, such that the only structural variation is the horizontal translation of the semi-circle along the top of the box.

The correct correspondence is established by fixing the points at the corners of the box and at the two ends of the semi-circle. This then produces a correct model with the expected single mode of variation, as is shown in the figure. A naïve, but incorrect, attempt at defining a correspondence would be to define it according to the distance along the shape. A single point is

[1] Remember that, as defined previously in Chap. 2, $\mathbf{x}^{(i)}$ denotes the ith shape point on an entire shape $\mathbf{x} = \{\mathbf{x}^{(i)} : i = 1, \dots n_P\}$, whereas $\mathbf{x}_i$ denotes the ith *entire* shape in a collection of such shapes. We hope the brackets will aid the reader in keeping in mind the two different meanings.

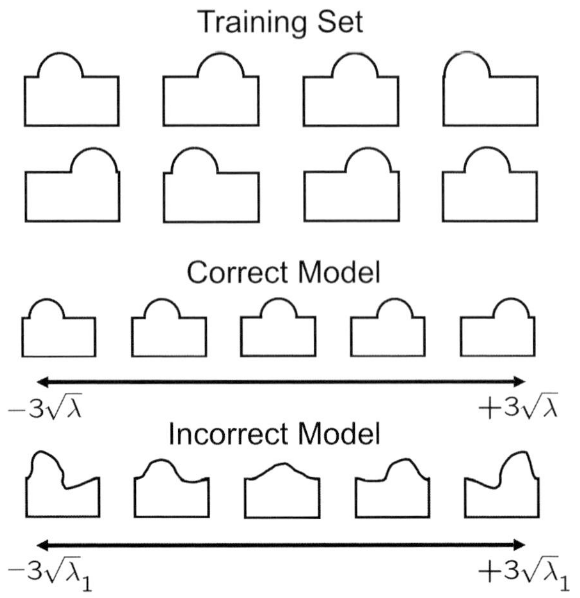

Fig. 4.1 Top: A training set of synthetic box-bump shapes. **Middle:** The model built using the correct correspondence, which has just one mode of variation. **Bottom:** The first mode of a model built using an incorrect correspondence. For both models, examples generated by varying the first mode within the range $\pm 3\sqrt{\lambda_1}$ are shown.

fixed on the top left-hand corner of the box and the correspondence of the rest of the shape is assigned by spacing points equally around each shape. This is usually referred to as *arc-length parameterisation.* The first mode of the model built according to this incorrect correspondence is also shown in the figure. It is obvious that this is a very poor model – even when using just the first mode, it creates examples totally unlike those seen in the training set (that is, it is not specific).

However, if we calculate the trace of the model covariance for both models, we find that the correct model yields a value of 0.4973, whereas the value for the incorrect model is 0.3591. That is, this objective function prefers the inferior model, which is a very obvious failure.

We hence see that using this simple definition of physical distance between shapes as a measure of similarity is not reliable, even though it is intuitive, and simple to implement.

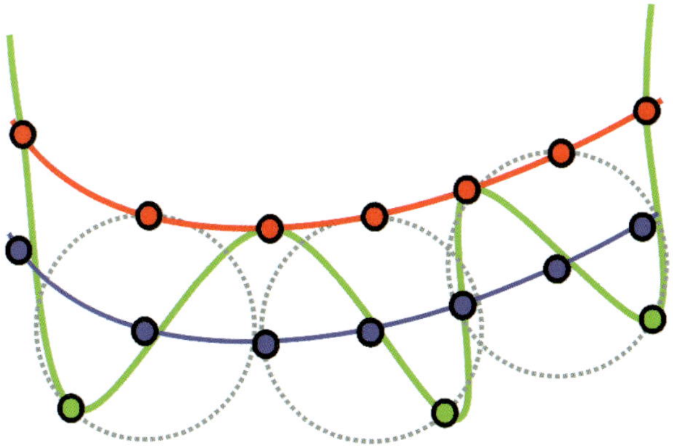

Fig. 4.2 Two shapes (red and green) are each to be compared with another shape (blue). The Euclidean point-to-point distance from the red or the green shape to the blue shape is the same (as indicated by the construction lines in grey), even though in terms of intuitive ideas of shape similarity, we would want the red shape to be closer than the green shape.

4.1.2 Bending Energy

The previous objective function considered the physical distance between actual shapes as a measure of similarity. As noted above, there are cases where this Euclidean distance fails. But there are other problems with Euclidean distance, in that by considering only pairwise distances between points, it neglects much of what we intuitively think of as shape similarity or dissimilarity. An example is shown in Fig. 4.2, where the red and green shapes are equal Euclidean distances from the blue shape, yet the red example is actually a much better fit in terms of *shape* similarity. And adding more shape points doesn't help, since then the red example is actually further away than the green example.

In order to distinguish cases that Euclidean distance cannot, an obvious approach is to consider the physical idea of deforming one shape so that it matches the other exactly. If we assign an energy to this deformation, then the deformation energy becomes our measure of shape similarity, with low energies indicating high similarity, and vice versa.

Let us consider a pair of *aligned* shapes as two sets of points $\{\mathbf{x}^{(i)}, \mathbf{y}^{(i)}\}$, $i = 1, \dots n_P$, lying in $\mathbb{R}^d$. For example, we could consider one-dimensional shapes lying in $\mathbb{R}^2$ or $\mathbb{R}^3$, or two-dimensional shape surfaces lying in $\mathbb{R}^3$. The whole space is then deformed so that a general point $\mathbf{x} \in \mathbb{R}^d$ moves to:

$$\mathbf{x} \mapsto \mathbf{x} + \mathbf{u}(\mathbf{x}),$$

where $\mathbf{u}(\mathbf{x})$ is the deformation field. We suppose that the points of the shape $\{\mathbf{x}^{(i)}\}$ are carried along with the deformation, so that they move so as to coincide exactly with the points of the other shape $\{\mathbf{y}^{(i)}\}$. That is:

$$\mathbf{x}^{(i)} + \mathbf{u}(\mathbf{x}^{(i)}) = \mathbf{y}^{(i)}.$$

An objective function for the bending energy of the deformation field can be written in the form:

$$\mathcal{L} = \int_{\mathbb{R}^d} d\mathbf{x}\ (L\mathbf{u}(\mathbf{x})) \cdot (L\mathbf{u}(\mathbf{x})) + \sum_{i=1}^{n_P} l_i \left\| \mathbf{u}(\mathbf{x}^{(i)}) + \mathbf{x}^{(i)} - \mathbf{y}^{(i)} \right\|^2. \tag{4.4}$$

L is some scalar differential operator, that defines the bending energy density of the deformation field.[2] The $\{l_i\}$ are a set of Lagrange multipliers, which ensure that the shapes match after the deformation. In general, this is just the formulation of a spline interpolant, where the exact form of the differential operator L determines the exact nature of the spline.[3]

For example, we could take $L = \nabla^2$, which would give (according to the boundary conditions that were also imposed on the deformation field), either the biharmonic thin-plate [62, 11] or the biharmonic clamped-plate [9, 114] spline interpolant of the deformation of the shape points.

There are deformation fields $\mathbf{u}^{(0)}(\mathbf{x})$ for which $L\mathbf{u}^{(0)}(\mathbf{x}) \equiv \mathbf{0}$, which hence make no contribution to the bending energy. For example, if $L = \nabla^2$, these transformations are just the linear transformations. In general, these deformation fields correspond to alignment of the shapes. As stated previously, we assume, for the sake of simplicity, that the shapes have already been aligned.

The general problem can be solved using the method of Green's functions, as discussed in Appendix A. The optimal matching deformation field is computed by taking the functional derivative $\frac{\delta \mathcal{L}}{\delta \mathbf{u}(\mathbf{x})}$ and equating it to zero. The resulting optimal value of the energy is just a function of the two original shapes, and for our aligned shapes, can be written in the form:

---

[2] Note that this differential operator L is not the same as that referred to in Appendix A. They are related by the fact that we can pass derivatives across within the bending-energy integral, so that:

$$\int_{\mathbb{R}^d} d\mathbf{x}\ (L\mathbf{u}(\mathbf{x})) \cdot (L\mathbf{u}(\mathbf{x})) = \int_{\mathbb{R}^d} d\mathbf{x}\ \mathbf{u}(\mathbf{x}) \cdot (L^\dagger L\mathbf{u}(\mathbf{x})) + \text{boundary terms},$$

where $L^\dagger$ is the Lagrange dual of L. This then gives the related (self-dual) differential operator $\mathfrak{L} = L^\dagger L$ as used therein.

[3] If the matching is not exact, but there is some trade-off between the energy of the deformation field, and the closeness of the match, then we have a smoothing rather than an interpolating spline. Mathematically, it is the difference between the formal optimisation of the bending energy given, where the Lagrange multipliers are also variables to be optimised over, and the optimisation of the same expression, but where the $\{l_i\}$ take definite, *finite* values. The exact values chosen determine the relative trade-off between bending energy and degree of match at each point.

$$\mathbf{G}_{ij} \doteq G(\mathbf{x}^{(i)}, \mathbf{x}^{(j)}),$$

$$\mathcal{L}_{\text{opt}} = \sum_{i,j=1}^{n_P} \mathbf{u}(\mathbf{x}^{(i)})\mathbf{G}_{ij}^{-1}\mathbf{u}(\mathbf{x}^{(j)}) \equiv \sum_{i,j=1}^{n_P} \sum_{\mu=1}^{d} u_\mu(\mathbf{x}^{(i)})\mathbf{G}_{ij}^{-1} u_\mu(\mathbf{x}^{(j)}),$$

where $G(\cdot, \cdot)$ is the appropriate Green's function. Note that the matrix $\mathbf{G}$, hence $\mathbf{G}^{-1}$ depends only on the positions of the points on only one of the shapes. This means that in general this bending-energy similarity measure is not *symmetric*, in that the energy differs depending on which shape we define as the fixed shape, and which the deforming shape. Although it is of course possible to define a symmetric version by repeating the procedure with the shapes swapped.

The use of such bending energies to match shapes and to define correspondence has been considered by several authors (e.g., [12, 142]). The necessary computation of the inverse of the Green's function matrix $\mathbf{G}^{-1}$ involves the inversion of an $n_P \times n_P$ matrix, which is of order n_P^3. There are approximate methods [61] which can reduce this to $(0.1 n_P)^3$, but this still means that the method is unsuitable if we wish to consider the continuum limit $n_P \to \infty$.

Another problem occurs if we wish to consider the correspondence across a group of shapes. Let us suppose that we had just three shapes A, B, and C. We could match A to B and C to B, and hence infer a correspondence between A and C. However, it is not guaranteed that this inferred correspondence would be consistent with the answer we would obtain if we matched A to C directly. And this problem potentially occurs when we wish to generalize any pairwise algorithm to the groupwise case in this manner.

Let us return to the Euclidean distance method we considered previously. We see that one problem is that it only considers proximity measured at discrete points as a measure of similarity. If we are going to compute point-based measures of similarity, it could be argued that what we should be comparing is the *local* shape in the region of each point, rather than just the distance between points. Such a method would be able to differentiate between the cases considered in Fig. 4.2. We will now consider a few such methods.

4.1.3 Curvature

Let us consider a point on a shape, and the shape in the vicinity of that point. To lowest order we have just the position of this point, as was considered in the case of Euclidean distance. To next order, we have the gradient of the shape at that point. And taking the next order gives quantities formed from the second-derivatives of the shape, which are the various measures of curvature.

Curvature is a shape descriptor with an intuitive appeal. It has the advantage that we can construct scalar measures of curvature. Consider a curve in two dimensions $\mathbb{R}^2$, where the curve has a parametric description in terms of Cartesian coordinates $(x(t), y(t))$. The curvature at a point is given by:

$$\kappa \doteq \frac{1}{R} = \frac{|x_t y_{tt} - y_t x_{tt}|}{(x_t^2 + y_t^2)^{\frac{3}{2}}},$$

where R is the radius of curvature, and x_t, x_{tt} denote the derivatives $\frac{dx(t)}{dt}$ and $\frac{d^2x(t)}{dt^2}$, respectively.

For shape surfaces, we obviously obtain different radii of curvature at a point depending on the direction we chose. The maximum and minimum values of the curvature are called the *principal curvatures*, and we can take either the arithmetic (the *mean curvature*), or the geometric (the square-root of the *Gaussian curvature*) mean of the principal curvatures as our single curvature measure.

A simple pointwise comparison of curvatures between two shapes A and B can then be made using a sum-of-squares objective function:

$$\mathcal{L} = \sum_{j=1}^{n_P} (\kappa_A^{(j)} - \kappa_B^{(j)})^2, \tag{4.5}$$

where $\kappa_A^{(j)}$ is the curvature measured at point j on shape A. Such an objective function obviously tends to establish correspondence between points of similar curvature. Curvature can also be used as one component of more complex objective functions (e.g., [24, 118, 73, 97]).

For some classes of object, such as the synthetic shapes in Fig. 4.1, curvature tends to be similar at corresponding points, but this is not true for all classes of shape. Computing curvature can also be computationally problematic, since it requires the estimation of second derivatives on a shape. If we have a noisy shape, high-curvature points can arise, purely through noise, and this can lead to these points having a disproportionate effect on the final correspondence.

4.1.4 Shape Context

Curvature at a point is a very local description of shape. Belongie et al. [7] developed the idea of *shape context*. The basic idea is that this is some description of the rest of the shape, from the viewpoint of the particular point in question. This work was based on the approach of pairwise geometric histograms, developed by Thacker et al. [176] in 1995.

For each point in turn, the relative vector position of all other points on the same shape is computed. This set of vectors is converted into log-polar

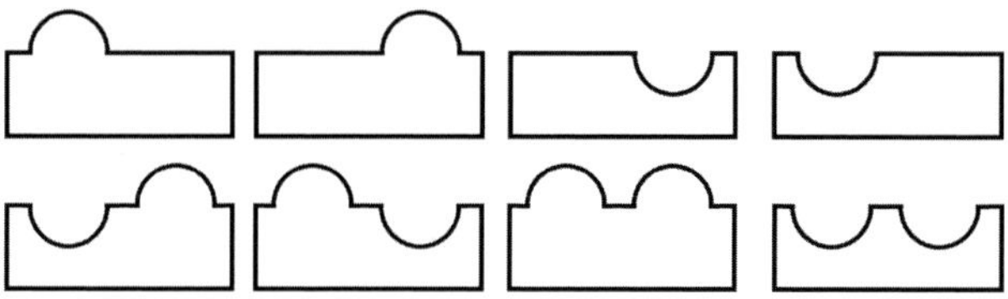

Fig. 4.3 Example double box-bump shapes. If we consider pairwise matches between shapes on the top row, it could be interpreted as a single bump moving and flipping. Whereas if we also include groupwise information from the shapes on the bottom row, we see that a more correct interpretation is of two bumps, which can appear, disappear, and flip.

coordinates, and the distribution is recorded using a histogram. The similarity of a given pair of points, one on each shape, is computed by comparing the histograms at those points. The histograms are compared using the χ-squared statistic:

$$\mathcal{L} = \sum_{j=1}^{n_P} \sum_{k=1}^{n_{bins}} \frac{\left(h_A^{(j)}(k) - h_B^{(j)}(k)\right)^2}{h_A^{(j)}(k) + h_B^{(j)}(k)}, \tag{4.6}$$

where $h_A^{(j)}(k)$ is the occupancy count of the kth bin of the histogram at point j on shape A.

Many other shape-based objective functions have also been proposed (e.g., [135, 74, 80]), using different measures of similarity. But none produce correspondences that can be considered as correct for all classes of object.

We have already mentioned problems with trying to generate groupwise correspondence using objective functions that are defined for generating correspondence for a pair of shapes. The naïve approach of defining groupwise correspondence via repeated-pairwise correspondence has an obvious problem, in that it is only using a limited amount of the total available information. From a theoretical point of view, the match between any pair of shapes really needs to be determined within the wider context of the group. It is easy to imagine situations where matching is ambiguous just in a pairwise context, whereas introducing information from the entire group can help to remove this ambiguity. Consider the simple example illustrated in Fig. 4.3. If we use pairwise matches between shapes on the top row, we could get a pairwise correspondence that would lead to the *groupwise* interpretation of a single bump moving and inverting. If we then tried to match a shape from the top row to one on the bottom row, we could have a possible ambiguity. From the point of view of curvature matching say, we cannot distinguish between a convex and a concave bump. Whereas if we also included information on shapes from the bottom row, we would instead obtain the correct groupwise interpretation and correspondence in terms of two bumps, that could appear, disappear, or invert.

We hence progress to the discussion of fully groupwise objective functions.

4.2 Model-Based Objective Functions

As we stated at the start of this chapter, we have divided objective functions into those which measure *similarity* between shapes, such as those discussed above, and those that also consider the statistics of the *dissimilarity* across a set of shapes.

The simplest way to quantify the statistics of dissimilarity is by considering the second-order statistics of the training set, which is described by either the $dn_P \times dn_P$ covariance matrix $\mathbf{D}$ (2.16) or the $n_S \times n_S$ covariance matrix $\widetilde{\mathbf{D}}$ (2.98):

$$D_{\mu\nu} \doteq \sum_{i=1}^{n_S} (\mathbf{x}_i - \bar{\mathbf{x}})_\mu (\mathbf{x}_i - \bar{\mathbf{x}})_\nu, \quad \widetilde{D}_{ij} \doteq (\mathbf{x}_i - \bar{\mathbf{x}}) \cdot (\mathbf{x}_j - \bar{\mathbf{x}}).$$

We first consider a simple objective function built from the determinant of the covariance matrix. We then consider a slightly more sophisticated treatment, which uses a model built from the covariance matrix in a statistical technique known as bootstrapping. Finally, we consider an information-theoretic objective function based on the minimum description length (MDL) principle. This is presented in some detail, along with various approximations to the MDL objective function.

4.2.1 The Determinant of the Model Covariance

The first groupwise objective function was proposed by Kotcheff and Taylor [101]. It is commonly referred to as the determinant of the covariance matrix, $\mathbf{D}$, except this description has one major flaw, in that the determinant of the covariance matrix is in general zero, due to the presence of zero eigenvalues. What is actually meant is:

$$\mathcal{L}_{\text{det}} = \log(\det(\mathbf{D})) = \log\left(\prod_{a=1}^{n_m} \lambda_a\right) = \sum_{a=1}^{n_m} \log(\lambda_a), \tag{4.7}$$

but only including the set of eigenvalues $\{\lambda_a\}$ that are *non-zero* eigenvalues of $\mathbf{D}$. This objective function can be thought of as measuring the volume that the training set occupies in shape space, and it hence tends to favour compact models.

In Appendix B, we calculate the gradient of this objective function with respect to variation of the training shapes (B.24), which gives the result:

$$\frac{\delta \mathcal{L}_{\text{det}}}{\delta \mathbf{x}_i} = \sum_{a=1}^{n_m} \frac{1}{\lambda_a} \frac{\delta \lambda_a}{\delta \mathbf{x}_i} = \sum_{a=1}^{n_m} \frac{2b_a^{(i)}}{\lambda_a} \mathbf{n}^{(a)}. \tag{4.8}$$

This gradient has the property that it is in the *same* direction as the gradient of the Mahalanobis distance at the position of the shape being perturbed.[4] It will hence tend to shrink the modes with smaller variance before it shrinks the modes with larger variance, having the net effect of concentrating the variance into the larger modes.

Compare this with the gradient of the trace of the covariance matrix:

$$\mathcal{L}_{\text{Tr}} \doteq \sum_{a=1}^{n_m} \lambda_a \;\Rightarrow\; \frac{\delta \mathcal{L}_{\text{Tr}}}{\delta \mathbf{x}_i} = \sum_{a=1}^{n_m} 2 b_a^{(i)} \mathbf{n}^{(a)}.$$

It is the additional factor of $\frac{1}{\lambda_a}$ that makes the difference between the gradient of the Euclidean distance for the case of the trace (Sect. 4.1.1), versus the gradient of the Mahalanobis distance for the determinant. The Mahalanobis distance is superior to the Euclidean distance, in that it is a distance that incorporates the knowledge about the distribution of the training set. However, inspection of (4.7) shows that the objective function has a degenerate minimum when any eigenvalue approaches zero. To overcome this, Kotcheff and Taylor added a small regularization constant ϵ:

$$\mathcal{L}_{\text{det}} = \log(\det(\mathbf{D} + \epsilon \mathbb{I})) - n_S \log \epsilon = \sum_{a=1}^{n_S} \left(\log(\lambda_a + \epsilon) - \log \epsilon\right), \qquad (4.9)$$

where $\mathbb{I}$ is the identity matrix. A non-zero value of ϵ can be seen as a measure of uncertainty as to whether an eigenvalue λ_a can actually be taken to be zero. It can hence be linked to an estimate of the noise on our training shapes, the idea being that if there is noise on the training data, eigenvalues that should be zero will actually be non-zero, and this size of the smallest eigenvalues will be determined by the size of the noise.

In summary, this objective function has an intuitive appeal, but there is no rigorous justification for its choice. It explicitly favours compact models, but there is no particular reason to suppose that it will favour models with other desirable properties.

As regards simple objective functions based on the covariance matrix, the sum and product of the eigenvalues (the trace and determinant of the covariance matrix) just about exhausts our options. We hence proceed to considering more sophisticated objective functions, based not on the covariance matrix per se, but on the statistical model built using this covariance matrix.

[4] Kotcheff and Taylor stated this result, but their proof was flawed; see Appendix B for a detailed analysis.

4.2.2 Measuring Model Properties by Bootstrapping

Properties of our statistical model of the data can be estimated by bootstrapping. The idea is to build a model pdf from the current correspondence, and then use this pdf to stochastically generate new shape instances. A detailed comparison between the generated shapes and the original training data can then be performed.

We will look at how bootstrapping can be used to measure two properties that are essential to a good model: specificity and generalization ability.[5]

A specific model should only generate instances of the object class that are similar to those in the training set – thus specificity is crucial if the model is to be used in applications such as image segmentation. The generalization ability of a model measures its ability to represent unseen instances of the class of object modelled – this is a fundamental property as it allows a model to learn the characteristics of a class of object from a limited training set. If a model is over-fitted to the training set, it will be unable to generalize to unseen examples.

Let us now look at the precise details of how each property can be estimated in practice.

4.2.2.1 Specificity

Specificity can be assessed by generating a population of instances using the model and comparing them to the members of the training set. Suppose we generate M examples from the model pdf $\{\mathbf{y}_A : a = 1, \ldots M\}$, using only n_m modes of the model. We define a quantitative measure of specificity:

$$\hat{S}(n_m) \doteq \frac{1}{M} \sum_{A=1}^{M} \min_i \|\mathbf{y}_A - \mathbf{x}_i\|. \tag{4.10}$$

That is, for each member of the sample set $\{\mathbf{y}_A\}$, we find the distance to the nearest element of the training set. Obviously, the more specific a model, the smaller these distances.

4.2.2.2 Generalization Ability

The generalization ability of a model is measured from the training set using leave-one-out reconstruction. A model is built using all but one member of the training set and then fitted to the excluded example. The accuracy to

[5] In Chap. 9, we give a detailed analysis of these measures in the context of model evaluation. More sophisticated measures are also considered there, but in the present context, we will consider just the naïve measures.

Algorithm 4.1 : Leave-One-Out Generalization Ability of a Model.

- For $n_m = 1, \ldots n_s - 2$:
  - For $i = 1, \ldots n_s$:
    1. Build the model $(\bar{\mathbf{x}}^{(i)}, \mathbf{N}^{(i)} = \{\mathbf{n}^{(a)} : a = 1 \ldots n_m\})$ from the training set with $\mathbf{x}_i$ removed.
    2. Calculate the parameter vector for shape i: $\mathbf{b^{(i)}} = (\mathbf{N^{(i)}})^{\mathbf{T}}(\mathbf{x_i} - \bar{\mathbf{x}}^{(\mathbf{i})})$.
    3. Reconstruct the unseen shape using n_m shape parameters:

    $$\tilde{\mathbf{x}}_i(n_m) = \bar{\mathbf{x}}^{(i)} + \sum_{a=1}^{n_m} b_a^{(i)} \mathbf{n}^{(\mathbf{a})}.$$

    4. Calculate the sum of squares approximation error:

    $$\epsilon_i^2(n_m) = \|\mathbf{x}_i - \tilde{\mathbf{x}}_i(n_m)\|^2.$$

  - Repeat
  - Calculate the mean squared error: $G(n_m) = \frac{1}{n_s} \sum_{i=1}^{n_s} \epsilon_i^2(n_m)$.
- Repeat

which the model can describe the unseen example is measured and the process is repeated, excluding each example in turn. The approximation error is averaged over the complete set of trials. The generalization ability is again measured as a function of the number of shape parameters, n_m, used in the reconstructions. The basic procedure is given in Algorithm 4.1.

Although these measures give an intuitive indication of model performance, they suffer from three problems. The first is that each measure is a function of the number of modes retained in the model. However, if any number except the complete set of modes is used then the optimisation can 'cheat' by hiding variation in the excluded modes. A bigger problem is that optimising specificity will not necessarily produce a general model and vice versa. For example, it is simple to see that the most specific model possible for a given set of training data is just a sum of δ-function pdfs at each data point. But qualitatively, such a model has no generalization ability whatsoever! This limitation could be overcome by combining the two measures into a single objective function as a weighted sum. But is not obvious how each measure should be weighted, nor is it clear that this weighting should be fixed throughout the optimisation. The final problem with these measures is the computational cost. Both measures are costly to calculate, but generalization ability is particularly expensive since it involves re-building the model n_s times.

Because of the limitations of all the model-based objective functions discussed so far, a new approach was taken [48, 53],[6] using ideas from information theory, as will be discussed in the next section.

4.3 An Information Theoretic Objective Function

We seek a principled basis for choosing an objective function that will directly favour models with good generalization ability, specificity, and compactness. The ability of a model to generalise *whilst being specific* depends on its ability to interpolate and, to some extent, extrapolate the training set. In order to achieve these properties, we apply the principle of Occam's razor, which can be roughly paraphrased as the statement that the simplest description of the training set will interpolate/extrapolate the best.

This notion of the simplicity of the description can be quantified using ideas from information theory, and in particular, the principle of minimum description length (MDL) [143, 144].

The basic idea is that we consider the problem of transmitting our entire training set as a coded message to a receiver. We suppose that the person receiving the message has a basic codebook, which enables them to decode the message, and hence reconstruct the training set. The length of this message is the *description length* of the training set. The point of this approach is that it takes *incommensurate* data terms, and by encoding them, produces a *commensurate* representation of the data, in that everything is reduced to a simple message length measured in bits.

We could just send the raw values of the coordinate positions of each point on each shape. This however does not make any use of the *similarity* across the set of shapes. A better encoding would be to send the mean shape, and then send, for each shape, the deviations for each point from the corresponding point on the mean shape. However, this still does not make use of the fact that there are correlations between points on a given shape, and also correlations between different shapes. These correlations are captured by the PCA principal axes that were defined previously in Sect. 2.1.3. This then gives a further-improved scheme, where we send the mean shape, the principal modes of variation (that is, the eigenvectors $\mathbf{n}^{(a)}$), and finally, the parameter vectors for each shape. This is obviously advantageous, since in most cases, the dimensionality of the parameter space $\mathbb{R}^{n_m}$ is much less than the dimensionality of the original shape space $\mathbb{R}^{dn_P}$. However, the final point to note is that the parameter vectors are not evenly distributed across parameter space, but are approximately distributed according to our model pdf $p(\mathbf{b})$. We hence should take account of this model pdf when constructing our encoding of the parameter vectors – basically, the idea is that regions of

[6] Note that the use of MDL had already been introduced into the computer vision community in a different context, for example, see [108, 66].

parameter space which have a high probability density should be specifiable using short messages, whereas regions which have a lower probability density can be specified using longer messages, so as to give an optimum message length when averaged over the entire training set. Such a scheme will obviously be adversely affected if the pdf used in encoding is actually a poor fit to the data.

This type of scheme effectively encodes the training set of shapes using the model. What we then have to send to our receiver is a message consisting of two parts. First, the parameters of the model, enabling the receiver to reconstruct the model, and second, the training set encoded using this model. We hence have a total message length which consists of two parts:

$$\boxed{\mathcal{L}_{\text{total}} = \mathcal{L}_{\text{params}} + \mathcal{L}_{\text{data}},} \tag{4.11}$$

which is the two-part coding formulation of MDL [105]. Before we go into specific details, it can already be seen that this scheme potentially enables interesting trade-offs between model complexity, and the degree to which the model fits the data.

For example, consider the case where our data lies on some straight line in data space. A well-fitting model would describe the exact position of this line. The remaining part of the message would just be a single number for each data point, describing the position of the data point along this line. A poorly-fitting model would be one, say, that described not this line, but some sub-space that included the line. This would mean that we would need more than one parameter to describe the position of each data point in this sub-space. However, if all the data lay on some line that was *not* straight, we would have a possible trade-off between the complexity of the model (do we describe the exact, curved line, or just the sub-space in which it lies), versus the degree of fit of the model to the data (the line is a better fit, but a more complex model, whereas the sub-space is a simpler model, but a poorer fit to the actual data). In this context, MDL can be used to solve the model selection problem (e.g., [83]), by comparing the best achievable message length across different classes of model.

In our application, we will restrict ourselves to a single class of model (e.g., Sect. 2.2.1, multivariate Gaussian models). What will vary is the *groupwise correspondence* across the data, which, as we have shown previously, effects both the dimensionality of the model (for example, the case illustrated in Fig. 4.1), and the degree of fit between the data and the model. It is hence a suitable case for application of the MDL principle.

In the next sections, we will derive a simple form for the description length of data encoded using a multivariate Gaussian. For such a model, we first have to send the mean shape (the origin of the PCA coordinate system), then the set of PCA directions $\{\mathbf{n}^{(a)}\}$. Once we have described this PCA coordinate system, the remaining data we have to send is just the positions of each data point along each axis. The positions along each axis are distributed as a one-dimensional *centred* Gaussian, where the distributions along each axis are now uncorrelated. The Gaussians are centred since the origin is the mean shape.

Before we can derive this description length, we first have to give the codeword lengths for a few basic operations as follows.

4.3.1 Shannon Codeword Length and Shannon Entropy

Let us first consider transmitting a number $0 < x < 1$ to some precision δ as a binary string. If $\delta = \frac{1}{2^k}$, then the codeword length is given by:

$$l(x;\delta) = -\log_2(\delta) \text{ bits},$$

where $\log_2$ denotes logarithms to base 2. We would like a meaning of message length that applies whether we encode using a binary string, or a decimal string. We hence define our message length as:

$$l(x;\delta) = -\log\delta,\ \ 0 < x < 1, \tag{4.12}$$

where log denotes the natural logarithm, and the length is now measured in *nats* or *nits* as opposed to bits. In the general case, transmission of a number within some range of width R, to a precision δ, requires a codeword of length:

$$\boxed{l(x;R,\delta) = -\log\frac{\delta}{R},\ \ y < x < y + R,} \tag{4.13}$$

which we can derive from the previous result by a simple scaling.

The next case we require is the codeword length for encoding *unbounded* integers. An integer $n = 2^k$ requires k bits. We hence take the codeword for an integer as:

$$l_{\text{int}}(n) = 1 + \log n,$$

where the 1 has been added so that the message length for sending 1 is one nat, rather than zero. This is not the full result for integers as given by Rissanen [143], using the iterated logarithm, but it is a continuous and tractable approximate form, suitable for our purposes.

The next case is transmitting the occurrence of some event from a finite set of such events. Consider a situation where we have a random process which has some set of distinct possible outcomes, where the αth outcome occurs with probability P_α, $\sum_\alpha P_\alpha = 1$. We transmit the fact of the occurrence of an event of type α by transmitting a codeword. We take as our full set of possible codewords the real numbers $0 < x < 1$. To specify event α, we define that the number x has to lie within some portion of the real line of length x_α. The idea here is that to get the shortest possible message length on average, frequent events are indicated by short codewords, whereas less frequent events can have longer codewords.

For example, suppose we have three events A, B, C. We could encode A by numbers $x < \frac{1}{2}$, $x_A = \frac{1}{2}$, B by numbers $0.5 < x < 0.75$, and C by numbers $0.75 < x < 1$ ($x_B = x_C = \frac{1}{4}$). In binary, sending 0.0 is enough to indicate that A has occurred (less the leading zero and point gives us one bit). Sending 0.1 indicates that it is not A, but we need to send 0.10 to unambiguously indicate B, whereas 0.11 indicates C.[7] And $-\log_2 x_a = -\log_2(0.5) = 1$ bit, whereas $-\log_2 x_B = -\log_2(0.25) = 2$ bits.

Hence for some number that lies within the range of length x_α, we only need to send it to a precision of $\delta = x_\alpha$ to be able to verify that it does indeed lie within this range. This gives, using the above result, a codeword length of $l_\alpha = -\log x_\alpha$. We can hence write the total message length for transmitting a sequence of N total events as:

$$\mathcal{L} = -\sum_\alpha N P_\alpha \log x_\alpha + C\left(1 - \sum_\alpha x_\alpha\right),$$

where C is a Lagrange multiplier, ensuring that all numbers $0 < x < 1$ are assigned to an event. We can find the optimum codeword assignment as follows:

$$\frac{\partial \mathcal{L}}{\partial x_\alpha} = \frac{N P_\alpha}{x_\alpha} - C = 0 \Rightarrow x_\alpha \propto P_\alpha \Rightarrow x_\alpha = P_\alpha, \tag{4.14}$$

where we have also used the normalization constraint. Hence for a set of events that occur with probabilities P_α, the optimal Shannon codeword length for event α is [160]:

$$\boxed{l_\alpha = -\log P_\alpha.} \tag{4.15}$$

The mean codeword length is:

$$\bar{l} = -\sum_\alpha P_\alpha \log P_\alpha,$$

which is just the Shannon entropy. Note that we do not need to know the actual codewords, we just need to know the optimum possible lengths for a mutually unambiguous set of codewords.

We can now consider transmitting values x that are distributed according to some model pdf $p(x)$, where they are transmitted to some fixed precision Δ. We will assume that we have already transmitted the parameters of the

[7] Note that the possible confusion here between our use of precision, and other terms such as *accuracy*, is the distinction between *truncation* and *rounding*. Hence sending a number to be correct to the *nearest* integer requires two digits, since 0.9 and 1.4 will both be rounded to 1, whereas 1.9 rounds to 2. Hence something decoded by rounding as 1 actually could be 1 ± 0.5, where 0.5 is referred to as the *accuracy*. But when using truncation, we just need one digit. The width of the possible *range* is the same in each case.

pdf, enabling the receiver to construct $p(x)$, and that the precision $\varDelta$ we are using has also been transmitted.

The set of all possible values of x to precision $\varDelta$ hence forms an infinite sequence of bins of width $\varDelta$, with bin positions $\{\hat{x}_\alpha\}$. The probability associated with bin α is given by:

$$P_\alpha \doteq \int_{\hat{x}_\alpha}^{\hat{x}_\alpha+\varDelta} p(x)dx. \tag{4.16}$$

If we can calculate these bin probabilities (which are of course normalized since the pdf $p(x)$ is normalized), we can use this result to compute the description length, since from (4.15), the optimum codeword length for transmission of a number that lies in bin α is just $l_\alpha = -\log P_\alpha$.

We will now consider the specific case where our model $p(\mathbf{b})$ is a multivariate Gaussian.

4.3.2 Description Length for a Multivariate Gaussian Model

As was explained at the start of this section, the description length for a multivariate Gaussian can be written as the sum of the description lengths for a series of one-dimensional centred Gaussians.

The origin of our coordinate system is given by the mean shape. The message length for transmission of the mean shape will depend on the number of shape points, the dimensionality of our shapes, and the precision to which we describe their positions. However, all of these quantities remain unchanged as we vary the groupwise correspondence, hence we can treat this term as a constant contribution to our description length.

We require exact reconstruction of our shapes, to a precision $\varDelta$, hence we have to retain all possible modes so that $n_m = n_S - 1$. The number of points n_P on each shape has to be taken so that the polygonal shapes are an adequate representation of our input shapes, hence n_P is also fixed. The description length for our set of n_m axis directions $\{\mathbf{n}^{(a)}\}$ is hence also fixed, being some function of n_S, the point precision $\varDelta$, and the number of points n_P, and the dimensionality of the space occupied by our original shapes d.

The point precision $\varDelta$ can be determined by quantizing the point positions of our original training shapes. Comparison of the original shape and the quantized version allows a maximum permissable value of $\varDelta$ to be determined. For example, for shape boundaries obtained from image data, $\varDelta$ will typically be of the order of the pixel-size.

There is one remaining parameter to be determined, which is the effective scale of our shapes R. Let us suppose that all of our aligned training shapes can be enclosed by some sphere or circle of radius r. Then a given point on any shape can move at most a distance $2r$ between shapes. This means that the maximum separation between shape vectors, is:

$$R = 2r\sqrt{n_P}.$$

This is also the lowest upper-bound on the range of the projections of the shape vectors onto the PCA axes.

We can hence write our total description length in the form:

$$\mathcal{L}_{\text{total}} = f(R, \Delta, n_S) + \sum_{a=1}^{n_m}\left(\mathcal{L}^{(a)}_{\text{params}} + \mathcal{L}^{(a)}_{\text{data}}\right) = f(R, \Delta, n_S) + \sum_{a=1}^{n_m}\mathcal{L}^{(a)}. \tag{4.17}$$

The function $f(R, \Delta, n_S)$ is then the (fixed) description length for the mean shape, the PCA directions etc. For each mode a, $\mathcal{L}^{(a)}_{\text{params}}$ is the description length for the parameters of the ath centred Gaussian in the direction $\mathbf{n}^{(a)}$, and $\mathcal{L}^{(a)}_{\text{data}}$ is the description length for the data in this direction (that is, the set of *quantized*[8] shape parameter values (2.34) $\{\hat{b}^{(i)}_a : i = 1, \ldots n_S\}$, $b^{(i)}_a \doteq (\mathbf{x}_i - \bar{\mathbf{x}}) \cdot \mathbf{n}^{(a)}$).

Let us first consider the parameter term $\mathcal{L}^{(a)}_{\text{params}}$. Each centred Gaussian has one parameter, the variance σ_a^2, or the width σ_a. If this value is encoded to some precision δ_a, then the actual number is the quantized version of this:

$$\hat{\sigma}_a = n\delta_a, \quad n \in \mathbb{N}.$$

The strict upper bound on the maximum separation of the data points projected onto the PCA axis R means that we have a strict upper limit on the variance:

$$\boxed{\sigma_{\max} = \frac{R}{2}.}$$

We also choose to set a lower bound on the modelled variance $\sigma_{\min}$. We hence have to consider the separately the cases $\sigma_a > \sigma_{\min}$ and $\sigma_a < \sigma_{\min}$.

The Case $\boldsymbol{\sigma_a > \sigma_{\min}}$

We will first consider the case where $\sigma_a > \sigma_{\min}$. To encode $\hat{\sigma}_a$, we hence have a value to precision δ_a, within the range $\sigma_{\max} - \sigma_{\min}$. This hence has a codeword length (4.13):

$$l(\hat{\sigma}_a) = l(\hat{\sigma}_a; \sigma_{\max} - \sigma_{\min}, \delta_a) = \log\frac{\sigma_{\max} - \sigma_{\min}}{\delta_a}.$$

---

[8] We will use b to denote a continuum value, and $\hat{b}$ to denote the corresponding quantized value.

The receiver cannot decode the value of $\hat{\sigma}_a$ unless they know the value of δ_a, we hence have to include the description length for δ_a. If we assume that δ_a is of the form $2^{\pm k}$, $k \in \mathbb{N}$, this then gives a description length:

$$l(\delta_a) \approx 1 + |\log \delta_a|,$$

where the additional nat codes for the sign. Putting these together gives:

$$\boxed{\mathcal{L}^{(a)}_{\text{params}} = 1 + \log \frac{\sigma_{\text{max}} - \sigma_{\text{min}}}{\delta_a} + |\log \delta_a|.} \tag{4.18}$$

Let us now consider the encoding of the data $\{\hat{b}_a^{(i)} : i = 1, \dots n_S\}$, which has been quantized to a precision Δ as determined from the original data. For a centred Gaussian of width $\hat{\sigma}$, the bin probability (4.16) for a bin at position $\hat{b}$ is given by:

$$P(\hat{b}) \doteq \int_{\hat{b}}^{\hat{b}+\Delta} \frac{1}{\sqrt{2\pi\hat{\sigma}^2}} \exp\left(-\frac{y^2}{2\hat{\sigma}^2}\right) dy \approx \frac{\Delta}{\hat{\sigma}\sqrt{2\pi}} \exp\left(-\frac{1}{2\hat{\sigma}^2}\hat{b}^2\right).$$

It can be shown numerically that this is a very good approximation (to within 99% of the correct value), for all values $\sigma > 2\Delta$. We hence choose to take our minimum modelled variance as:

$$\boxed{\sigma_{\text{min}} = 2\Delta.}$$

We find that for PCA direction $\mathbf{n}^{(a)}$:

$$l(\hat{b}_a) = -\log P(\hat{b}_a) = -\log \Delta + \log \hat{\sigma}_a + \frac{1}{2}\log 2\pi + \frac{1}{2\hat{\sigma}_a^2}\hat{b}_a^2.$$

Note that in Sect. 2.2.1, an optimum value of σ_a was determined using the Maximum Likelihood criterion, giving:

$$\sigma_a^2 = \frac{1}{n_S}\sum_{i=1}^{n_S}(\mathbf{x}_i - \bar{\mathbf{x}}) \cdot \mathbf{n}^{(a)} = \frac{1}{n_S}\sum_{i=1}^{n_S}(b_a^{(i)})^2 = \frac{\lambda_a}{n_S}.$$

If we instead use the MDL principle, and optimise:

$$\sum_{i=1}^{n_S} l(\hat{b}_a^{(i)}) = n_S \log \sigma_a + \frac{1}{2\sigma_a^2}\lambda_a + \text{constant},$$
$$\Rightarrow \frac{\partial}{\partial \sigma_a}\sum_{i=1}^{n_S} l(\hat{b}_a^{(i)}) = \frac{n_S}{\sigma_a} - \frac{\lambda_a}{\sigma_a^3},$$
$$\Rightarrow \sigma_a^2 = \frac{\lambda_a}{n_S},$$

we hence obtain the same result. Note that if we had made a different choice for encoding σ_a, we would obtain a slightly different result (e.g., see [115]). However, the differences are not significant in the limit of large numbers of shape examples, and we will hence retain the above as our optimal estimate of the *unquantized* value σ_a of the variance of the quantized data. Substituting back in, we obtain:

$$\mathcal{L}_{\text{data}}^{(a)} \doteq \sum_{i=1}^{n_S} l(\hat{b}_a^{(i)}) = -n_S \log \Delta + n_S \log \hat{\sigma}_a + \frac{n_S}{2}\log 2\pi + \frac{n_S}{2}\left(\frac{\sigma_a^2}{\hat{\sigma}_a^2}\right).$$

This result does not simplify trivially, since the variance of the quantized data $\sigma_a^2 \doteq \frac{1}{n_S}\sum_i (\hat{b}_a^{(i)})^2$ is not exactly equal to the value of the *quantized* variance $\hat{\sigma}_a^2$. In general, they differ by some amount d_a where:

$$\hat{\sigma}_a = \sigma_a + d_a, \ |d_a| \leq \frac{\delta_a}{2},$$

where δ_a is the precision of $\hat{\sigma}_a$ as above.

The exact value of d_a depends on the data. We can however estimate its likely effects. If we assume a flat distribution for d_a over its allowed range, we can hence calculate the ensemble average of quantities involving d_a. We can then make the approximation of replacing such quantities by the ensemble average. That is:

$$f(\hat{\sigma}_a) \approx \mathbb{E}\left[f(\sigma_a + d_a)\right],$$

where $\mathbb{E}\left[\cdot\right]$ denotes the average over the distribution of d_a:

$$\mathbb{E}\left[f(\sigma_a + d_a)\right] \doteq \frac{1}{\delta_a}\int_{-\frac{\delta_a}{2}}^{+\frac{\delta_a}{2}} f(\sigma_a + y)dy.$$

Consider the terms in the message length that depend on $\hat{\sigma}_a$ or on δ_a:

$$\mathcal{L}^{(a)} = |\log \delta_a| - \log \delta_a + n_S \log \hat{\sigma_a} + \frac{n_S}{2}\left(\frac{\sigma_a^2}{\hat{\sigma}_a^2}\right) + \dots.$$

We hence compute:

$$\begin{aligned}
\mathbb{E}\left[\log \hat{\sigma}_a\right] &= \log \sigma_a + \frac{1}{\delta_a} \int\limits_{-\frac{\delta_a}{2}}^{+\frac{\delta_a}{2}} \log\left(1 + \frac{y}{\sigma_a}\right) dy, \\
&= \log \sigma_a + \frac{1}{2} \log\left(1 - \frac{\delta_a^2}{4\sigma_a^2}\right) - 1 + \frac{\sigma_a}{\delta_a} \log\left(\frac{2\sigma_a + \delta_a}{2\sigma_a - \delta_a}\right), \\
&= \log \sigma_a - \frac{\delta_a^2}{24\sigma_a^2} + O\left(\frac{\delta_a^4}{\sigma_a^4}\right). \\
\mathbb{E}\left[\frac{1}{\hat{\sigma}_a^2}\right] &= \frac{1}{\sigma_a^2}\left[1 + \frac{\delta_a^2}{4\sigma_a^2} + O\left(\frac{\delta_a^4}{\sigma_a^4}\right)\right]
\end{aligned}$$

Hence:

$$\mathcal{L}^{(a)} \approx |\log \delta_a| - \log \delta_a + \frac{n_S \delta_a^2}{12\sigma_a^2} + \ldots,$$

where we have only written the terms that depend on δ_a. We can hence find an optimum value of δ_a by setting the derivative to zero. Because of the modulus term, we have to consider separately the cases $\delta_a < 1$ and $\delta_a > 1$. We hence find the optimum parameter precision is given by:

$$\boxed{\delta_a^* = \delta^*(\sigma_a, n_S) = \min\left(1, \sigma_a \sqrt{\frac{12}{n_S}}\right).} \tag{4.19}$$

It has been shown numerically [115] that this estimate of the optimum parameter precision is reliable, provided the dataset is not too small.

Putting all this together, we get the final result for the optimum description length for direction $\mathbf{n}^{(a)}$, for the case $\sigma_a > \sigma_{\text{min}}$, is:

$$\begin{aligned}
\mathcal{L}_1^{(a)} = \mathcal{L}_1(\sigma_a, n_S, R, \Delta) = 1 - n_S \log \Delta &+ \frac{n_S}{2} \log 2\pi + \frac{n_S}{2} \\
&+ \log\left(\frac{\sigma_{\text{max}} - \sigma_{\text{min}}}{\delta_a^*}\right) + |\log \delta_a^*| \\
&+ n_S \log \sigma_a + \frac{n_S (\delta_a^*)^2}{12\sigma_a^2}.
\end{aligned} \tag{4.20}$$

The Case $\sigma_a < \sigma_{\text{min}}$

If $\sigma_a < \sigma_{\text{min}}$, but the data lies in more than one bin, we decide to encode the data using a pdf of width σ_{min}. We also set the parameter precision to:

$$\delta_a = \delta^*(\sigma_{\text{min}}, n_S).$$

An analogous derivation to that given above then yields the result:

$$\mathcal{L}_2^{(a)} = \mathcal{L}_2(\sigma_{\min}, n_S, R, \Delta) = 1 - n_S \log \Delta + \frac{n_S}{2} \log 2\pi$$
$$+ \log\left(\frac{\sigma_{\max} - \sigma_{\min}}{\delta_a}\right) + |\log \delta_a| + n_S \log \sigma_{\min}$$
$$- \frac{n_S \delta_a^2}{24\sigma_{\min}^2} + \frac{n_S \sigma_a^2}{2\sigma_{\min}^2}\left(1 + \frac{\delta_a^2}{4\sigma_{\min}^2}\right). \quad (4.21)$$

As a check, we note that $\mathcal{L}_1^{(a)}$ and $\mathcal{L}_2^{(a)}$ agree at $\sigma_a = \sigma_{\min}$.

The final case to consider is when $\sigma_a < \sigma_{\min}$, but the data all lie in one bin. This means that since the data is also centred, all the data lie in the central bin at the origin. We hence do not have to transmit any information in this case.

The obvious advantage of this MDL objective function is that the objective function is well-behaved, even in the limit $\sigma_a \mapsto 0$, that is, $\lambda_a \mapsto 0$. This is in direct contrast to the determinant of the covariance matrix, which required explicit regularization to deal with small eigenvalues (4.9).

The only free parameters are the number of shape points n_P and the data precision Δ. The number of shape points does not enter explicitly into the calculation, we can hence take the limit $n_P \mapsto \infty$ and use the integral definition of the covariance matrix (2.124). As regards the data precision, the quantization of point positions can lead to discontinuity in the objective function close to convergence. This can be overcome by averaging the value of the objective function over a range of values of Δ. The required integral can be solved by numerical integration.

4.3.3 Approximations to MDL

The full expression for the description length of a training set derived above (4.20) and (4.21) is complex, but this is necessary to deal with the general case. In order to gain further insight into the objective function, Davies et al. looked at a limiting case [53].

We consider the dual limit of infinitely small data precision ($\Delta \mapsto 0$), and an infinitely large dataset ($n_S \mapsto \infty$). The full MDL objective function can be *approximated* by:

$$\mathcal{L}_1(\sigma, n_S, R, \Delta) \approx g(R, \Delta, n_S) + (n_S - 2)\log \sigma, \quad (4.22)$$
$$\mathcal{L}_2(\sigma, n_S, R, \Delta) \approx g(R, \Delta, n_S) + (n_S - 2)\log \sigma_{\min} + \frac{n_S + 3}{2}\left(\frac{\sigma^2}{\sigma_{\min}^2} - 1\right), \quad (4.23)$$

where $g(R, \Delta, n_S)$ is some fixed function. The first point to note is that we have continuity as $\sigma \mapsto \sigma_{\min}$ from below. Furthermore, we can see that in this dual limit, the part of the objective function that depends on the $\{\sigma_a\}$ contains terms similar to the determinant of the covariance matrix (4.9). However, the MDL objective function is always well-defined, even in the limit $\lambda_a \mapsto 0$, $\sigma_a \mapsto 0$, where such a direction makes no contribution to the objective function. Whereas in the form used previously [101] it would make an infinitely large contribution, without the addition of artificial correction terms.

Thodberg [177] simplified this approximation further by assuming that the third case never occurs – that is, the range of the data in any direction is always greater than the quantization parameter, Δ. The $f(R, \Delta, n_S)$ and $g(R, \Delta, n_S)$ terms can now be ignored, since they are constant for a given training set. This then gives for the total description length:

$$\mathcal{L}_{\text{total}} \approx \sum_{\sigma_p \geq \sigma_{\min}} (n_s - 2) \log \sigma_p + \sum_{\sigma_q < \sigma_{\min}} \left[(n_s - 2) \log \sigma_{\min} + \frac{(n_s + 3)}{2} \left(\left(\frac{\sigma_q}{\sigma_{\min}} \right)^2 - 1 \right) \right]. \tag{4.24}$$

Dividing by $n_s - 2$ and using the approximation $(n_s + 3)/(n_s - 2) \approx 1$:

$$\mathcal{L}_{\text{total}} \approx \sum_{\sigma_p \geq \sigma_{\min}} \log \sigma_p + \sum_{\sigma_q < \sigma_{\min}} \left[\log \sigma_{\min} + \frac{1}{2} \left(\left(\frac{\sigma_q}{\sigma_{\min}} \right)^2 - 1 \right) \right]. \tag{4.25}$$

By adding suitable constants, and scaling (subtracting $\log \sigma_{\min}$ from each term, multiplying by 2, then adding 1 to each term), we get the transformed expression:

$$\mathcal{L}_{\text{total}} \approx \sum_{\sigma_p \geq \sigma_{\min}} \left(\log \left(\frac{\sigma_p^2}{\sigma_{\min}^2} \right) + 1 \right) + \sum_{\sigma_q < \sigma_{\min}} \frac{\sigma_q^2}{\sigma_{\min}^2}. \tag{4.26}$$

If we substitute $\sigma_p^2 = n_P \lambda_p$, and similarly $\sigma_{\min}^2 = n_P \lambda_{\min}$, we obtain the final expression:

$$\mathcal{L}_{\text{total}} \approx \sum_{\lambda_p \geq \lambda_{\min}} \left(\log \left(\frac{\lambda_p}{\lambda_{\min}} \right) + 1 \right) + \sum_{\lambda_q < \lambda_{\min}} \frac{\lambda_q}{\lambda_{\min}}, \tag{4.27}$$

which is the form used by Thodberg. Considering (4.27), we see that this simplification is a trade off between two terms: one similar to the *determinant* of the data covariance matrix ($\sum \log \lambda_a$, (4.9)), and the other that is similar to the *trace* of the data covariance matrix ($\sum \lambda_a$, (4.3)). This trade-off works because as we noted earlier, the determinant has problems with small eigenvalues, whereas the trace does not. The parameter $\lambda_{\min}$ determines the point where we effectively switch between the determinant-type term and the trace-type term. However, the relative importance of the two terms is governed by the parameter, $\lambda_{\min}$, for which there is now no obvious choice of value, since we have removed the significance of the data quantization.

Thodberg and Olafsdottir [178] further modified this expression by the addition of an arbitrarily weighted term that penalises dissimilarity in curvature at corresponding points. Note that it would be possible to introduce curvature into MDL by transmitting both the point positions and the curvature at each point. This could be seen as an over-complete description of the polygonal shape. However, this is not the approach taken by Thodberg and Olafsdottir, who add an arbitrarily weighted term. As we have already said above, there is no reason to believe that corresponding points should lie on regions of similar curvature and it may, in fact, skew the description length measure. It also negates the theoretical elegance of the MDL approach, reducing the objective function to a sum of incommensurate terms.

4.3.4 Gradient of Simplified MDL Objective Functions

We now turn to the question of incorporating an MDL objective function into an optimisation framework. Optimisation algorithms can often perform better if they are given gradient information about the objective function, as we will see in detail in Chap. 7.

We will suppose that the parameterisation of the ith shape is controlled by some vector of parameters $\boldsymbol{\alpha}^{(i)}$, where the individual parameters are given by $\{\alpha_A^{(i)} : A = 1, \ldots M\}$.[9] The optimisation algorithm needs the gradient of the objective function with respect to each of the parameters for each shape.

A naïve approach would be to calculate the gradient numerically, using a simple finite-difference scheme. The partial derivative with respect to the Ath parameter for the ith shape is approximated by:

$$\frac{\partial \mathcal{L}}{\partial \alpha_A^{(i)}} \approx \frac{\mathcal{L}\left(\alpha_A^{(i)} + \Delta\alpha_A^{(i)}\right) - \mathcal{L}\left(\alpha_A^{(i)}\right)}{\Delta\alpha_A^{(i)}}, \tag{4.28}$$

where $\Delta\alpha_A^{(i)}$ is some suitably-small perturbation of $\alpha_A^{(i)}$. However, as with any approach of this type, truncation errors and roundoff errors are introduced [139]. A more serious drawback is the high computational cost of using such a scheme: the whole model must be rebuilt for the estimation of every component of the gradient – this involves re-building the covariance matrix and calculating its eigen-decomposition $n_s \times M$ times.

However, part of the Jacobian of the gradient can be computed analytically in closed form. Let us suppose we have an objective function which is only a

---

[9] We will talk about re-parameterisation in greater detail in Chaps. 5 and 6. For now, the reader just needs to know that the relevant variables that we are trying to optimise over are the $\boldsymbol{\alpha}^{(i)}$ for each shape $i = 1, \ldots n_S$.

function of the eigenvalues of the covariance matrix of the data (such as the determinant of Kotcheff and Taylor [101], or some of the approximations to MDL given above), so that:

$$\mathcal{L} = \mathcal{L}(\{\lambda_a\}).$$

The Jacobian matrix for the gradient of the objective function is then given by:

$$\frac{\partial \mathcal{L}}{\partial \alpha_A^{(i)}} = \sum_{a=1}^{n_m} \frac{\partial \mathcal{L}}{\partial \lambda_a} \frac{\partial \lambda_a}{\partial \alpha_A^{(i)}}.$$

The computation of $\frac{\partial \mathcal{L}}{\partial \lambda_a}$ is trivial. The computation of $\frac{\partial \lambda_a}{\partial \alpha_A^{(i)}}$ is more complicated. Remember that the eigenvalues $\{\lambda_a\}$ are the eigenvalues of the covariance matrix of the data. For finite-dimensional shape representations, the $dn_P \times dn_P$ covariance matrix $\mathbf{D}$ can be used (2.16). Whereas for the infinite-dimensional, continuous representation of shape, the alternative $n_S \times n_S$ covariance matrix $\widetilde{\mathbf{D}}$ (2.98) is needed. The eigenvalues are the same in each case, but the computation is different. We consider first the finite-dimensional case.

For the finite-dimensional shape representation, the ith shape is given by the shape vector $\mathbf{x}_i$ with elements $\{\mathbf{x}_{i\mu} : \mu = 1, \ldots dn_P\}$. For the corresponding $dn_P \times dn_P$ covariance matrix $\mathbf{D}$ with elements $D_{\mu\nu}$, using the chain-rule for derivatives, the computation becomes:

$$\frac{\partial \lambda_a}{\partial \alpha_A^{(i)}} = \sum_{\mu,\nu,\eta=1}^{dn_P} \frac{\partial \lambda_a}{\partial D_{\mu\nu}} \frac{\partial D_{\mu\nu}}{\partial x_{i\eta}} \frac{\partial x_{i\eta}}{\partial \alpha_A^{(i)}} = \sum_{\eta=1}^{dn_P} \frac{\partial \lambda_a}{\partial x_{i\eta}} \frac{\partial x_{i\eta}}{\partial \alpha_A^{(i)}}.$$

The computation of $\frac{\partial \lambda_a}{\partial x_{i\eta}}$ from the point of view of PCA and the covariance matrix $\mathbf{D}$ is given in Appendix B, with the result (B.24):

$$\boxed{\frac{\partial \lambda_a}{\partial x_{i\eta}} = 2b_a^{(i)} n_\eta^{(a)},}$$

where $\mathbf{n}^{(a)}$ is the ath eigenvector of $\mathbf{D}$, and $b_a^{(i)}$ is the projection of the ith shape vector onto the vector $\mathbf{n}^{(a)}$ (the parameter vector (2.34)). This result was first introduced by Ericsson and Åström [65], who, using the results in [131], derived it using the singular value decomposition (SVD) of the data matrix. The relation between the PCA and the SVD approach is also given in Appendix B.

The final term in the computation of the full Jacobian is the variation of the ith shape with respect to the re-parameterisation parameters for that shape, $\frac{\partial x_{i\eta}}{\partial \alpha_A^{(i)}}$, and this obviously depends on the exact details of the re-parameterisation. This can be calculated numerically, as we will see in Sect. 7.1.3.

For the case of the infinite-dimensional shape representation, or the alternative covariance matrix $\widetilde{\mathbf{D}}$, the corresponding computation is:

$$\frac{\partial \lambda_a}{\partial \alpha_A^{(i)}} = \sum_{j,k=1}^{n_S} \frac{\partial \lambda_a}{\partial \widetilde{D}_{jk}} \frac{\partial \widetilde{D}_{jk}}{\partial \alpha_A^{(i)}}.$$

In Appendix B, we show that (B.25):

$$\boxed{\frac{\partial \lambda_a}{\partial \widetilde{D}_{ij}} = \frac{\widetilde{n}_i^{(a)} \widetilde{n}_j^{(a)}}{\|\widetilde{\mathbf{n}}^{(a)}\|^2},}$$

where $\widetilde{\mathbf{n}}^{(a)}$ is the ath eigenvector of $\widetilde{\mathbf{D}}$. This result was first derived by Hladůvka and Bühler [94], who extended the SVD approach to this case. In Appendix B, this result is derived from the point of view of PCA, and the link to the SVD approach of Hladůvka and Bühler is also explained.

The next term is:

$$\frac{\partial \widetilde{D}_{jk}}{\partial \alpha_A^{(i)}} = \int \frac{\delta \widetilde{D}_{jk}}{\delta \mathbf{S}_i(\mathbf{x})} \frac{\delta \mathbf{S}_i(\mathbf{x})}{\delta \alpha_A^{(i)}} dA(\mathbf{x}).$$

From Appendix B:

$$\widetilde{\mathbf{S}}_i(\mathbf{x}) \doteq \mathbf{S}_i(\mathbf{x}) - \bar{\mathbf{S}}(\mathbf{x}),$$

$$\frac{\delta \widetilde{D}_{jk}}{\delta \mathbf{S}_i(\mathbf{x})} = \frac{1}{A n_S} \left[(n_S \delta_{ij} - 1)\widetilde{\mathbf{S}}_k(\mathbf{x}) + (n_S \delta_{ik} - 1)\widetilde{\mathbf{S}}_j(\mathbf{x}) \right].$$

So that, as in the finite case, the only term left to calculate is $\dfrac{\delta \mathbf{S}_i(\mathbf{x})}{\delta \alpha_A^{(i)}}$, the gradient of the shape function with respect to the control parameter of the re-parameterisation. Once we have this term, the integrals over the surface of the mean shape can be calculated numerically, and the Jacobian of the gradient of the objective function can be evaluated.

Hladůvka and Bühler [94] give a closed-form solution for this final term, but it is restricted to a specific transformation model (which will be described later in Sect. 5.1.4.2). We will see later in Sect. 7.1.3 of Chap. 7 that numerical methods can be used to calculate this term for any representation, albeit with some small loss of accuracy.

4.4 Concluding Remarks

In this chapter, we have reviewed several objective functions. The remaining question is: which is the best one to use in practice for groupwise model-building?

As we have already explained, the pairwise objective functions are not suitable for building a groupwise model. However, they can usefully be employed to initialize the groupwise correspondence. We achieve this by performing a series of pairwise optimisations between each training shape and a fixed reference shape. Such a series of pairwise optimisations is much quicker than a groupwise optimisation, and can hence serve as an efficient initialization stage for the more computationally intensive groupwise stages of the optimisation. The most appropriate choice of pairwise objective function depends on the precise features of the class of shapes being modelled.

The bootstrapped objective functions (generalization and specificity) are too computationally demanding to be used in practice. This leaves the determinant of the covariance matrix, the full MDL objective function and the various MDL approximations as the only viable model-based objective functions.

Let us begin by considering Thodberg's simplification of MDL (4.27). Some arbitrary assumptions and approximations were made in order to arrive at this simplification and the net result is an objective function that is not really a measure of description length at all. On the other hand, this simplification is simpler to compute than the full MDL objective function, and its gradient is readily computed.

The determinant-based objective function of Kotcheff and Taylor (see Sect. 4.2.1) seems to offer a more appealing choice: its gradient can be computed, and it is simpler than the gradient of Thodberg's approximation. However, the determinant only measures the compactness of the training set in shape space and does not explicitly measure other desirable model properties (such as generalization and specificity). It has, however, been shown to produce very similar models to the proper MDL objective function for a range of biomedical objects [172].

In conclusion, the objective function of choice is the full MDL objective function, but it comes at a high computational cost. In practice, the various approximations to it, or the determinant, can be employed to further refine our initial correspondence (that found using a *pairwise* scheme, as above), since the calculable gradient allows efficient evolution. The full MDL objective function is only employed when we are close to the optimum.

Chapter 5
Re-parameterisation of Open and Closed Curves

In the previous chapter, we described various objective functions that measure the quality of a model built from a given set of correspondences. Our aim is to find the correspondence that gives the optimum value of the chosen objective function, and in order to achieve this, we need a parameterised method of manipulating groupwise correspondence.

In Chap. 2, we introduced parameterised, infinite-dimensional representations of shape. To summarize, a shape S_i from the training set of n_S shapes is represented by a parametric shape-function (2.117) $\mathbf{S}_i(\cdot)$. The shape-function is defined by considering an initial one-to-one mapping $\mathfrak{X}_i$ from the parameter space X to the shape S_i thus:

$$X \stackrel{\mathfrak{X}_i}{\longmapsto} S_i, \ \mathbf{x} \stackrel{\mathfrak{X}_i}{\longmapsto} \mathbf{S}_i(\mathbf{x}). \tag{5.1}$$

The mapping $\mathfrak{X}_i$ hence associates a parameter value $\mathbf{x}$ to each point on the ith shape, the position of that point on the shape being the value of the shape function $\mathbf{S}_i(\mathbf{x})$.

Correspondence between shapes is then defined at points of the same parameter value $\mathbf{x}$, so that:

$$\mathbf{S}_i(\mathbf{x}) \sim \mathbf{S}_j(\mathbf{x}), \tag{5.2}$$

where $\sim$ denotes the dense point-to-point correspondence. In the present chapter, we focus on the case of shapes in two dimensions; that is, the parameter space X is one-dimensional, and we will use u as the parameter value, rather than the vector-valued parameter $\mathbf{x}$. The parameter space X is then quite simple, since we have only two choices for the topology of one-part shapes, either the open line, or the circle. For shapes topologically equivalent to an open line, the parameter space X is a segment of the real line $u \in [0, 1]$. For the circle case, we take $u \in [0, 1]$ as before (or equivalently, $u \in [0, 2\pi]$ as in the case of polar coordinates), but add the constraint that $\mathbf{S}_i(0) \equiv \mathbf{S}_i(1)$, which closes the two ends of the shape into a circle.

R. Davies et al., *Statistical Models of Shape*,
DOI: 10.1007/978-1-84800-138-1_5,

In Chap. 3 we introduced the idea of manipulating correspondence by re-parameterising each example shape. Let ϕ_i be a re-parameterisation function for the ith shape, which is a diffeomorphism of the line/circle as appropriate. The re-parameterisation is then given by:

$$u \xmapsto{\phi_i} u' \doteq \phi_i(u),\ \ \phi_i : [0,1] \longmapsto [0,1]. \tag{5.3}$$

This mapping then also acts on the shape-function $\mathbf{S}_i(\cdot)$, so that:

$$\mathbf{S}_i(\cdot) \xmapsto{\phi_i} \mathbf{S}'_i(\cdot), \boxed{\mathbf{S}'_i(u') \equiv \mathbf{S}'_i(\phi_i(u)) \doteq \mathbf{S}_i(u).} \tag{5.4}$$

That is, the point under consideration on the actual shape ($\mathbf{S}_i(u)$) does not change, but the parameter value for that point ($u \mapsto u'$), and the shape-function describing the shape ($\mathbf{S}_i(\cdot) \mapsto \mathbf{S}'_i(\cdot)$) both change. This means that when compared to another shape S_j (which is not being re-parameterised at the moment), the correspondence changes:

$$\mathbf{S}_j(u) \sim \mathbf{S}_i(u) \xmapsto{\phi_i} \mathbf{S}_j(u) \sim \mathbf{S}'_i(u) \ \Rightarrow\ \mathbf{S}_j(u) \sim \mathbf{S}_i(\phi_i^{-1}(u)). \tag{5.5}$$

In practice, each shape $\{S_i : i = 1, \ldots n_S\}$ has its own re-parameterisation function ϕ_i, with the correspondence changing under simultaneous re-parameterisation in an obvious fashion.

To utilize this representation of re-parameterisation within an optimisation framework, we must be able to represent and manipulate the set of re-parameterisation functions $\{\phi_i : i = 1, \ldots n_S\}$.

We assume that all the shapes in our training set have the correct relative orientation. That is, increasing the parameter value traverses any of the shapes in the same direction.[1] A suitable diffeomorphic mapping for re-parameterisation then has to retain this relative orientation, hence we need only consider that part of the diffeomorphism group that is continuous with the identity. For the case where the topology is that of the line, this is isomorphic to the set of all monotonically-increasing differentiable functions of the unit line.[2] For oriented shapes with circular topology, we could consider a general orientation-preserving diffeomorphism. However, it is sometimes simpler to restrict ourselves to the subgroup of diffeomorphisms that leave one point unchanged on each shape. If we label this point as the closure point for the line

[1] This can be considered as part of the general initial task of bringing the shapes into roughly the correct relative alignment. So, for instance, if we had a training set of hand outlines, consisting of examples of both left and right hands, the simplest way to achieve this would be to reflect the shapes of the right hands, so that all the hands can then be overlaid with the digits in rough correspondence. Then an initial parameterisation that traced first the thumb then each of the fingers would obviously be a consistent parameterisation across the set.

[2] Whereas a re-parameterisation that changed the orientation would be represented by a monotonically *decreasing* function, and would in effect swap the start and ends points of the curve.

($u = 0 = 1$), then the diffeomorphic transformations that leave this point stationary are the transformations of the line. Altering the corresponding point that is held fixed on all the shapes then widens the set of transformations, and any transformation we require can be obtained by concatenating such transformations.

The problem is then one of constructing parameterised sets of monotonic functions on the line segment $[0, 1]$, which can be applied to either open or closed line topologies (although with the proviso that the re-parameterisation is then not necessarily differentiable on the circle at the point where the ends join). We also consider specifically the case of circular topology, for re-parameterisations that are everywhere differentiable.

In this chapter, we consider several ways to build such sets of functions.

5.1 Open Curves

We consider first the case of open curves, which have the simplest topology. As explained above, for oriented open curves, consistent parameterisation is trivial, and the legal re-parameterisation functions are just the set of monotonically increasing functions over the unit line. In what follows, we shorten this to just monotonic.

5.1.1 Piecewise-Linear Re-parameterisation

The simplest way to construct a monotonic function from a finite set of parameters is by considering linear interpolation, which builds a piecewise-linear monotonic function [101].

Suppose we have a set of ordered node points $\{p_\alpha : \alpha = 1, \dots n\}$ spaced along the real line at positions $\{u_\alpha\}$. If we define the function values $\{\phi(u_\alpha)\}$ at each node, this then defines the function at any intermediate point by linear interpolation :

$$\phi(u) \doteq \phi(u_\alpha) + (\phi(u_{\alpha+1}) - \phi(u_\alpha)) \frac{u - u_\alpha}{u_{\alpha+1} - u_\alpha}, \quad u_\alpha \leq u \leq u_{\alpha+1}, \tag{5.6}$$

where p_α and $p_{\alpha+1}$ are the left and right nearest-neighbour nodes to the point at u.

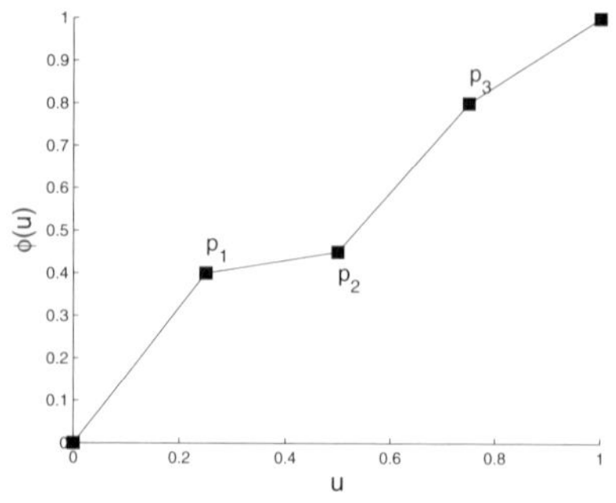

Fig. 5.1 An example of the piecewise-linear re-parameterisation, using three nodes with parameter values $\{u_\alpha\} = \{0.4, 0.45, 0.8\}$.

To ensure that $\phi(u)$ is diffeomorphic,[3] both the original node positions $\{u_\alpha\}$ and their new positions $\{\phi(u_\alpha)\}$ must be in the same order:

$$\begin{aligned} & 0 \leq u_\alpha < u_{\alpha+1} \ldots < u_n \leq 1, \\ \Rightarrow \quad & 0 \leq \phi(u_\alpha) < \phi(u_{\alpha+1}) \ldots \leq \phi(u_n) \leq 1. \end{aligned} \tag{5.7}$$

Typically, we start with nodes equally spaced on the shape, so that the initial parameterisation $\{u_\alpha\}$ is just a path-length parameterisation of the shapes, with corresponding points at equal fractional path-length distances along their respective shapes. The piecewise-linear representation is illustrated in Fig. 5.1.

It should be noted here that ensuring a homeomorphism is particularly simple for one-dimensional shapes in two or three dimensions, since it reduces to a simple case of retaining the relative ordering of points. The same consideration does not, however, hold for two-dimensional shapes in three dimensions, with a two-dimensional parameter space, where the homeomorphism constraint is considerably more complicated, as will be seen later.

5.1.2 Recursive Piecewise-Linear Re-parameterisation

If we require a denser re-parameterisation of the shapes, we could just create more equi-distant nodes. But a more interesting approach is to apply the initial piecewise linear approach in a recursive manner. This will also illustrate an alternative approach to the ordering constraint required for homeomor-

[3] $\phi(u)$ is not differentiable at the control points hence it is not diffeomorphic in a strict sense, but homeomorphic. Homeomorphic mappings are continuous, one-to-one mappings with a continuous inverse, whereas diffeomorphisms add the constraint that the mapping and its inverse must be differentiable (to some order), rather than just continuous.

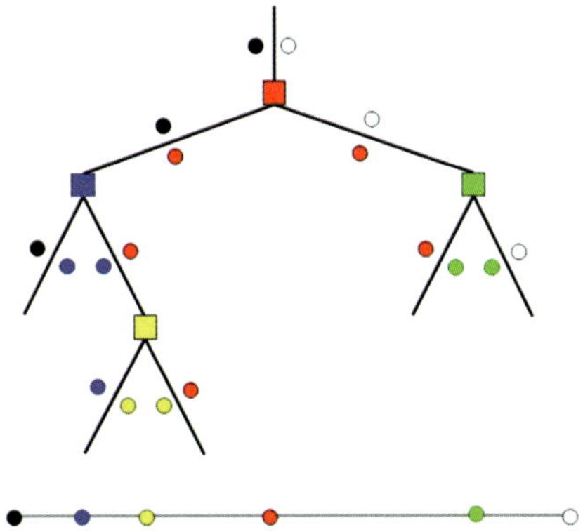

Fig. 5.2 A tree graph showing, read from top to bottom, recursive subdivision of a line segment. The actual line segment is shown at the bottom, with nodes indicated by coloured circles. The edges of the graph represent line segments, with the pair of nodes (coloured circles) that define that line segment indicated next to it. The coloured squares which terminate edges indicate addition of a node (coded by colour) within that segment.

phisms, an approach that is fruitful since it can be generalized, and applied to shapes in three dimensions as well as shapes in two dimensions.

Consider an initial piecewise linear re-parameterisation function $\phi(u)$ defined on some set of n nodes. If we consider the part of the function between a pair of neighbouring nodes, the function in this region is just a straight line. This is directly analogous to the straight line we have if we consider the identity re-parameterisation ($\phi(u) \equiv u$) for the entire line ($u \in [0,1]$), before we constructed the re-parameterisation based on n nodes. So, just as we can construct a re-parameterisation of the line segment $[0,1]$ based on a set of nodes, so we can also increase the density of this re-parameterisation by applying this construction recursively onto segments of the line defined by the sets of nodes created at the previous level.

Let us take the simplest case where we add only one node each time we apply the process. For the addition of just a single node, the position of that node can be totally described by giving the identities of the two higher-level nodes that describe the segment into which the node is going to be placed (the *parent* nodes, p_α and p_β, say), plus the fractional distance $\tau_{\alpha\beta}$ of the new *daughter* node p_γ along that segment, where:

$$u_\gamma \doteq u_\alpha + \tau_{\alpha\beta}(u_\beta - u_\alpha), \;\; u_\beta > u_\alpha. \tag{5.8}$$

Each new segment can then be further divided, by adding new daughter nodes.

The final set of nodes can then be described by a simple branching tree structure as shown in Fig. 5.2. The nodes are represented by coloured circles, and the edges of the graph correspond to line segments, where the circles to each side of the graph edge represent the nodes defining that line segment.

A square represents the action of adding a new node inside the line segment represented by the graph edge that it terminates. We hence generate two new line segments, creating two new branches of the graph, with each of the two nodes that defined the original line segment now helping to define one of the new line segments, along with the new node.

This process is repeated, the final collection of line segments and nodes being the terminal branches of the completed tree.

This may seem an unwarranted obfuscation of the original simple concept of a set of ordered nodes. However, consider now how the homeomorphism condition constrains the allowed movements of the nodes. For a node with descendants, the homeomorphism constraint is satisfied with respect to the descendants provided they move proportionately as the ancestor moves. With respect to the parents of a node, movement of the daughter node is allowed provided that the node remains between the parents.

Another way to view this is that movement is allowed provided the connectivity structure of the nodes is preserved. Rather than the connectivity given by ordering of the nodes, we instead use a representation of connectivity based on the parent-daughter relationship, with links running from each daughter to both parents. We see that movement of any node is allowed, provided that the directed line segments linking it to its parents retain their orientation, and the any descendant nodes move proportionately. For example, if a child node moved to the right of its righthand parent, the line segment linking it to this parent would flip direction, which is not permitted.

This concept can be easily generalized to higher dimensions. For shapes in three dimensions, the nodes are now connected to form a triangulated mesh, the triangles of the mesh being the direct counterparts of the line segments in two dimensions. Triangles of this mesh can be sub-divided by placing a new child node within the triangle, so that each child node has three parents. As for the line segments, the original triangle is now replaced by three new triangles. The homeomorphism constraint is satisfied provided that the triangles defined by the parent-child relationship do not flip. This process is explained in detail in the next chapter, which concentrates specifically on shapes in three dimensions.

Although it is simple to compute, the piecewise-linear representation is not differentiable at the node positions. This is sufficient for some cases, but not all. In the following sections, we discuss how to construct parametric representations of re-parameterisation functions that are differentiable (at least $\mathcal{C}^1$) everywhere.

5.1.3 Localized Re-parameterisation

The recursive linear re-parameterisation is localized in the sense that the addition of a daughter node affects only points within its parent segment. We now consider differentiable methods of re-parameterisation that are explicitly localized.

An intuitive way of viewing the localized re-parameterisation of a curve is to introduce a motion field, $h(u)$, that is only non-zero in a local bounded region. Without loss of generality, let us consider a transformation localized to the interval $[-1, 1]$, where:

$$\phi(u) = \begin{cases} u + h(u)p & \text{if } -1 \leq u \leq 1 \\ u & \text{otherwise} \end{cases}, \tag{5.9}$$

hence the re-parameterisation is just the identity outside of this region. The motion field $h(u)$ is taken to be a differentiable function, and p is the *amplitude* of the motion field. Continuity at the edges of the region mean that $h(-1) = h(1) = 0$. To ensure differentiability at the edges of the region, we have the further constraint that $h'(-1) = h'(1) = 0$. A homeomorphic (i.e., monotonic) re-parameterisation places constraints on p, as detailed in the following Theorem.

Theorem 5.1. Homeomorphic Localized Re-parameterisation. *For a motion field as defined in (5.9) with the constraints:*

$$h(-1) = h(1) = 0 \ \& \ h'(-1) = h'(1) = 0, \tag{5.10}$$

the corresponding re-parameterisation is homeomorphic provided that:

$$|p| < \frac{1}{|h'(u)|} \ \forall \ -1 \leq u \leq 1. \tag{5.11}$$

Proof. From (5.9):

$$\phi(u) = u + ph(u).$$

Let us consider two points with an infinitesimal separation: u and $u + \Delta u$, where $0 < \Delta u \ll 1$. Then:

$$\phi(u + \Delta u) = \phi(u) + \Delta u + ph'(u)\Delta u + O\left((\Delta u)^2\right).$$

We defined Δu such that $u + \Delta u > u$. The transformation is hence homeomorphic provided that the image points are in the same order. Hence:

$$\phi(u + \Delta u) > \phi(u) \ \Rightarrow \ \Delta u\,(1 + ph'(u)) > 0 \ \Rightarrow \ 1 + ph'(u) > 0.$$

$$\therefore \left.\begin{array}{l} \text{If } h'(u) < 0 \ \Rightarrow \ p < \frac{1}{-h'(u)} \\ \text{If } h'(u) > 0 \ \Rightarrow \ p > -\frac{1}{h'(u)} \end{array}\right\} \ \boxed{|p| < \frac{1}{|h'(u)|}.} \tag{5.12}$$

Note that this is effectively just the one-dimensional equivalent of evaluating the determinant of the Jacobian for the transformation, and constraining it to be positive. □

An obvious choice of function for representing the motion field, $h(u)$, is a portion of a Gaussian [89], so that:

$$h(u) = \frac{1}{\sigma\sqrt{2\pi}} \exp\left(-\frac{u^2}{2\sigma^2}\right), \tag{5.13}$$

where σ controls the width of the Gaussian. Using (5.11), we see that the transformation is homeomorphic within the region if the motion of the centre is bounded:

$$\begin{aligned} \text{If: } \sigma \leq 1,\ & |p| < \sigma^2\sqrt{2\pi}\exp(\tfrac{1}{2}). \\ \text{else if: } \sigma > 1,\ & |p| < \sigma^3\sqrt{2\pi}\exp(\tfrac{1}{2\sigma^2}). \end{aligned} \tag{5.14}$$

To maintain continuity at the borders of the region, we must add a constant term, so that:

$$h(u) = \frac{1}{\sigma\sqrt{2\pi}} \exp\left(-\frac{u^2}{2\sigma^2}\right) - B \tag{5.15}$$

$$= \frac{1}{\sigma\sqrt{2\pi}} \exp\left(-\frac{u^2}{2\sigma^2}\right) - \frac{1}{\sigma\sqrt{2\pi}} \exp\left(-\frac{1}{2\sigma^2}\right), \tag{5.16}$$

so that $h(1) = h(-1) = 0$. However, this function does not have a zero derivative at $u = \pm 1$, but:

$$h'(\pm 1) = \mp \frac{1}{2\sigma^3\sqrt{2\pi}} \exp\left(-\frac{1}{2\sigma^2}\right). \tag{5.17}$$

This mismatch obviously decreases the smaller the width σ we choose – that is, the further along the tails of the Gaussian we make the cut.[4] However, given the nature of the Gaussian, it can never be completely removed.

An alternative function is one based on the bi-harmonic clamped-plate spline (see [184] and the Appendix), where:

$$\begin{aligned} h(u) &= 1 - u^2 + u^2 \ln(u^2), \\ h(\pm 1) &\equiv 0, h'(\pm 1) \equiv 0. \end{aligned} \tag{5.18}$$

The constraints of continuity and differentiability at the boundary of the region are hence met automatically (see Fig. 5.3). The homeomorphism constraint (5.11) gives the amplitude constraint $|p| < \frac{e}{4}$.

If we consider the Taylor series expansion of (5.18) about $u^2 = 1$, and retain only the first non-zero term, we obtain a simple polynomial function which is also suitable for our purposes [30]:

[4] For surfaces, Heimann et al. [89] cut off the Gaussian at 3σ, which in our case, when mapped to the range $[-1, +1]$, is equivalent to setting $\sigma = \frac{1}{3}$.

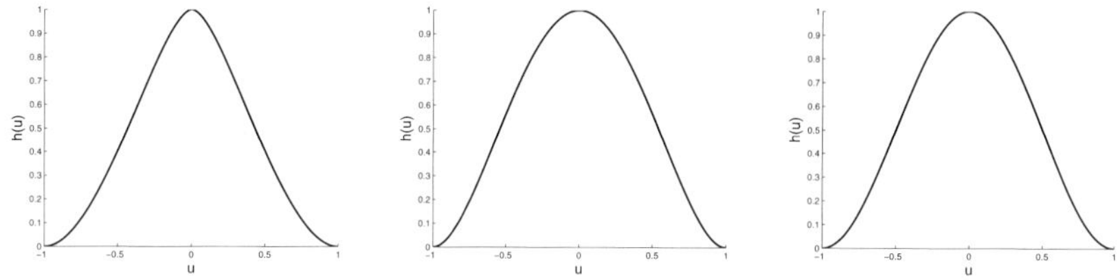

Fig. 5.3 Three functions suitable for the local displacement scheme, plotted over the interval $[-1, 1]$. **Left**: clamped plate spline (5.18), **Centre**: the polynomial representation (5.19), **Right**: the trigonometric representation (5.20).

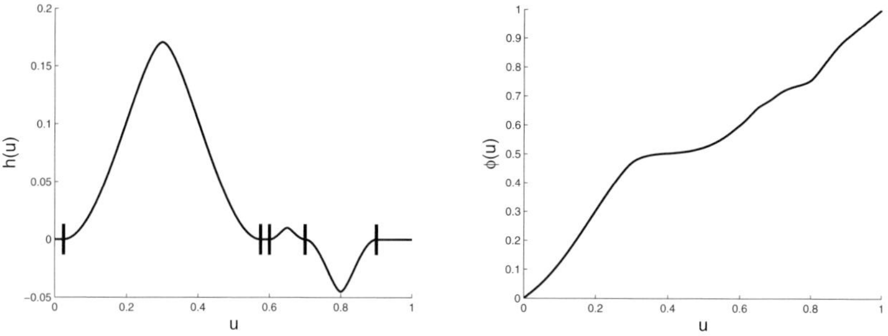

Fig. 5.4 A re-parameterisation function formed using a combination of local transformations. Three clamped plate splines of widths $\{w_k\} = \{0.55, 0.10, 0.20\}$ were placed at positions $\{a_k\} = \{0.30, 0.64, 0.80\}$. The centrepoints of each region were moved to new *relative* positions of $\{p_k\} = \{0.62, 0.20, -0.45\}$. **Left:** The motion field, $h(u)$, showing the territories of the local region. **Right:** The re-parameterisation function $\phi(u)$.

$$h(u) = (1 - u^2)^2. \tag{5.19}$$

This function (see Fig. 5.3) inherits the continuity and differentiability properties of the clamped-plate spline at the boundary, but it is slightly simpler to compute. The maximum allowed displacement is $|p| < \frac{8}{3\sqrt{3}}$.

We can also construct suitable functions using trigonometric functions. For example:

$$h(u) = \frac{1}{2}\left(1 + \cos(\pi u)\right), \;\; h(\pm 1) = 0, \;\; h'(\pm 1) = 0, \tag{5.20}$$

as is plotted in Fig. 5.3. The homeomorphism constraint yields: $|p| < \frac{2}{\pi}$.

In practice, several such local transformations are applied successively. Each transformation has a position of the centre a_k, a width w_k (which is the size of the actual region that is then mapped to the interval $u \in [-1, 1]$), plus the amplitude of the deformation p_k. Figure 5.4 shows an example of such a deformation, using three clamped-plate splines, of varying widths, positions, and amplitudes.

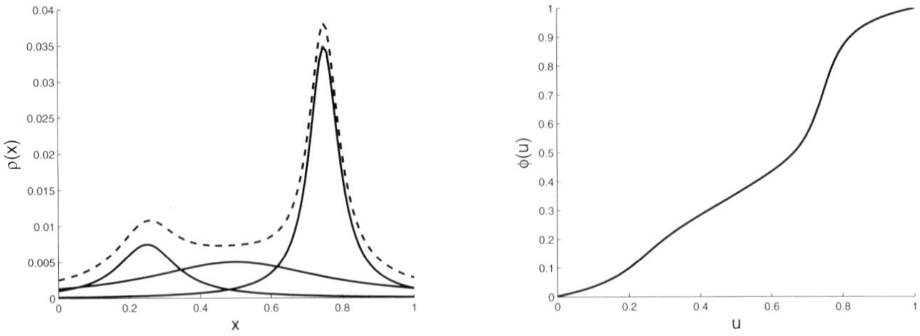

Fig. 5.5 An illustration of the integral representation of re-parameterisation. **Left:** three kernel functions (solid lines) and its sum (dashed line). **Right:** the integral of the summed kernel function plus the identity.

5.1.4 Kernel-Based Representation of Re-parameterisation

We now move on to consider more general differentiable methods of re-parameterisation. We first concentrate on the case of open curves. These re-parameterisations can also then be applied to closed curves, except that the re-parameterisation is then only differentiable everywhere except at the point where the two ends are joined. In a later section (Sect. 5.2), we consider re-parameterisation for closed curves which are differentiable everywhere.

To recap, we wish to construct functions $\phi(u),\ u \in [0,1]$ that are monotonic, with $\phi(0) = 0$ and $\phi(1) = 1$, and also differentiable.

One simple way of achieving this is where $\phi(u)$ is the cumulative distribution function (cdf) of some normalized probability density function (pdf) $\rho(x)$ [51]:

$$\phi(u) = \int_0^u \rho(x)dx, \quad \int_0^1 \rho(x)dx = 1, \ \ \rho(x) \geq 0 \ \forall \ x \ \in \ [0,1]. \tag{5.21}$$

Building a parametric representation of $\phi(\cdot)$ then reduces to the problem of building a parametric representation of a general pdf $\rho(x)$. As in kernel density estimation (that we considered earlier in Sect. 2.1), a general pdf can be constructed from a weighted sum of basic kernels. The positions, widths and heights of the individual kernels then form the parametric representation of $\rho(x)$, hence of $\phi(u)$ – the idea is illustrated in Fig. 5.5. In fact, the piecewise-linear representation (5.6) considered earlier can be cast in this form, where the kernel is a square bump of varying width and position.

This suggests that to obtain a differentiable function, we need to use smooth, differentiable kernel functions. One obvious choice would be to use a combination of Gaussian kernel functions [51]. However, this is not to-

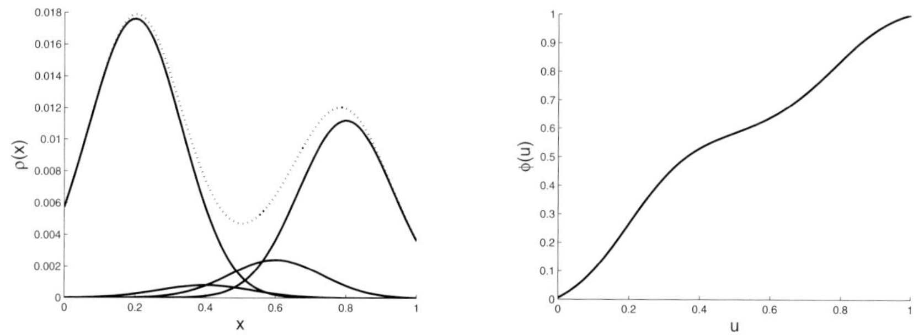

Fig. 5.6 An illustration of a re-parameterisation constructed from a set of Gaussian kernels. The parameters of the re-parameterisation are: $\{a_k\} = \{0.2, 0.4, 0.6, 0.8\}$, $\{w_k\} = \{\frac{2}{15}, \frac{2}{15}, \frac{2}{15}, \frac{2}{15}\}$, and $\{A_k\} = \{0.550, 0.025, 0.075, 0.350\}$. **Left:** the individual kernel functions (solid lines) and their sum (dashed line). **Right:** the integral of the sum of kernel functions plus the identity.

tally straightforward since we require $\rho(x)$ to be normalized over the range $0 \leq x \leq 1$, rather than over the entire real line. We hence define the Gaussian-kernel pdf:

$$\rho(x) = \frac{1}{N}\left[1 + \sum_{k=1}^{n_k} \frac{A_k}{\sigma_k\sqrt{2\pi}} \exp\left(-\frac{1}{2\sigma_k^2}(x - a_k)^2\right)\right], \tag{5.22}$$

where we have n_k Gaussian kernels, with centres $\{0 \leq a_k \leq 1\}$, widths $\{\sigma_k\}$ and amplitudes $\{A_k\}$. The normalization factor N is given by:

$$N = 1 + \sum_{k=1}^{n_k} \frac{A_k}{2}\mathrm{erf}\left(\frac{1 - a_k}{\sigma_k\sqrt{2}}\right) + \sum_{k=1}^{n_k} \frac{A_k}{2}\mathrm{erf}\left(\frac{a_k}{\sigma_k\sqrt{2}}\right), \quad \mathrm{erf}(x) \doteq \frac{2}{\sqrt{\pi}}\int_0^x e^{-r^2}\,dr. \tag{5.23}$$

An example of a Gaussian kernel re-parameterisation is shown in Fig. 5.6. The function is defined so that if all the kernel heights are zero ($A_k = 0$), the final $\phi(u)$ is just a straight line.

As the Gaussian width σ_k approach zero, a Gaussian approaches the Dirac δ-function. Hence in principle, such a mixture of Gaussians can approximate any pdf to any arbitrarily small degree of accuracy.

There are however other possible choices of kernels. One example is the Cauchy function, as follows.

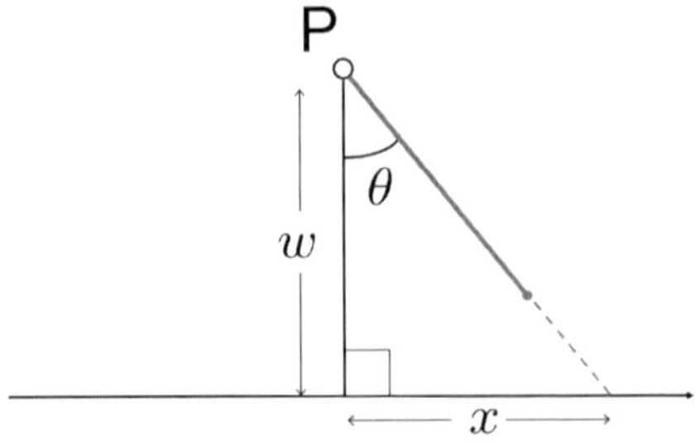

Fig. 5.7 Geometric interpretation of the Cauchy distribution. A pivot P is placed a height w above the plane, with a straight rod suspended from the pivot. The horizontal distance x is determined by the angle θ. If θ has a flat distribution between $-\frac{\pi}{2}$ and $\frac{\pi}{2}$, then x will be distributed according to the Cauchy pdf $f(x; w, 0)$ as defined in (5.24).

5.1.4.1 Cauchy Kernels

The Cauchy distribution[5] [111] leads to a unimodal, symmetric probability density function, defined over the whole real line:

$$f(x; w, a) = \frac{1}{\pi} \frac{w}{w^2 + (x-a)^2}, \quad -\infty < x < \infty, \;\; w \geq 0, \tag{5.24}$$

where w is the width of the Cauchy (the scale parameter) and a is the position of the centre (the location parameter). It has a simple geometric interpretation, as is shown in Fig. 5.7.

In this geometric interpretation, the Cauchy distribution is generated from a uniform distribution via a simple transformation of variables. This means that it is simple to calculate the integral of the Cauchy pdf. Because of the existence of this analytic form, Davies et al. [51] chose to use a sum of Cauchy pdfs to represent a general density $\rho(x)$. In order to calculate the cumulative distribution function, we need the integral of the pdf (derived from the geometric interpretation, or [79], page 81, 2.172):

$$g(u; w, a) = \int_0^u f(x; w, a) dx = \frac{1}{\pi} \left[\arctan\left(\frac{u-a}{w}\right) + \arctan\left(\frac{a}{w}\right) \right]. \tag{5.25}$$

Note that as for Gaussians, the Cauchy pdf approaches the Dirac δ-function in the limit, hence can be used to represent an arbitrary density $\rho(x)$ to any required degree of accuracy. As for a sum of Gaussians, we use a sum of n_k Cauchy kernels plus a constant term, so that:

---

[5] Also referred to in the literature as the *Lorentzian distribution* or the *Lorentz distribution*.

$$\rho(x) = \frac{1}{N}\left[1 + \sum_{k=1}^{n_k} A_k f(x; w_k, a_k,)\right], \quad 0 \leq x \leq 1, \tag{5.26}$$

$$B \doteq \sum_{k=1}^{n_k} \frac{A_k}{\pi} \arctan\left(\frac{a_k}{w_k}\right), \tag{5.27}$$

$$N \doteq 1 + B + \sum_{k=1}^{n_k} \frac{A_k}{\pi} \arctan\left(\frac{1-a_k}{w_k}\right), \tag{5.28}$$

where A_k is the magnitude of the kth kernel. The cdf, hence $\phi(u)$, is then given by:

$$\phi(u) = \int_0^u \rho(x)dx, \quad 0 \leq x \leq 1$$

$$\Rightarrow \quad \boxed{\phi(u) = \frac{1}{N}\left[u + B + \sum_{k=1}^{n_k} \frac{A_k}{\pi} \arctan\left(\frac{u-a_k}{w_k}\right)\right].} \tag{5.29}$$

An example of such a re-parameterisation is shown in Fig. 5.8. As before, the parameters $\{A_k\}$ are such that if $A_k = 0 \ \forall \ k$, the re-parameterisation is just the identity re-parameterisation.

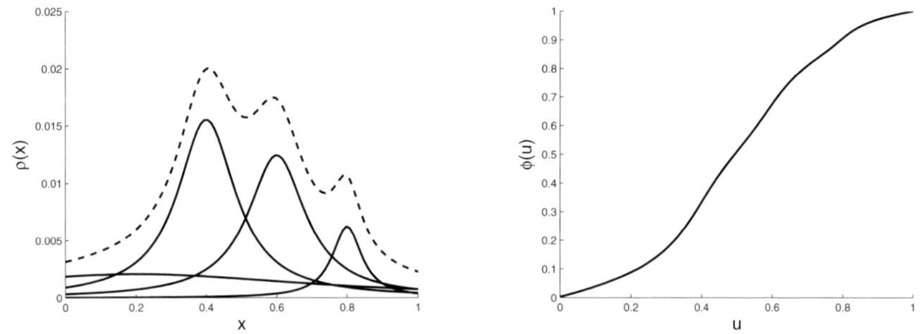

Fig. 5.8 An illustration of a re-parameterisation constructed from a set of Cauchy kernels. The parameters of the re-parameterisation are: $\{a_k\} = \{0.20, 0.40, 0.60, 0.80\}$, $\{w_k\} = \{0.60, 0.10, 0.10, 0.05\}$, and $\{A_k\} = \{0.40, 0.50, 0.40, 0.10\}$. **Left:** the kernel functions (solid lines) and its sum (dashed line). **Right:** the integral of the kernel function plus the identity.

5.1.4.2 Polynomial Re-parameterisation

The cdf approach to constructing a diffeomorphic re-parameterisation function borrows ideas from kernel density estimation (Sect. 2.1), and in par-

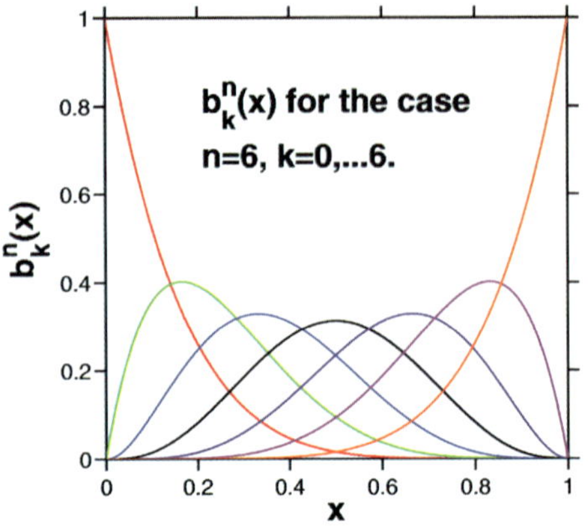

Fig. 5.9 The 6th-order Bernstein polynomials.

ticular borrows the idea of using a symmetric, monotonic[6] kernel function. However, since we are *not* trying to estimate an actual density, we need not retain this kernel-based link between data points and the estimated pdf. In particular, we can build up our pdf from simpler, polynomial functions, with the advantage that differentiation and integration are much simpler for polynomial functions that they are for those kernel-based functions we considered previously.

Hladůvka and Bühler [94] show that weighted combinations of Bernstein polynomials can be used as a basis for representing $\rho(x)$ in (5.21). The kth Bernstein polynomial of degree n is given by:

$$b_k^n(x) = \ {}^n\mathrm{C}_k x^k (1-x)^{n-k}, \ \ n \geq 0, \ \ k = 0, \ldots, n, \tag{5.30}$$

where ${}^n\mathrm{C}_k$ is the usual binomial coefficient. The Bernstein basis polynomials $\{b_k^n(x) : k = 0, \ldots, n\}$ of degree n form a basis for polynomial functions of degree n.

If we consider the unit interval $[0, 1]$, we see that $b_k^n(x)$ is non-negative. For $0 < k < n$, they have a maximum:

$$\frac{d}{dx} b_k^n(x) = \ {}^n\mathrm{C}_k x^{k-1} (1-x)^{n-k-1} (k - nx) \Rightarrow \frac{d}{dx} b_k^n(x) = 0 \rightarrow x = \frac{k}{n}. \tag{5.31}$$

We hence see that for a fixed order n, the maxima of the set:

$$\{b_k^n : k = 1, \ldots, n-1\}$$

are equally spaced along the interval between 0 and 1. See Fig. 5.9 for examples.

---

[6] That is, each *half* of the symmetric function is strictly monotonic.

If we integrate each basis polynomial over the range $[0, 1]$:

$$\int_0^1 b_k^n(x)dx = {}^n\mathrm{C}_k \int_0^1 x^k(1-x)^{n-k}dx = \frac{1}{n+1}. \tag{5.32}$$

We hence can construct a normalized pdf as the weighted sum of Bernstein polynomials of order n:

$$\rho_n(x) \doteq \frac{n+1}{\sum_{j=0}^{n} w_j} \sum_{k=0}^{n} w_k b_k^n(x), \;\; w_k \;\geq\; 0, \tag{5.33}$$

where $\{w_k\}$ are the weights. The re-parameterisation is then the cdf of $\rho(x)$. In order to calculate this, we define[7]:

$$B_k^n(u) \doteq \int_0^u (n+1)b_k^n(x)dx, \tag{5.34}$$

hence:

$$\phi(u) \doteq \phi(u;\{w_j\}) = \frac{1}{\sum_{j=0}^{n} w_j} \sum_{k=0}^{n} w_k B_k^n(u). \tag{5.35}$$

This form makes the dependance on the parameters $\{w_k\}$ explicit, since the functions $B_k^n(\cdot)$ do not depend on these weights. This means that the $B_k^n(u)$ can be pre-computed, which is computationally advantageous.

Note that unlike the kernel cases considered previously, the identity re-parameterisation is obtained when $w_1 = w_2 = \ldots = w_n > 0$.

Although this form has been derived for re-parameterisation of the unit line, we can re-parameterise the unit circle by joining the ends as before. However, for the case of the bump-like Bernstein polynomials $(0 < k < n)$, we have an additional property, that:

$$b_k^n(0) = b_k^n(1) \;\Rightarrow\; B_k^{\prime n}(0) = B_k^{\prime n}(1) \;\Rightarrow \phi'(0) = \phi'(1), \tag{5.36}$$

hence if we restrict ourselves to these bump-like polynomials, $\phi(u)$ is also differentiable at the join, hence differentiable on the entire circle.

The explicit dependance on the parameters $\{w_k\}$ also means that we can differentiate the re-parameterisation function with respect to these parameters thus:

[7] Note that we are using a slightly different notation to that used by Hladůvka and Bühler [94].

$$\frac{\partial\phi(u;\{w_k\})}{\partial w_h} = -\frac{1}{\left(\sum\limits_{j=0}^{n} w_j\right)^2}\sum_{k=0}^{n} w_k B_k^n(u) + \frac{1}{\left(\sum\limits_{j=0}^{n} w_j\right)} B_h^n(u)$$

$$= \frac{1}{\left(\sum\limits_{j=0}^{n} w_j\right)^2}\sum_{k=0}^{n} \left(B_h^n(u) - B_k^n(u)\right). \qquad (5.37)$$

This form is then suitable for use in gradient descent methods of optimisation, which is the reason why Hladůvka and Bühler [94] introduced this polynomial representation of re-parameterisation.

The Bernstein representation also lends itself naturally to a multi-resolution approach. If we consider $\{b_k^n(u)\}$ and $\{b_k^{2n}(u)\}$, we see that the maxima (5.31) of the latter are twice as dense as the maxima of the former, the new additional maxima lying halfway between the old maxima.

However, unlike the kernel approaches we considered previously, the positions (and associated widths) of these maxima are not adjustable, which means that the polynomial approach does not have quite the same degree of flexibility as the kernel methods, although it does have considerable computational advantages.

Having considered various approaches to re-parameterising the unit line, we now consider specifically the case of the unit circle.

5.2 Differentiable Re-parameterisations for Closed Curves

As noted before, for closed curves ($S(0) \equiv S(1)$), $\phi(u)$ must be a homeomorphism of the unit circle. The formalism for open curves can be used by just joining the curves at $u = 0 = 1$, but then $\phi(u)$ is in general not differentiable at this point. Making $\phi(u)$ differentiable at $u = 0 = 1$ using the cdf formalism then requires that $\rho(0) = \rho(1)$, which is not generally true for the parametric pdfs we have considered so far (although it is true for the case of the Bernstein polynomial representation).

The solution is to use kernel functions defined around a circle, rather than along the real line, so that the kernel is smooth and everywhere differentiable.

5.2.1 Wrapped Kernel Re-parameterisation for Closed Curves

There are distributions that can be naturally defined directly on the circle (such as the von Mises distribution [189]), which have significance as regards the field of statistics. But in our case, the statistical properties of such distributions (fascinating as they are [111, 69]), is not the issue. We require a flexible approach which can generate differentiable distributions on the circle, and ideally, we would like these distributions to be unimodal, symmetric, of variable width/scale, and also have tractable cdfs. One approach that can achieve this is to create distributions on the circle from suitable distributions on the real line.

There are two basic ways to transfer pdfs from the real line to the circle. The most intuitive is by wrapping [111], an idea that goes back to at least 1917 [154]. Imagine a unit circle rolling along the entirety of the real line. Each time a particular point on the circumference touches the real line, it gathers the value of the pdf associated with that point on the real line. The final wrapped pdf at that point on the circle is then calculated by summing the infinite number of contributions that the point has collected. For a probability density function $f(x),\ x \in \mathbb{R}$, the wrapped pdf is:

$$\sum_{p=-\infty}^{\infty} f(x + 2\pi p). \tag{5.38}$$

The second approach is by mapping, where a one-to-one mapping is constructed from the entire real line to the circle, and the distribution transferred across.

The problem with the wrapping approach is that, in general, the expression for the pdf on the circle involves an infinite summation which is not expressible in closed form. However, for the particular case of the Cauchy distribution, wrapping and the particular mapping we consider here produce identical distributions [116]. We hence describe here the *mapping* approach, and use this to derive the final form of the *wrapped* or *circular* Cauchy distribution [107].

We map the entire real line to the unit circle as follows. The coordinate x on the extended real line[8] maps to the Cartesian coordinates (X, Y) thus [116]:

$$X(x) \doteq \frac{1 - x^2}{1 + x^2},\ Y(x) \doteq \frac{2x}{1 + x^2} \quad \Rightarrow\ X^2 + Y^2 \equiv 1, \tag{5.39}$$

or equivalently, in complex notation:

$$Z(x) \doteq \frac{1 + ix}{1 - ix} = X(x) + iY(x). \tag{5.40}$$

[8] Basically, we have the real numbers plus the point at infinity, to close the circle.

We hence see that the point so defined lies on the unit circle. We can then use the usual polar angle θ, where $X \doteq \cos\theta$, $Y \doteq \sin\theta$. The associated inverse transformation is:

$$x(\theta) = \frac{1}{\sin\theta}(1 - \cos\theta) = \tan\left(\frac{\theta}{2}\right). \tag{5.41}$$

For a pdf $f(x)$ on the real line, the corresponding mapped pdf $\tilde{f}(\theta)$ on the circle is:

$$\tilde{f}(\theta) \doteq \frac{dx}{d\theta} f(x(\theta)). \tag{5.42}$$

We can also relate the cdfs for the pdfs, since:

$$\int_0^\theta \tilde{f}(\psi)d\psi \equiv \int_0^{x(\theta)} f(y)dy. \tag{5.43}$$

Our first instinct might be to take $f(x)$ to be a Gaussian pdf, to obtain the *mapped Gaussian distribution*, with pdf:

$$\begin{aligned} f(x) &\doteq \frac{1}{\sigma\sqrt{2\pi}} \exp\left(-\frac{1}{2\sigma^2}x^2\right), \\ \tilde{f}(\theta) &= \frac{1}{2\sigma\sqrt{2\pi}}\left(1 + \tan^2\frac{\theta}{2}\right)\exp\left(-\frac{1}{2\sigma^2}\tan^2\frac{\theta}{2}\right). \end{aligned} \tag{5.44}$$

However, if we look at the derivative of $\tilde{f}(\theta)$, we see that this mapped distribution is only unimodal for the case $\sigma \leq \frac{1}{\sqrt{2}}$. Hence we cannot explore the full range, from a Dirac δ-function to a flat distribution, by using this kernel. See Fig. 5.10 for examples.

If we perform this mapping for the Cauchy distribution (5.24), and take $f(x) = f(x; w, 0)$, what we obtain is the *wrapped* Cauchy distribution [107, 111],[9] with the density:

$$\begin{aligned} \tilde{f}(\theta; \Omega, 0) &= \frac{1}{2\pi}\frac{1 - \Omega^2}{1 + \Omega^2 - 2\Omega\cos\theta}, \\ \text{where:} \quad \Omega &\doteq \frac{1-w}{1+w}, \quad 0 \leq \Omega \leq 1, \end{aligned} \tag{5.45}$$

where Ω is the angular width parameter. If we compute the first derivative, we see that $\tilde{f}(\theta; \Omega, 0)$ has a single peak at $\theta = 0$ for all allowed values of Ω. As $\Omega \to 0$, we see that $\tilde{f}(\theta; \Omega, 0) \to \frac{1}{2\pi}$, which is a uniform distribution. Conversely, if $\Omega = 1 - \epsilon,\ \epsilon \ll 1$, then $\tilde{f}(\theta; \Omega, 0) \to \frac{\epsilon}{2\pi(1-\cos\theta)}$, which ap-

[9] Also referred to in the literature as the *circular* Cauchy distribution.

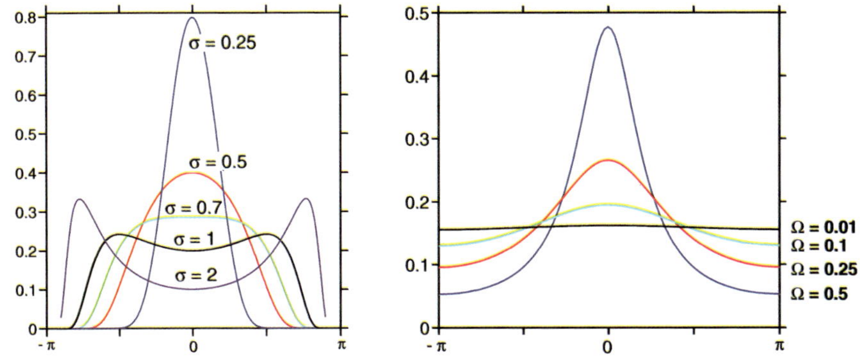

Fig. 5.10 Left: Mapped Gaussian distribution for values of $\sigma = 0.25, 0.5, 0.7, 1$, and 2. **Right:** Mapped Cauchy distribution for values of $\Omega = 0.01, 0.1, 0.25$, and 0.5. For reasons of clarity, functions are plotted over the range $[-\pi, \pi]$ rather than $[0, 2\pi]$.

proaches the Dirac δ-function. Unlike the mapped Gaussian distribution, the mapped/wrapped Cauchy distribution hence explores the full required range (see Fig. 5.10 for examples).

This pdf $\tilde{f}(\theta; \Omega, 0)$ also has a closed-form indefinite integral. Using the relation to the unmapped Cauchy distribution, the cumulative distribution for the wrapped Cauchy is:

$$\begin{aligned}\tilde{g}(\theta; \Omega, 0) &\doteq \int_0^\theta \tilde{f}(\psi; \Omega, 0) d\psi, \\ &= \int_0^u f(x; w, 0) dx = g(u; w, 0), \\ \text{where:} \qquad & u \doteq \tan \frac{\theta}{2}, \;\; w \doteq \frac{1-\Omega}{1+\Omega}. \\ \Rightarrow \tilde{g}(\theta; \Omega, 0) &= \frac{1}{2\pi} \arctan\left(\frac{(1-\Omega^2)\sin\theta}{(1+\Omega^2)\cos\theta - 2\Omega} \right),\end{aligned} \tag{5.46}$$

where the inverse tangent is defined from 0 to 2π. Note that this cdf has the property that:

$$\tilde{g}(\theta; \Omega, 0) = 1 - \tilde{g}(-\theta; \Omega, 0). \tag{5.47}$$

It hence follows that the cumulative distribution for a wrapped Cauchy function of width Ω centred at a is:

$$\tilde{g}(\theta; \Omega, a) = \tilde{g}(a, \Omega, 0) + \tilde{g}(\theta - a, \Omega, 0) - \frac{1}{2}[1 - \text{sign}(\theta - a)], \tag{5.48}$$

where:

$$\mathrm{sign}(x) = \begin{cases} -1 \text{ if } x < 1, \\ 1 \quad \text{otherwise.} \end{cases} \tag{5.49}$$

To represent $\rho(\theta)$ (which we take for the moment to be an un-normalized pdf), we use a set of wrapped Cauchy kernels plus a constant term:

$$\rho(\theta) = \frac{1}{2\pi} + \sum_{k=1}^{n_k} A_k \tilde{f}(\theta; \Omega_k, a_k) \quad 0 \leq \theta \leq 2\pi, \tag{5.50}$$

where A_k is the magnitude of the kth kernel. The cdf $\phi(u)$ is then obtained by integrating $\rho(\theta)$. Using the previous result, we find:

$$\phi(u) \doteq \int_0^{2\pi u} \rho(\theta) d\theta = u + \sum_{k=1}^{n_k} A_k \, \tilde{g}(2\pi u; \Omega_k, a_k), \quad 0 \leq u \leq 1. \tag{5.51}$$

It is then simple to normalize this cdf, to obtain the final expression:

$$\phi(u) = \frac{1}{N} \left[u + \sum_{k=1}^{n_k} A_k \tilde{g}(2\pi u; \Omega_k, a_k) \right] \quad \text{where} \quad N = 1 + \sum_{k=1}^{n_k} A_k. \tag{5.52}$$

As before, the parameters are such that $A_k = 0 \;\forall\; k$ gives the identity re-parameterisation.

Note that, for closed curves, the position of the origin, $\phi(0)$, is still a free parameter. In effect, this additional degree of freedom just corresponds to a uniform rotation of the parameter space for each shape. Optimisation with respect to this degree of freedom will be considered in a later chapter.

5.3 Use in Optimisation

This chapter has looked at several representations of re-parameterisation that are suitable for use in an optimisation framework, using the techniques that will be described in Chap. 7.

Since we are re-parameterising all of the shapes in our training set, and since each parametric re-parameterisation function can involve many parameters, it is obvious that we are dealing with a large-scale optimisation problem. It is therefore advantageous if we can lower the dimensionality of our search space, by reducing the number of free parameters.

If we consider some of the representations given here, such as the kernel-based methods, it can be seen that a typical re-parameterisation function will involve many such kernels, and that each kernel is specified by a set of three parameters, which are basically the position parameter, the width parameter, and the amplitude parameter. For a multiscale or coarse-to-fine optimisation strategy, we need to be able to specify the scales and positions a priori. It hence makes sense to fix the *positions* and *widths* of each kernel, and only search over the amplitude parameter. This obviously reduces the dimensionality of the search space by a factor of three.

Similar considerations hold for other representations of re-parameterisation. In general, we refer to those variables which are held fixed as the *auxiliary* parameters. Table 5.1 lists the auxiliary parameters for each method.

We initialize the search algorithm at the identity re-parameterisation $\phi(u) = u$, and the parameter values required to represent the identity are also listed in the table.

Table 5.1 A summary of the representations of re-parameterisation described in this chapter.

Representation	Optimisation Parameters	Initial Parameter Values	Constraints	Auxiliary Variables
Piecewise-linear (Sect. 5.1.1)	Positions $\{\phi(u_\alpha)\}$ of nodes $\{p_\alpha\}$	$\{\phi(u_\alpha)\} = \{u_\alpha\} = \{\frac{1}{n_\alpha+1}, \frac{2}{n_\alpha+1}, \dots \frac{n_\alpha}{n_\alpha+1}\}$	$\phi(u_\alpha) < \phi(u_{\alpha+1})$	
Recursive piecewise-linear (Sect. 5.1.2)	Node fractional distances, $\{\tau_{\alpha\beta}\}$	$\{\tau_{\alpha\beta}\}$ within allowed range	$0 < \tau_{\alpha\beta} < 1$	
Clamped-plate spline (Sect. 5.1.3)	Node displacements, $\{p_k\}$	$\{p_k\} = \{0, \dots, 0\}$	$\|p_k\| \leq \frac{e}{4}$	The width $\{w_k\}$ and centre point $\{a_k\}$ of each local region
Bounded polynomial (Sect. 5.1.3)	Node displacements, $\{p_k\}$	$\{p_k\} = \{0, \dots, 0\}$	$\|p_k\| < \frac{8}{3\sqrt{3}}$	The width $\{w_k\}$ and centre point $\{a_k\}$ of each local region
Kernel-based representations (Sect. 5.1.4.1,Sect. 5.2.1)	Kernel magnitudes, $\{A_k\}$	$\{A_k\} = \{0, \dots, 0\}$	$A_k \geq 0$	The width $\{\Omega_k\}$ and position, $\{a_k\}$, of each kernel
Bernstein polynomials (Sect. 5.1.4.2)	Weights, $\{w_k : k = 1, \dots n-1\}$	$\{w_k = \beta\}, \beta \geq 0$	$w_k \geq 0$	Polynomial order n

Chapter 6
Parameterisation and Re-parameterisation of Surfaces

In the previous chapter, we showed how correspondence across a set of curves can be manipulated, by parameterising and then re-parameterising the curves. In this chapter, we move up a dimension, and consider the case of shapes in three dimensions – that is, surfaces.

The general approach is the same. However, surfaces present additional challenges and complications. Consider first the initial parameterisation of a shape. For a curve, this is trivially simple, we just have to be able to traverse the curve without retracing our steps, and assign increasing parameter values as we go. Re-parameterisation is also relatively straightforward, since any monotonic function preserves the ordering along a curve and hence is a valid re-parameterisation. No such ordering exists on a surface. This means that the initial parameterisation has to be explicitly constructed, and complicates the construction of re-parameterisation functions.

To recap, suppose we have a shape S. We also have a topological primitive X with the same topology as the shape. The initial parameterisation then proceeds by finding (by some method) a one-to-one mapping $\mathfrak{X}$ between the topological primitive X and the shape S. For the case of curves we considered in the last chapter, the topological primitive corresponding to an open curve is just the open line segment, whereas for closed curves, X is the unit circle. By parameterising the topological primitive, we hence parameterise the original shape. The parameterisation of the topological primitive is often fairly straightforward, so that by a slight abuse of notation, we will use X to represent both the topological primitive and the resulting parameter space, and $\mathfrak{X}$ to represent both the mapping from the topological primitive to the shape, and the mapping from the parameter space to the shape. For our parameterised shape S, we then have:

$$X \overset{\mathfrak{X}}{\longmapsto} S, \; \mathbf{x} \overset{\mathfrak{X}}{\longmapsto} \mathbf{S}(\mathbf{x}), \tag{6.1}$$

where $\mathbf{x}$ is the parameter value of a point on the shape S, and $\mathbf{S}(\cdot)$ is a vector-valued shape function representing the entire shape S:

R. Davies et al., *Statistical Models of Shape*,

DOI: 10.1007/978-1-84800-138-1_6,

$$\mathbf{S}(\mathbf{x}) \doteq (S^x(\mathbf{x}), S^y(\mathbf{x}), S^z(\mathbf{x})) \in \mathbb{R}^3. \tag{6.2}$$

To parameterise a set of shapes $\{S_i : i = 1, \ldots n_S\}$, all with the same topology, we then have to construct a mapping $\mathfrak{X}_i$ from the common parameter space X to each shape in the set. Correspondence between shapes is then defined by common parameter value $\mathbf{x}$, so that:

$$\begin{aligned} X \overset{\mathfrak{X}_i}{\longmapsto} S_i, \ \mathbf{x} \overset{\mathfrak{X}_i}{\longmapsto} \mathbf{S}_i(\mathbf{x}), \\ X \overset{\mathfrak{X}_j}{\longmapsto} S_j, \ \mathbf{x} \overset{\mathfrak{X}_j}{\longmapsto} \mathbf{S}_j(\mathbf{x}), \\ \therefore\ \mathbf{S}_i(\mathbf{x}) \sim \mathbf{S}_j(\mathbf{x}), \end{aligned} \tag{6.3}$$

where $\sim$ denotes correspondence. Correspondence is then manipulated by re-parameterising the topological primitive:

$$\begin{aligned} \mathbf{x} \overset{\phi_i}{\longmapsto} \phi_i(\mathbf{x}) \ \& \ \mathbf{S}_i \overset{\phi_i}{\longmapsto} \mathbf{S}'_i, \ \text{where } \mathbf{S}'_i(\phi_i(\mathbf{x})) \doteq \mathbf{S}_i(\mathbf{x}) \ \forall\ i = 1, \ldots n_S, \\ \therefore\ \mathbf{S}_i(\mathbf{x}) \sim \mathbf{S}_j(\mathbf{x}) \overset{\phi_i,\phi_j}{\longmapsto} \mathbf{S}'_i(\mathbf{x}) \sim \mathbf{S}'_j(\mathbf{x}), \end{aligned} \tag{6.4}$$

where ϕ_i is the re-parameterisation function for the ith shape.

In this chapter, we discuss how to construct such parameterisations and re-parameterisations for surfaces. We concentrate first on the cases of open surfaces (with a single boundary), and closed surfaces with spherical topology. The extension to the case of other topologies (e.g., toroidal) will also be discussed.

6.1 Surface Parameterisation

Our input surfaces $\{S_i\}$ are typically represented as triangulated meshes. For a single shape S, a triangulated mesh consists of a set of triangles $\{\mathbf{t}^\alpha\}$, where $\mathbf{v}^\alpha$ represents the vertices of the triangle $\mathbf{t}^\alpha$, so that:

$$\mathbf{v}^\alpha \doteq \{\mathbf{v}^{\alpha a} \in \mathbb{R}^3 : a = 1, 2, 3\}. \tag{6.5}$$

The triangles are of course joined, so that it will sometimes be convenient to refer to the entire set of non-coincident nodes $\{\mathbf{v}^A\}$, where the triangulation is constructed by assigned nodes to triangles thus:

$$\mathbf{v}^\alpha = \{\mathbf{v}^{\alpha 1}, \mathbf{v}^{\alpha 2}, \mathbf{v}^{\alpha 3}\} = \{\mathbf{v}^A, \mathbf{v}^B, \mathbf{v}^C\}, \tag{6.6}$$

for some set of values $\{A, B, C\}$. A given node $\mathbf{v}^A$ will typically be a vertex of several triangles, where $\mathbf{v}^A \in \mathbf{t}^\alpha \ \Rightarrow\ \mathbf{v}^A \in \mathbf{v}^\alpha$. We then also define the set of neighbours N_A of a vertex $\mathbf{v}^A$, where:

$$N_A \doteq \{\mathbf{v}^B \neq \mathbf{v}^A : \mathbf{v}^A \in \mathbf{t}^\alpha,\ \ \mathbf{v}^B \in \mathbf{t}^\alpha\},$$
$$\text{or } N_A \doteq \{B : \mathbf{v}^B \neq \mathbf{v}^A, \mathbf{v}^A \in \mathbf{t}^\alpha,\ \ \mathbf{v}^B \in \mathbf{t}^\alpha\} \tag{6.7}$$

as convenient.

The continuous mapping $\mathfrak{X}$ between the shape S and the parameter space X is typically constructed by first defining the mapping between the vertices of the mesh $\{\mathbf{v}^A\}$ and their associated set of parameter values $\{\mathbf{x}^A \in X\}$. The full mapping from X to S is then constructed by interpolating the values of $\{\mathbf{x}^A\}$ (hence $\{\mathbf{v}^A\}$) in between the vertices, using the common triangulation $\{\mathbf{t}^\alpha\}$.

Mesh parameterisation has been a field of intense research activity over the last decade – see [70] or [162] for a review. Most approaches have focused on finding a parameterisation that minimises some measure of distortion between the triangulated mesh on the shape, and the triangulated mesh in parameter space. Typical optimisation criteria include angle distortion (e.g., approximate [96] or exact [3, 85] conformal mappings, where a conformal mapping is one that preserves all angles), area distortion (e.g., [196]), changes in edge length, or a combination of these (see [162] for a review).

For the purposes of statistical shape modelling, it has been found that useful criteria are that the parameterisation should be consistent across a set of examples, and that the distortion of area should be minimal. The set of parameterisations define the initial correspondence from which optimisation of correspondence by re-parameterisation begins. An inconsistent set of parameterisations can hence lead to a poor initial correspondence, which then hinders the optimisation algorithm's ability to find the global minimum, as well as leading to a long convergence time. Area distortion leads to under- or over-sampling of the surfaces, giving a poor representation of the shape. This is because sampling is performed evenly over the parameter space, rather than directly on the shapes (see Sect. 7.3.3).

Many published methods of parameterisation (see review articles [70, 162]) are sufficient for certain classes of objects. Here, we describe a general method of parameterisation, which is suitable for building statistical shape models of many classes of objects. An initial discrete parameterisation is first achieved, based on the work in [71] and [14]. This discrete parameterisation is then interpolated (using barycentric interpolation) to create a continuous parameterisation. This parameterisation can then be further modified, to address the criteria of area distortion and inconsistency mentioned above, and a final parameterisation reached by optimisation [54] as will be described below.

6.1.1 Initial Parameterisation for Open Surfaces

We first consider the initial parameterisation of open surfaces, a parameterisation which can then be further refined as will be discussed in Sects. 6.1.4 and 6.1.5.

A simply connected[1] open surface with a single boundary (i.e., no holes or handles), is topologically equivalent to the unit disc in the plane. For the purposes of calculation, it is more convenient to use the unit square, so that:

$$X = \{\mathbf{x}\}, \mathbf{x} = (x, y), 0 \leq x \leq 1, 0 \leq y \leq 1. \tag{6.8}$$

The first stage of the parameterisation is mapping the nodes $\{\mathbf{v}^A\}$ of the triangulated mesh. This starts by mapping all the nodes at the boundary of the mesh to the boundary of the unit square. The spacing between nodes on the square is adjusted so that the relative spacing between neighbouring boundary nodes is preserved. Additional nodes then need to be placed at the corners of the unit square to ensure the entire domain is covered, with corresponding nodes added on the boundary of the shape.

For the remaining non-boundary nodes, let us consider first adjusting the position of just one node, $\mathbf{x}^A$, say, leaving the positions of all other nodes fixed. The position of the moving node can be written as a linear combination of the positions of its neighbours:

$$\mathbf{x}^A = \sum_{B \in N_A} w_{AB} \mathbf{x}^B, \quad \sum_{B \in N_A} w_{AB} = 1, \tag{6.9}$$

with weights $\{w_{AB}\}$. In general, the weights can be taken to depend on the distances between nodes on the triangulated shape, so that:

$$w_{AB} = f\left(\|\mathbf{v}^A - \mathbf{v}^B\|, \{\|\mathbf{v}^A - \mathbf{v}^C\| : C \in N_A\}\right). \tag{6.10}$$

The function f can then be adjusted so that nodes which are closest together on the shape, remain closest in parameter space, and so on. The simplest approach is to assume that all the weights $\{w_{AB}\}$ have equal value $w_{AB} = \frac{1}{\text{size}(N_A)}$, so that a node $\mathbf{x}^A$ is placed at the mean position of its neighbours.

If we now consider moving all the nodes, (6.9) then defines a set of equations. Let M be the set of indices of the movable, non-boundary nodes, and F be the set of indices of the fixed boundary nodes. The position vectors of the movable nodes are collected into a vector of vectors $\mathbf{X} = \{\mathbf{x}^A : A \in M\}$. We then construct a matrix of scalars $\mathbf{P}$, where:

$$\begin{gathered}\mathbf{P} = \{P_{AB} : A \in M, B \in M\}, \\ P_{AA} = 1. \text{ If } B \in N_A, P_{AB} = -w_{AB}, \text{ else } P_{AB} = 0.\end{gathered} \tag{6.11}$$

[1] *Simply connected* means that any path between any pair of points can be continuously deformed into any other such path.

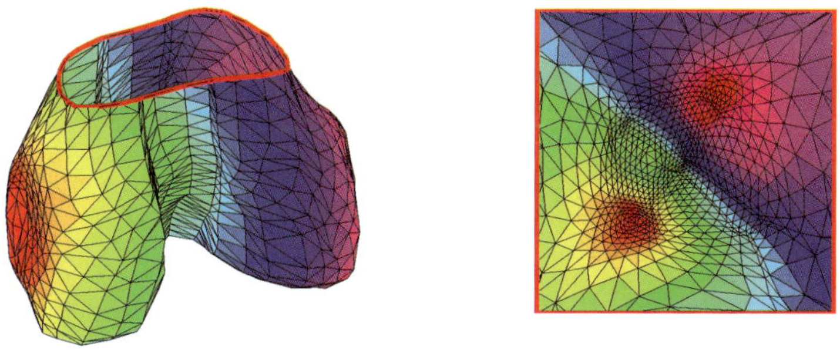

Fig. 6.1 Initial parameterisation of an open surface, using the method in (6.13). **Left**: A view of the triangulated open surface in $\mathbb{R}^3$, representing the distal end of a human femur. **Right**: The triangulated mesh mapped to the parameter space of the unit square (6.8). The colours denotes the correspondence between the meshes, and the thick red curve represents the boundary of the surface.

A second vector of vectors $\mathbf{Y}$ is constructed, where:

$$\mathbf{Y}_A = \sum_{B \in F \& B \in N_A} w_{AB} \mathbf{x}^B. \tag{6.12}$$

The set of equations can then be written compactly in matrix form as:

$$\mathbf{PX} = \mathbf{Y}, \tag{6.13}$$

which is solved for the variable $\mathbf{X}$, the collection of positions of the movable nodes. There are various algorithms that can be used to solve this equation (see Chap. 2 of [139]), but the conjugate gradient method works well.

An example of an open surface parameterised using this method is given in Fig. 6.1. It can be seen from the figure that the mapping is correct,and that none of the triangles have flipped. Note that on the original shape, the triangles of the mesh are approximately of equal area. This is not the case on the parameter space, where triangles in the proximity of the boundary tend to be much larger than triangles further from the boundary. Refining this parameterisation to minimise the area distortion is dealt with in a later section.

6.1.2 Initial Parameterisation for Closed Surfaces

In this section, we consider building the initial parameterisation for a triangulated closed surface with the topology of a sphere. As for the case of triangulated open surfaces, we first construct the mapping for the nodes of the mesh, and then interpolate to find the continuous mapping.

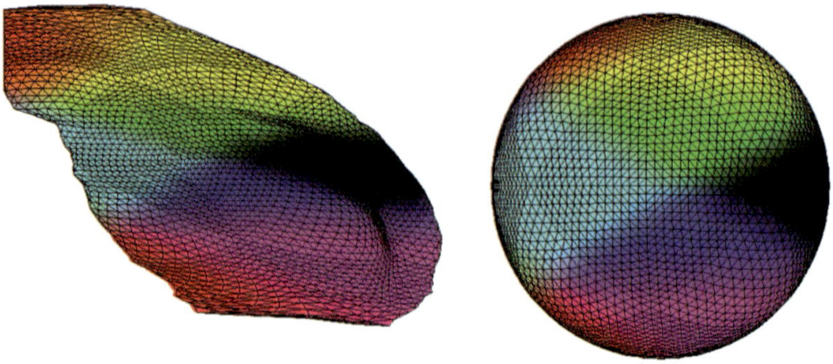

Fig. 6.2 Initial parameterisation of a closed surface. **Left:** Triangulated mesh for a closed surface, representing the anterior horn of a human brain ventricle. **Right:** The triangulated mesh in the parameter space $\mathbb{S}^2$. The colours denote the correspondence between the meshes.

The parameter space for our closed shapes is then just the unit sphere. The coordinates of a general point $\mathbf{x}$ in the parameter space are then the usual polar coordinates (θ, ψ), with $0 \leq \theta \leq \pi$, $0 \leq \psi < 2\pi$.

The initial parameterisation can be constructed using a similar method to that in (6.9). However, unlike open surfaces, there are no boundary nodes, so we have to impose other boundary conditions in order to obtain a unique solution.

One way to do this is as follows. We pick two nodes on the surface which lie at opposite extremes of the shape – for example, we could pick the nodes with the minimum and maximum values of the z coordinate. We will denote these nodes by $\mathbf{v}_{NP}$ and $\mathbf{v}_{SP}$, which stands for north pole and south pole, respectively. These are mapped to the poles of the unit sphere, with $\theta_{NP} = \pi$, and $\theta_{SP} = 0$.

Setting appropriate boundary conditions for the ψ coordinate is more difficult as it is periodic. We can accommodate this by holding fixed a Greenwich meridian ($\psi = 0$). The nodes which map to this meridian are chosen as those nodes which lie on the shortest path through the shape mesh between $\mathbf{v}_{NP}$ and $\mathbf{v}_{SP}$, and this shortest path can be calculated by using Dijkstra's algorithm [60].

This then provides sufficient boundary conditions to yield a unique solution, using an analogous construction to that given for the case of open surfaces. An example of such an initial spherical parameterisation is given in Fig. 6.2.

6.1.3 Defining a Continuous Parameterisation

So far, we have shown how to map triangulated meshes from shape surfaces to the corresponding topological primitive, from the triangulated mesh $(\{\mathbf{v}^A\}, \{\mathbf{t}^\alpha\})$ to the mapped mesh $(\{\mathbf{x}^A\}, \{\mathbf{t}^\alpha\})$, and how this mapping can be controlled to respect various properties of the original mesh, and preserve its connectivity. We now show how such a mapping can be interpolated, so as to create a continuous mapping between the shape and parameter space.

Let $\mathbf{v}$ be a general point on the shape surface. Suppose this point lies within some triangle $\mathbf{t}^\alpha$. The position of this point can then be described in terms of the positions of the vertices of this triangle $\{\mathbf{v}^{\alpha a}\}$ by the use of barycentric coordinates $(a^\alpha, b^\alpha, c^\alpha)$. Let $\text{Area}(\mathbf{u}, \mathbf{v}, \mathbf{w})$ denote the area of the triangle formed by the three points $\mathbf{u}, \mathbf{v}$, and $\mathbf{w}$. Then the barycentric areal [45] coordinates are defined thus:

$$a^\alpha(\mathbf{v}) \doteq \frac{\text{Area}(\mathbf{v}, \mathbf{v}^{\alpha 1}, \mathbf{v}^{\alpha 2})}{\text{Area}(\mathbf{v}^{\alpha 1}, \mathbf{v}^{\alpha 2}, \mathbf{v}^{\alpha 3})} \tag{6.14}$$

$$b^\alpha(\mathbf{v}) \doteq \frac{\text{Area}(\mathbf{v}, \mathbf{v}^{\alpha 1}, \mathbf{v}^{\alpha 3})}{\text{Area}(\mathbf{v}^{\alpha 1}, \mathbf{v}^{\alpha 2}, \mathbf{v}^{\alpha 3})} \tag{6.15}$$

$$c^\alpha(\mathbf{v}) \doteq \frac{\text{Area}(\mathbf{v}, \mathbf{v}^{\alpha 2}, \mathbf{v}^{\alpha 3})}{\text{Area}(\mathbf{v}^{\alpha 1}, \mathbf{v}^{\alpha 2}, \mathbf{v}^{\alpha 3})}. \tag{6.16}$$

The position of the general point is recovered from the barycentric coordinates via:

$$\mathbf{v} \equiv a^\alpha \mathbf{v}^{\alpha 1} + b^\alpha \mathbf{v}^{\alpha 2} + c^\alpha \mathbf{v}^{\alpha 3}. \tag{6.17}$$

The barycentric coordinates for the triangle inside which the point $\mathbf{v}$ lies satisfy:

$$0 \leq a^\alpha \leq 1,\ 0 \leq b^\alpha \leq 1,\ 0 \leq c^\alpha \leq 1 \ \&\ a^\alpha + b^\alpha + c^\alpha = 1. \tag{6.18}$$

The triangle that contains $\mathbf{v}$ can hence be found using an exhaustive search, then the barycentric coordinates are then used in an obvious fashion to define the corresponding point $\mathbf{x}$ in the corresponding triangle on the topological primitive/parameter space X:

$$\mathbf{x}(\mathbf{v}) \doteq a^\alpha \mathbf{x}^{\alpha 1} + b^\alpha \mathbf{x}^{\alpha 2} + c^\alpha \mathbf{x}^{\alpha 3}. \tag{6.19}$$

The inverse transformation $\mathbf{x} \stackrel{\mathfrak{X}}{\longmapsto} \mathbf{v}(\mathbf{x})$ is defined analogously.

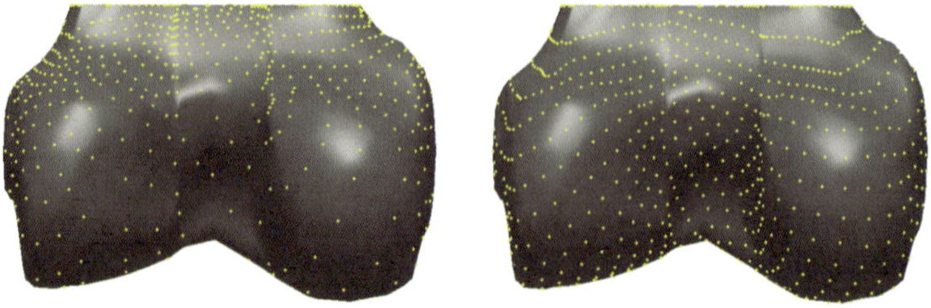

Fig. 6.3 An example of a distal femur sampled according to its parameterisation before (**Left**) and after (**Right**) correcting for areal distortion.

6.1.4 Removing Area Distortion

We have shown how to construct an initial continuous parameterisation for both open and closed surfaces. However, as can be seen from Fig. 6.1, this initial mapping can cause considerable areal distortion, with triangles that are a similar size on the shape mapping to triangles with very different areas in the parameter space.

The extent of areal distortion can be quantified by the following objective function. Using the previous notation, $\text{Area}(\mathbf{v}^{\alpha}) \equiv \text{Area}(\mathbf{v}^{\alpha 1}, \mathbf{v}^{\alpha 2}, \mathbf{v}^{\alpha 2})$, which is the area of triangle $\mathbf{t}^{\alpha}$ on the shape. The objective function can then be written as:

$$\mathcal{L} = \sum_{\alpha} \left(\frac{\text{Area}(\mathbf{v}^{\alpha})}{\sum\limits_{\beta} \text{Area}(\mathbf{v}^{\beta})} - \frac{\text{Area}(\mathbf{x}^{\alpha})}{\sum\limits_{\beta} \text{Area}(\mathbf{x}^{\beta})} \right)^2 . \tag{6.20}$$

We hence see that this algorithm compares the fractional area of a triangle on the shape with fractional area of the corresponding triangle on the parameter space. The node positions, $\{\mathbf{x}^A\}$, on the parameter space can be manipulated using any of the representations of re-parameterisations described later in this chapter, in order to minimise this objective function.

The benefit of correcting for areal distortion is illustrated in Fig. 6.3. The point is that from now on, we will be working with surfaces using this parameterisation. For instance, surfaces will be sampled according to the parameterisation. The figure shows a surface sampled according to a parameterisation before and after correction for areal distortion. It can be seen that the corrected parameterisation gives a much more even sampling of the shape surface that the uncorrected one.

Note that we need only perform this correction for a single, reference shape, since as will be shown in the next section, this can be propagated to all of the other examples in the training set of shapes.

6.1.5 Consistent Parameterisation

So far, we have just considered the problem of parameterising a single shape, and most methods of shape parameterisation only go this far. But for our application of building models, we have to parameterise an entire training set of shapes, not just a single example.

If the group of shapes is not taken into consideration, the parameterisations of individual shapes are unlikely to be consistent with other parameterisations in the group. The problem is demonstrated in Fig. 6.4, which shows the initial correspondence between a pair of surfaces, parameterised using the method described in Sect. 6.1.

Note that our training set of shapes is assumed to be aligned, so that we have a rough initial correspondence defined across the set. The idea is now to manipulate the parameterisations, so that they are consistent across the set, according to this initial definition of correspondence.

Suppose we take one shape as a reference. All the shapes in the set are mutually aligned, which hence defines an initial correspondence across the set. And an initial parameterisation has also been generated for each shape. For the initial shape, we also adjust its parameterisation for areal distortion, as in Fig. 6.3.

We then decimate the parameterisation of the reference shape. It was found that retaining about 20% of the original nodes was usually sufficient. Note that this does not mean losing the fidelity of the shape representation, since the original dense triangulation is maintained and manipulated. Decimation just reduces the number of degrees of freedom of the problem, and eases computation.

Let $\{\tilde{\mathbf{v}}^A_{ref}\}$ be the set of nodes of the decimated reference shape, and $\{\mathbf{v}^B_i\}$ the set of nodes for the *undecimated* triangulation of the ith shape in the set. A correspondence is then generated between each point of $\{\tilde{\mathbf{v}}^A_{ref}\}$, and its closest point (according to the point-to-point distances between our *aligned* set of training shapes). This then defines an index correspondence between the reference and each of the other shapes, where:

$$I_i(A) = B, \text{ where } B = \arg\min\left(\|\tilde{\mathbf{v}}^A_{ref} - \mathbf{v}^B_i\|\right). \tag{6.21}$$

We now manipulate the parameterisation of the ith shape, using this correspondence derived from the aligned shapes. Our objective function measures the degree to which the parameterisations are consistent with this closest-point correspondence defined above, which is quantified by:

$$\mathcal{L} \doteq \sum_A \|\tilde{\mathbf{x}}^A_{ref} - \mathbf{x}^{I_i(A)}_i\|^2. \tag{6.22}$$

To summarize, the nodes of the decimated triangulations are held fixed on the reference shape, and the full triangulation is held fixed on the ith shape. They

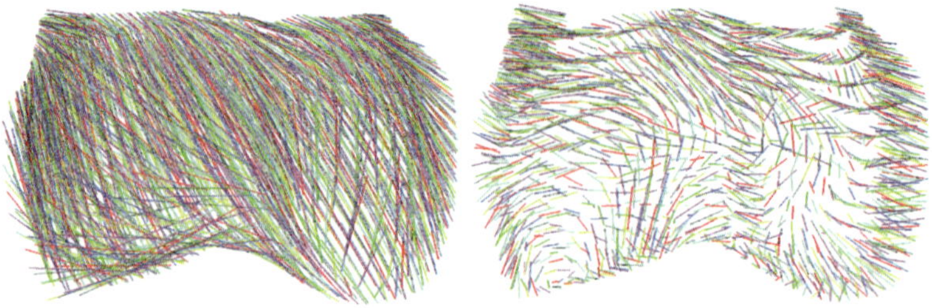

Fig. 6.4 The correspondence according to parameter value of a pair of distal femurs, where the coloured lines join corresponding points on the pair of shapes. **Left:** SPHARM parameterisation [14, 99, 76]; **Right:** groupwise consistent parameterisation.

are placed in correspondence so as to minimise the sum of squared distances between nodes on the shapes. We then manipulate the parameterisation of the ith shape (i.e., the points $\{\mathbf{x}_i^B\}$), so as to also try to minimise the sum of squared distances between corresponding points, but now in parameter space. The node positions $\{\mathbf{x}_i^B\}$ on the parameter space can be manipulated using any of the representations of re-parameterisation which will be described next in this chapter.

Figure 6.4 illustrates the effect of producing a consistent set of parameterisations. The correspondence illustrated here is that given by equal parameter value, rather than the closest-point referred to above. It can be seen that after optimisation, points on different shapes with equal parameter values are much closer together on the physical shapes than before the optimisation.

In this section, we have shown how to generate initial parameterisations for both closed and open surfaces. We have also shown how to construct objective functions which allow these initial parameterisations to be refined, generating parameterisations that minimise areal distortion, as well as parameterisations that are consistent across a set of shapes.

The final issue is how to manipulate these parameterisations, not just in order to refine the initial parameterisations as given above, but also to solve the final groupwise correspondence problem.

6.2 Re-parameterisation of Surfaces

In this section, we consider how to generate legal re-parameterisations for surfaces. As was shown in Chap. 3, and demonstrated in the previous chapter for the specific case of shapes in two dimensions (curves), such re-parameterisations involve generating homeomorphic[2] mappings of the parameter space into itself.

We need to recall here what we are actually trying to achieve, which is exploring the space of groupwise correspondences across a set of shapes, in order to locate the minimum of some groupwise objective function. Homeomorphisms then correspond to legal moves in this space.

As regards solving the optimisation problem itself, there is another consideration: such optimisation problems of groupwise or pairwise correspondence of either shapes or images [23] are in general ill-posed[3] in the Hadamard sense [86], and are not solvable without some form of regularization. From the point of view of practical computation, we require a finite-dimensional search space. There are several different approaches to such regularization. In what follows, we take the approach of explicitly constructing homeomorphic transformations that belong to some finite-dimensional space of parameterised transformations (that is, parametric re-parameterisation functions, which we can think of as *hard regularization*). The alternative approach, of *soft regularization* and non-parametric re-parameterisation functions, is considered in a later chapter (see Chap. 8).

6.2.1 Re-parameterisation of Open Surfaces

In a previous section, we describe how open surfaces could be mapped to the unit square, to provide an initial parameterisation.

The relevant re-parameterisation function is then a homeomorphic mapping of the unit square. Many such mappings have been defined in the field of non-rigid image registration (for example, [22, 149, 75]). The fluid-based representation of Christensen [22] (a *soft* regularization method) will be considered later (see Chap. 8). In the current section, we will consider just the *hard* regularization methods, based on sets of parameterised homeomorphic transformations. This can be considered as a generalization and an extension of the piecewise-linear and local representations of re-parameterisation defined previously for open curves (see Sect. 5.1).

6.2.1.1 Recursive Piecewise Linear Re-parameterisation

As noted in the previous chapter, the one-dimensional piecewise-linear representation of curve re-parameteristaion (in Sect. 5.1.1) has no natural extension to surface re-parameterisation since the explicit ordering constraints in

[2] In technical terms, a *homeomorphic* mapping (homeomorphism) is one which is one-to-one, and continuous in both directions. A *diffeomorphism* is slightly more restricted, in that the mapping is differentiable (to some order), rather than just continuous.

[3] A *well-posed* problem has a unique solution, which depends continuously on the data.

(5.7) cannot be applied on two-dimensional surfaces. The recursive piecewise-linear representation (Sect. 5.1.2), on the other hand, only uses an implicit ordering, allowing a straightforward extension to surface re-parameterisation. In Sect. 5.1.2, we also briefly discussed an extension that used recursive subdivision of triangles, which is of general applicability. However, for the case of open surfaces, the parameter space is a unit square, which suggests a variant based on rectangles.

The method proceeds as follows. Suppose we start with the set of four nodes that correspond to the corners of the entire unit square. The representation is refined by successively adding nodes in between those already present. Each new node subdivides the parameter space into rectangles, which are then further subdivided. In effect, we are building a simple tiling of the unit square, with rectangular tiles, the nodes at the corners of these rectangles forming a mesh.

Let us consider a single rectangle of this mesh, defined by the four nodes with positions:

$$\square^{\alpha\gamma\beta\delta} = \{\mathbf{p}_\alpha, \mathbf{p}_\gamma, \mathbf{p}_\beta, \mathbf{p}_\delta\}. \tag{6.23}$$

We will assume that we have Cartesian coordinates (x, y), with *unit* axis vectors $\hat{\mathbf{x}}, \hat{\mathbf{y}}$, which are aligned with the sides of the original unit square. This means that to uniquely specify a rectangle, we need only give the positions of a pair of diagonal nodes, rather than the positions of all four corners. Compressing notation somewhat, we then refer to the rectangle $\square^{\alpha\beta}$, with node positions:

$$\begin{aligned} &\square^{\alpha\beta} \doteq \{\mathbf{p}_\alpha, \mathbf{p}_\gamma, \mathbf{p}_\beta, \mathbf{p}_\delta\}, \\ \text{where: } &\mathbf{p}_\gamma = \mathbf{p}_\alpha + ((\mathbf{p}_\beta - \mathbf{p}_\alpha)\cdot\hat{\mathbf{y}})\,\hat{\mathbf{y}},\ \ \mathbf{p}_\delta = \mathbf{p}_\alpha + ((\mathbf{p}_\beta - \mathbf{p}_\alpha)\cdot\hat{\mathbf{x}})\,\hat{\mathbf{x}}. \end{aligned} \tag{6.24}$$

We then place a new daughter node $\mathbf{p}$ within this rectangle. Its position is totally specified by the fractional distances along each of the sides of the parent rectangle $\square^{\alpha\beta}$, so that:

$$\begin{aligned} \mathbf{p} &\doteq p_x\hat{\mathbf{x}} + p_y\hat{\mathbf{y}}, \\ p_x &\doteq (\mathbf{p}_\alpha\cdot\hat{\mathbf{x}}) + \tau_x^{\alpha\beta}\left((\mathbf{p}_\beta - \mathbf{p}_\alpha)\cdot\hat{\mathbf{x}}\right), \\ p_y &\doteq (\mathbf{p}_\alpha\cdot\hat{\mathbf{y}}) + \tau_y^{\alpha\beta}\left((\mathbf{p}_\beta - \mathbf{p}_\alpha)\cdot\hat{\mathbf{y}}\right). \qquad (6.25) \\ \boldsymbol{\tau}^{\alpha\beta} &\doteq (\tau_x^{\alpha\beta}, \tau_y^{\alpha\beta}). \qquad (6.26) \end{aligned}$$

The parent rectangle $\square^{\alpha\beta}$ is hence subdivided into four new rectangles by this daughter node. Note that this process also introduces subsidiary nodes (as shown in Fig. 6.5), so that each new rectangle possesses four nodes.

Each of these new rectangles may then be subdivided in their turn by the addition of further daughter nodes. The advantage of this scheme is that it produces a self-similar tiling, hence can be extended to any required depth of recursion without altering the essential properties of the tiling.

Legal re-parameterisations of the unit square are then created by moving the *daughter* nodes from their original positions, with the nodes at the corners of the original unit square itself remaining fixed. The movement of a daughter node remains valid as long as it remains within the rectangle defined by its parents. The subsidiary nodes attached to a daughter node move with it. And when the parents of a node move, the daughter moves proportionately, so that the transformation remains homeomorphic. This pattern of movement is then bilinearly interpolated to define the movement at any point within the original unit square, which yields the final homeomorphic re-parameterisation function.

Although the mesh we have created looks quite complicated (see Fig. 6.5), the parameters of the re-parameterisation function itself are quite simple. For each daughter node, the free parameters are the fractional distances $\boldsymbol{\tau} \doteq (\tau_x, \tau_y)$, where $0 < \tau_x < 1$, $0 < \tau_y < 1$. We hence see that for a mesh with n daughter nodes (the entire mesh contains these daughter nodes and all their associated subsidiary nodes, hence contains far more than n nodes), the full space of parameters for the re-parameterisation function generated by this set of daughter nodes is just the unit hypercube in $\mathbb{R}^{2n}$. The identity re-parameterisation function corresponds to the daughter nodes not moving from their initial positions, and their allowed initial positions correspond to any point within the unit hypercube.

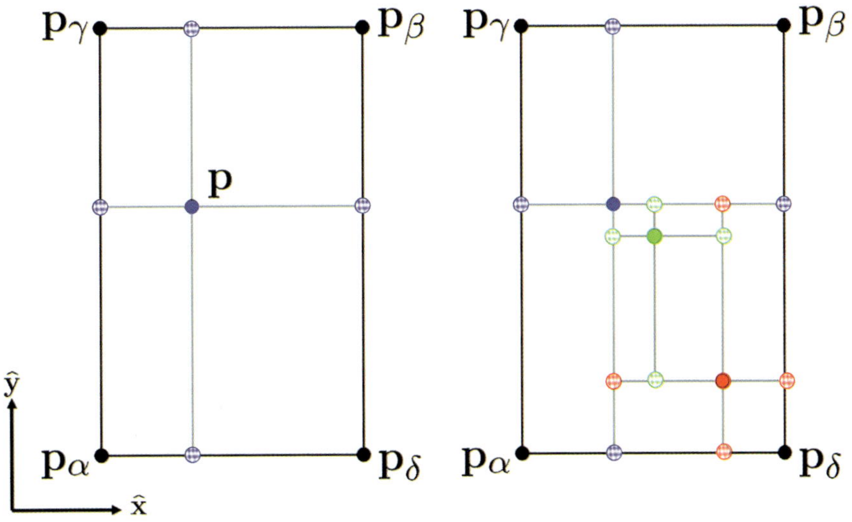

Fig. 6.5 Left: Subdivision of the parent rectangle $\square^{\alpha\beta}$ (black lines) by a daughter node $\mathbf{p}$ (blue circle). **Right:** Recursive subdivision, with daughter nodes displayed as circles filled with solid colour. The corresponding subsidiary nodes are the stippled filled circles of the same colour.

This representation is also *localized*, in the sense that the movement of any daughter node affects only the re-parameterisation of those points which lie within the associated parent rectangle. The method hence facilitates the use of a coarse-to-fine or a multiscale optimisation scheme.

The major drawback is that the re-parameterisation function itself is not differentiable at the points or edges of the mesh. This issue of differentiability will be addressed in the next section.

6.2.1.2 Localized Re-parameterisation

As we saw in the previous chapter, the piecewise-linear representation can be improved upon by instead using a composition of smoother, local transformations.

We could just consider a traditional spline-based representation of deformation, such as the free-form deformation image registration framework of Rueckert et al. [151], based on B-splines [57]. However, it is possible to use a simpler approach.

For the purposes of efficient computational optimisation, we would like a transformation that is localized in its effect. We would also like the scale of the affected region to be variable, to allow a coarse-to-fine or a multiscale optimisation approach. On the grounds of symmetry, the simplest region to consider is a circular disc, where the radius of the disc provides the scale.

Hence, without loss of generality, let us consider applying a transformation to the unit disk, centred at the origin. The simplest transformation can be described as a *localized translation*, where the centre of the disc translates, and all other points within the disc translate in the same direction, but with varying amplitude.

Let us parameterise this transformation by a vector $\mathbf{p}$, which represents the new position of the centre of the disc, hence the motion at the origin. The general form of a homeomorphic re-parameterisation function that can be built from such a transformation is as in the following theorem.

Theorem 6.1. Homeomorphic Local Translations.
Consider the strictly localized re-parameterisation function defined by:

$$\phi(\mathbf{x}) = \begin{cases} \mathbf{x} + h(||\mathbf{x}||)\mathbf{p} & \text{if } |\mathbf{x}| < 1 \\ \mathbf{x} & \text{otherwise} \end{cases}$$
$$\text{where } h(r) \geq 0 \ \forall \ r \geq 0,$$
$$h(0) = 1, \ \ h(1) = 0, \ \ \& \ h'(1) = 0. \tag{6.27}$$

$\phi(\mathbf{x})$ *is then homeomorphic provided that:*

$$\|\mathbf{p}\| < \frac{1}{|h'(r)|} \ \forall \ r. \tag{6.28}$$

Proof. Since all movement is in the direction of $\mathbf{p}$*, we do not have to check for folding due to movement of points whose initial separation contains some non-zero component perpendicular to this direction. It is hence sufficient to consider the relative movement of two points with initial positions* $\mathbf{x}$ *and* $\mathbf{x} + \epsilon\mathbf{p}$*, where* $0 < \epsilon \ll 1$*. Hence:*

$$\begin{aligned} \|\mathbf{x} + \epsilon\mathbf{p}\| &= \|\mathbf{x}\| + \frac{\epsilon\mathbf{p}\cdot\mathbf{x}}{\|\mathbf{x}\|} + O\left(\epsilon^2\right). \\ \therefore\ \phi(\mathbf{x}) &= \mathbf{x} + h(\|\mathbf{x}\|)\mathbf{p}, \\ \&\ \phi(\mathbf{x} + \epsilon\mathbf{p}) &= \phi(\mathbf{x}) + \epsilon\mathbf{p} + \epsilon\left(\frac{\mathbf{p}\cdot\mathbf{x}}{\|\mathbf{x}\|}\right)h'(\|\mathbf{x}\|)\mathbf{p} + O\left(\epsilon^2\right). \end{aligned} \tag{6.29}$$

The re-parameterisation is hence homeomorphic provided that:

$$\begin{aligned} \mathbf{p}\cdot\left(\phi(\mathbf{x}+\epsilon\mathbf{p}) - \phi(\mathbf{x})\right) > 0 \ &\Rightarrow\ 1 + \left(\frac{\mathbf{p}\cdot\mathbf{x}}{\|\mathbf{x}\|}\right)h'(\|\mathbf{x}\|) > 0, \\ &\Rightarrow\ p \doteq \|\mathbf{p}\| < \frac{1}{|h'(\|\mathbf{x}\|)|}. \end{aligned} \tag{6.30}$$

□

We saw two examples of similar functions in the previous chapter. The first (Sect. 5.18) is based on the clamped-plate spline (CPS) [112]. The biharmonic clamped-plate Green's function in two dimensions [9] leads to the function:

$$\textbf{CPS:}\quad h(r) = 1 - r^2 + r^2\ln(r^2),\ \text{constraint: } p < \frac{e}{4}, \tag{6.31}$$

The second is the bounded polynomial representation (Sect. 5.19):

$$\textbf{Polynomial:}\quad h(r) = \left(1 - r^2\right)^2,\ \text{constraint: } p < \frac{3\sqrt{3}}{8}, \tag{6.32}$$

We can also build such functions based on trigonometric functions, for example:

$$\textbf{Trigonometric:}\quad h(r) = \frac{1}{2}\left(1 + \cos(\pi r)\right),\ \text{constraint: } p < \frac{2}{\pi}. \tag{6.33}$$

Whichever form is chosen, several such local transformations can then be concatenated to produce a more general transformation (see Fig. 6.6 for examples). The combined transformation then depends on a set of control points, and control point motions, but also depends on the *order* in which the individual transformations are combined.

It is of course also possible to consider a more general spline interpolant, based on a set of control points, which does not depend on such an ordering.

We do not propose to consider such splines here, but just consider the simple case (sufficient for our purposes) of *localized* interpolants.

Consider a set of control points defined over a regular grid of positions. We say that the field of movement across the grid is interpolated from the movements defined at the control points. If the movement within each square only depends on the control points at the four corners of that square, then we have the *localized* interpolant that we require.

Let us consider a single such square, aligned with the coordinate axes. For ease of computation, we scale the square to unit size, with $0 \leq x \leq 1$, and $0 \leq y \leq 1$. Let $\mathbf{p}(x, y)$ denote the interpolated movement at the point (x, y), with corresponding re-parameterisation function:

$$\phi(x, y) \doteq (x, y) + \mathbf{p}(x, y).$$

We then consider localized interpolants of the form [41]:

$$\begin{aligned}\mathbf{p}(x, y) \doteq\ & h(x)h(y)\mathbf{p}(0,0) + h(1-x)h(y)\mathbf{p}(1,0) + \\ & h(x)h(1-y)\mathbf{p}(0,1) + h(1-x)h(1-y)\mathbf{p}(1,1),\end{aligned} \tag{6.34}$$

where $h(r),\ 0 \leq r \leq 1$ is the interpolation function. Written in this form, this is just a form of kernel smoothing of the motion field, although using (symmetric) kernels $k(x) \doteq h(|x|)$ of strictly compact support, to ensure the locality of our interpolant.

For sensible interpolation/smoothing, $h(r)$ should be a monotonically decreasing function. To obtain the correct values of the interpolated movement at the corners, we require that:

$$h(0) = 1,\ \ h(1) = 0. \tag{6.35}$$

The simplest function that satisfies this is the step-function:

$$h(r) = 1,\ \ r < \frac{1}{2},\ \ \text{else: } h(r) = 0, \tag{6.36}$$

which corresponds to nearest-neighbour interpolation.

The next-simplest is the linear function $h(r) = 1 - r$, and this kernel is then equivalent to *bilinear* interpolation. However, neither the nearest-neighbour or the bilinear interpolant is everywhere differentiable. To obtain a smoother interpolant, we hence need to choose a smoother kernel. First, imposing differentiability at the boundary of influence of a control point gives us the constraint that $h'(1) = 0$. Any of the functions considered above, the CPS (6.31), the polynomial (6.32), or the trigonometric (6.33), satisfy this and can be used to produce this smoothed-out version of bilinear interpolation. An example is shown in Fig. 6.6. Note however that in general, the conditions that need to be met so that the re-parameterisation given by such

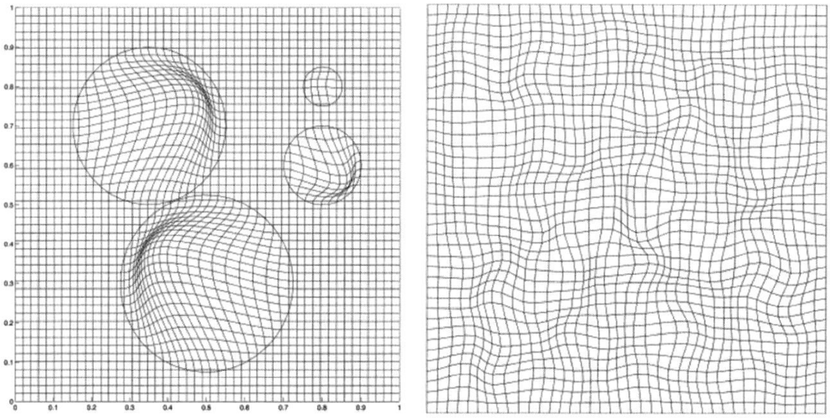

Fig. 6.6 Examples of local re-parameterisations. **Left**: A composition of four single-point clamped-plate spline transformations (6.31). **Right**: The clamped-plate smoothed interpolant (6.35), defined over a 10×10 grid of control points.

an interpolation is guaranteed to be homeomorphic are not straightforward, nor is their derivation.

We will now consider a further restriction on the allowed kernels. Bilinear interpolation has the important property that the interpolated values are constant if the control point values are constant. In terms of the kernel $h(r)$, this is equivalent to the requirement that:

$$h(r) + h(1 - r) = 1. \tag{6.37}$$

Of the functions considered above, only the trigonometric satisfies this constraint. A class of polynomial functions that satisfy this condition is:

$$\text{If } r \leq \frac{1}{2},\ \ h(r) = 1 - ar^n,\ \ \text{else } h(r) = a(1-r)^n,\ \ n \geq 1,\ \ a = 2^{n-1}.$$

If we consider just the class of kernels that satisfy (6.37), the interpolant becomes slightly simpler:

$$\begin{aligned} \mathbf{p}(x,y) \doteq\ & h(x)h(y)\mathbf{p}(0,0) + h(y)(1-h(x))\mathbf{p}(1,0) + \\ & h(x)\,(1-h(y))\mathbf{p}(0,1) + (1-h(x))(1-h(y))\mathbf{p}(1,1). \end{aligned} \tag{6.38}$$

Let us rewrite the node motions in terms of the mean motion:

$$\mathbf{p}(0,0) = \bar{\mathbf{p}} + \tilde{\mathbf{p}}(0,0),$$

where $\bar{\mathbf{p}}$ is the mean of the four motions. We then place bounds on the components of these residual motions at the nodes, so that:

$$-p \leq \tilde{\mathbf{p}}_x(0,0) \leq p, \text{ for some } p > 0. \tag{6.39}$$

By considering the Jacobian of the re-parameterisation function $\phi(x, y)$, and constraining this to remain *positive* (i.e., a valid homeomorphism), it is possible to derive a relation between the maximum allowed residual displacement p and the maximum value of $|h'(r)|$ (similar to that in (6.28)):

$$p \leq \frac{1}{\alpha |h'(r)|}, \tag{6.40}$$

where α is some number that can be determined (although only after a considerable amount of algebraic manipulation!).

6.2.2 Re-parameterisation of Closed Surfaces

We now move on to the re-parameterisation of closed surfaces. As we saw in Sect. 6.1.2, closed surfaces can be parameterised by mapping the surface of the shape onto the surface of a unit sphere. Points in the parameter space can then be represented using the usual spherical polar angular coordinates (θ, ϕ). Correspondence may now be manipulated by moving points around on the sphere, and to achieve this, we need to construct sets of homeomorphic mappings of the surface of the sphere.

Many of the techniques we have developed previously for re-parameterising open surfaces, or for re-parameterising open and closed curves can be adapted for this task, as we will show.

6.2.2.1 Recursive Piecewise-Linear Re-parameterisation

A recursive piecewise-linear representation of closed surface re-parameterisation can be constructed in a similar manner to open surfaces (described above in Sect. 6.2.1). However, given the differences between a unit square and a unit sphere, triangular meshes are preferred over the rectangular tiling that was used previously.

Let us consider first the task applied to a planar triangle $(\mathbf{p}_\alpha, \mathbf{p}_\beta, \mathbf{p}_\gamma)$ such as that shown in Fig. 6.7**(i)**. A daughter node P is created within the triangle, at position $\mathbf{p}$. Its position within the triangle is specified by the two fractional distances $\tau_1^{\alpha\beta\gamma}, \tau_2^{\alpha\beta\gamma}$. The line $\mathbf{r} - \mathbf{q}$ is constructed, passing through P and parallel to $\mathbf{p}_\beta - \mathbf{p}_\gamma$. The fractional distances are then that of P along the line $\mathbf{r} - \mathbf{q}$, and the point $\mathbf{q}$ along $\mathbf{p}_\gamma - \mathbf{p}_\alpha$, with $0 \leq \tau_{1,2}^{\alpha\beta\gamma} \leq 1$. We hence see that:

$$\mathbf{p} = (1 - \tau_1^{\alpha\beta\gamma})\mathbf{p}_\alpha + \tau_1^{\alpha\beta\gamma}\tau_2^{\alpha\beta\gamma}\mathbf{p}_\beta + \tau_1^{\alpha\beta\gamma}(1 - \tau_2^{\alpha\beta\gamma})\mathbf{p}_\gamma. \tag{6.41}$$

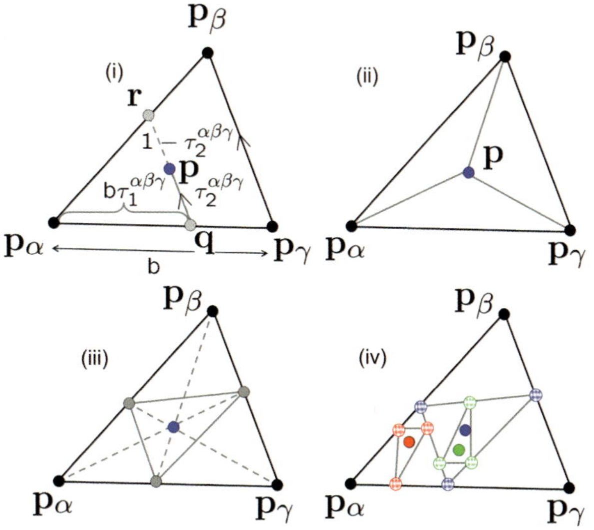

Fig. 6.7 Recursive piecewise-linear re-parameterisation of closed surfaces. **(i)** A daughter node (blue) placed within a planar triangle $(\mathbf{p}^\alpha, \mathbf{p}^\beta, \mathbf{p}^\gamma)$, with the fractional distances $\boldsymbol{\tau}^{\alpha\beta\gamma}$ indicated. **(ii)** A simple threefold subdivision generated by the daughter node. **(iii)** A fourfold subdivision generated by the daughter *control* node. The dotted grey lines are construction lines, the solid grey lines indicate the new edges added, with the new subsidiary nodes (grey filled circles). **(iv)** A recursive subdivision according to the fourfold scheme. The solid coloured circles are the various daughter control nodes, with stippled circles of the same colour indicating the added subsidiary nodes of the triangulation. Grey lines indicate the added edges.

We could use P to subdivide the parent triangle into three, as shown in Fig. 6.7 **(ii)**. However, this construction suffers from the problem that it is not self-similar. The sides of the parent triangle are retained as sides of the new daughter triangles, which therefore means that these triangles becoming increasingly sliver-like as the recursion proceeds, approaching a line in the infinite limit. A better subdivision is the fourfold subdivision shown in Fig. 6.7 **(iii)**, the construction as shown by the dashed lines in the diagram. This division is self-similar,[4] hence can be carried out to any required depth of recursion without changing the nature of the triangles. An example of such multi-level recursion is shown in Fig. 6.7 **(iv)**.

Deformations of these triangulated meshes proceed as before, where the motion of a daughter control node is legal as long as it remains within its parent triangle, and any daughter moves proportionately with its parent triangle. The subsidiary nodes of a daughter node move accordingly. As in the case of open surfaces (Sect. 6.2.1), for a triangulation with n daughter control nodes, the parameter space is the unit hypercube in $\mathbb{R}^{2n}$.

---

[4] The construction is obviously related to that used to produce the fractal called the *Sierpiński gasket* or *Sierpiński triangle* [164].

Now let us consider applying this procedure to points on the sphere. The construction is initialized by placing a triangulated polyhedron within the sphere (such as the inscribed tetrahedron or octahedron). The nodes of this polyhedron then form the initial triangulation, with connectivity as determined by the inscribed polyhedron. Since we are on the surface of the sphere, the entire mesh of *spherical* triangles covers the sphere, without any gaps or overlaps.

We now consider the *planar* triangle formed by the same three nodes. This is subdivided as shown previously. The new nodes are then projected back onto the surface of the sphere via:

$$\mathbf{r} \mapsto \frac{\mathbf{r}}{\|\mathbf{r}\|}, \tag{6.42}$$

where the origin is taken at the centre of the sphere. This hence produces a recursive, self-similar triangulation of the surface, whose movement, controlled as in the planar case, produces a homeomorphic re-parameterisation of the surface of the sphere.

As before, these recursive piecewise linear re-parameterisations are powerful, but possess the drawback that they are not differentiable along the edges of the triangulation. We hence move on to consider methods of re-parameterisation of the sphere which are differentiable almost everywhere.

6.2.2.2 Localized Re-parameterisation

In Sect. 6.2.1, we showed how to generate various parametric homeomorphic deformations of the unit disc. Hence a simple way of re-parameterising the sphere is to use these local transformations defined for open surfaces, but applied on *flattened* regions of the sphere.

To be specific, consider a region of the surface of the sphere, defined so as to be less than a distance w from some point C also on the surface. This sphere about C hence defines a cap on the surface of the unit sphere. In terms of this localized patch, we refer to C as the centre of the patch, with width w. Such a cap is obviously topologically equivalent to the disc, and can be mapped into the disc via an orthographic projection (as long as $w < \sqrt{2}$, to ensure that the cap is less than a hemisphere). A diagram illustrating the orthographic projection is shown in Fig. 6.8. The projection obviously means that points near the edge of a large cap will be compressed, and we found that a smaller limit of $w < \frac{1}{3}$ was suitable, and avoided extreme distortions.

Suppose the centre C has polar coordinates (θ_c, ψ_c), and a general point on the sphere (θ, ψ), which then maps to a point (x, y) in Cartesian coordinates on the tangent plane T_C to the sphere at C. It can be shown that [168, 193]:

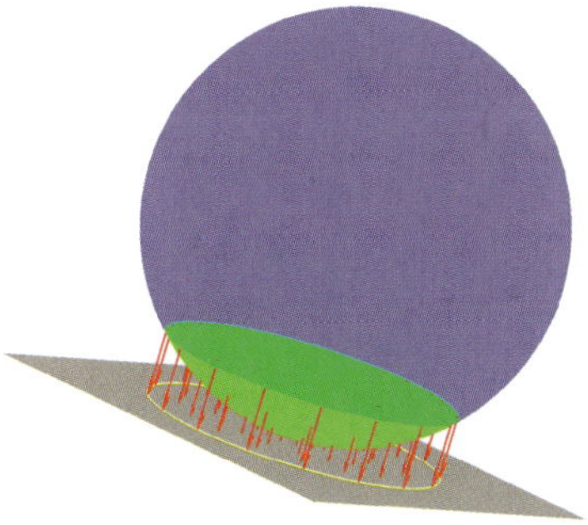

Fig. 6.8 An orthographic projection from the sphere onto the tangent plane.

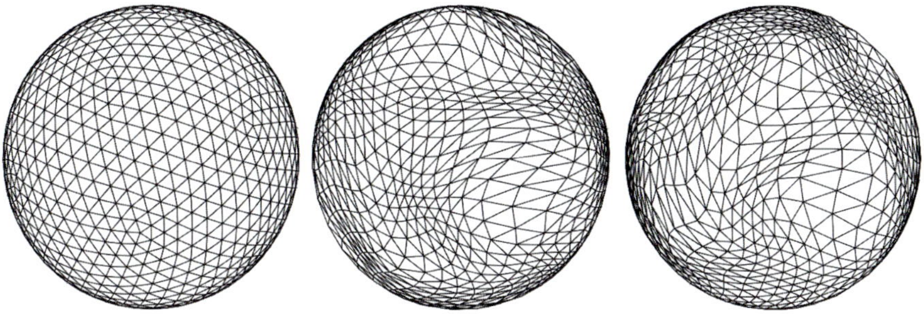

Fig. 6.9 Multiple local transformations of the sphere. **Left:** the original mesh; **Middle** and **Right:** the mesh after applying multiple local transformations.

$$\begin{aligned} x &= \cos\theta \sin(\psi - \psi_c) \\ y &= \cos\theta_c \sin\theta - \sin\theta_c \cos\theta \cos(\psi - \psi_c). \end{aligned} \tag{6.43}$$

Re-parameterisation is then performed on the disc (Sect. 6.2.1), using any function obeying the constraints of Theorem 6.1 for homeomorphic local transformations. The complete set of parameters for the transformation are then the position of the centre point C, the cap width w, and the amount of deformation of the centre point, $\mathbf{p}$ as in Theorem 6.1. After the transformation, points are projected back onto the sphere (see [193] for specific formulae and advice on implementation).

As for open surfaces, a number of local transformations can be combined to form a general transformation – see Fig. 6.9 for an example.

These transformations, as in the case of open surfaces, have the advantage that only a localized region is effected, which eases the computational load.

The width parameter means that they are also suitable for a multiscale or a coarse-to-fine optimisation strategy.

6.2.2.3 Cauchy Kernel Re-parameterisation

We can also adapt methods for re-parameterising closed curves (Sect. 5.2.1) to create methods for re-parameterising the sphere.

The basic idea is as follows. Consider an arbitrary point P on the sphere. This then defines a set of great circles, which intersect at P and at its antipodal point. Considering the arcs from P to the antipode, these can be re-parameterised using a re-parameterisation of a *open line.* And provided that this one-dimensional re-parameterisation varies smoothly around the sphere perpendicular to the arcs, we then have a re-parameterisation of the entire sphere.

We consider first a transformation which is symmetric around the axis. We then generalize this to include asymmetric transformations, and transformations which includes shearing.

6.2.2.4 Symmetric Theta Transformation

Let us consider an arbitrary point P on the unit sphere. For simplicity, assume that the spherical polar co-ordinates (θ, ψ) on the sphere have been redefined so that P corresponds to the point $\theta = 0$. Let us first consider a rotationally symmetric mapping that re-parameterises the θ coordinate:

$$\theta \mapsto f(\theta). \tag{6.44}$$

For the mapping to be homeomorphic and continuous with the identity, f must be a non-decreasing monotonic function over the range $0 \leq \theta \leq \pi$, with $f(0) = 0$ and $f(\pi) = \pi$. Hence any of the methods described in the previous chapter for re-parameterising *open* curves could be adapted for this purpose. However, we choose here the method described in Sect. 5.2.1 for closed curves, using the cumulative distribution of a wrapped Cauchy distribution and a constant term (5.46).[5] Then we have:

$$\begin{aligned} f(\theta) &= \int_0^\theta d\alpha \left[\frac{1}{(1+A)} \left(1 + A \left(\frac{1 - \Omega^2}{1 + \Omega^2 - 2\Omega \cos\alpha} \right) \right) \right] \\ &= \frac{1}{1+A} \left(\theta + A \arccos\left(\frac{(1+\Omega^2)\cos\theta - 2\Omega}{1 + \Omega^2 - 2\Omega\cos\theta} \right) \right), \end{aligned} \tag{6.45}$$

[5] Note that in (5.46), the cdf was written in terms of the arctan function, but that here we use a mathematically equivalent formulation in terms of the arccos function.

where $0 < \Omega < 1$) is the width parameter,[6] and A ($A \geq 0$) is the amplitude of the wrapped Cauchy. Note that here, we are taking the origin of the cdf to be the *centre* of the bump of the wrapped Cauchy, and integrating outwards. Hence points spread out around the bump position, with the centre $\theta = 0$ and the antipodal point $\theta = \pi$ fixed ($f(0) \equiv 0,\ f(\pi) \equiv \pi$).

This form of $f(\theta)$ is such that $A = 0$ corresponds to the identity re-parameterisation. In the limit of small values of the width parameter[6] $\Omega \to 0$:

$$f(\theta) = \theta + \frac{2\Omega A \sin\theta}{1+A} + O\left(\Omega^2\right), \tag{6.46}$$

which is the identity re-parameterisation plus a perturbation, so that the largest amount of movement is at the equator at $\theta = \frac{\pi}{2}$. Conversely, in the limit $\Omega \to 1$:

$$f(\theta) = \frac{1}{1+A}\left(\theta + A\pi\right) + O\left((1-\Omega)^2\right). \tag{6.47}$$

This corresponds to a *linear* function, which is the form of $f(\theta)$ for values of θ *greater* than some minimum, where smaller values of θ correspond to the initial steep rise given by integration over the sharp peak.

Equation (6.45) was constructed with the kernel centre P at the point $\theta = 0$. If the polar coordinates are defined with respect to an *arbitrary* axis, so that the kernel centre P has the position $\mathbf{a} \in \mathbb{R}^3$, then an arbitrary point $\mathbf{x} \in \mathbb{R}^3$ on the sphere moves to $\mathbf{x}'$ where:

$$\cos\theta \doteq \mathbf{a} \cdot \mathbf{x}, \tag{6.48}$$

$$\mathbf{x} \to \mathbf{x}' = \mathbf{a}\cos f(\theta) + \frac{\sin f(\theta)}{\sin\theta}\left(\mathbf{x} - \mathbf{a}\cos\theta\right). \tag{6.49}$$

In Fig. 6.10, we show an example of re-parameterising a surface using a single kernel. As can be seen from the figure, points spread out around the centre of the kernel, whilst compressing at the antipodal point.

6.2.2.5 Asymmetric Theta Transformations

The above transformation is symmetric in its effect about the centre of the kernel. We can also perform asymmetric transformations of the form:

$$\psi \ \mapsto\ \psi,\ \theta \mapsto f(\theta, \psi). \tag{6.50}$$

As before, let us redefine polar coordinates, so that the kernel centre P corresponds to the point $\theta = 0$. We take a second point $Q \neq P$, and define the ψ coordinate about P so that Q is on the meridian $\psi = 0$. An asymmetric transformation around the point P, towards a point Q can be achieved using

---

[6] Note that $\Omega \to 1$ gives a more highly-peaked distribution, whereas $\Omega \to 0$ gives a flatter distribution. See Fig. 5.10 for examples.

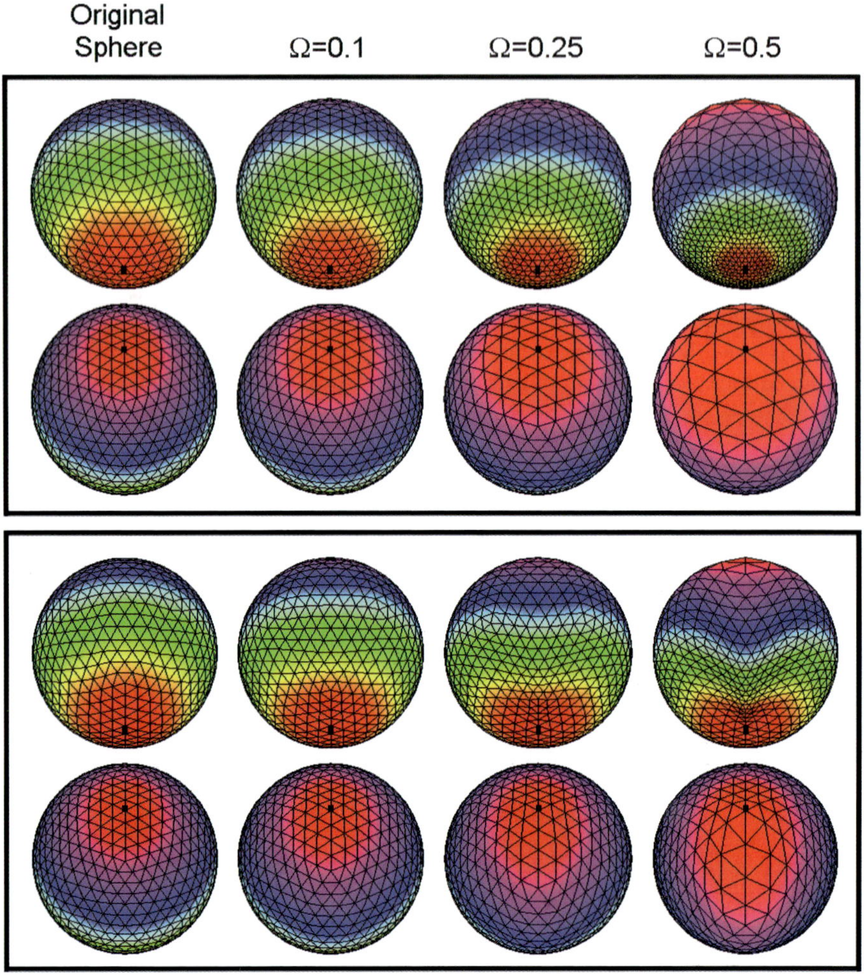

Fig. 6.10 A demonstration of a symmetric (**top** panel) and an asymmetric (**bottom** panel) theta transformation applied to the sphere. In each case, two views are presented. From the left, we have the original sphere, and the theta transformation (6.45) for the values of $\Omega = 0.1, 0.25$, and 0.5 and $A = 4$. In the bottom panel, for the asymmetric transformation (6.51), the same Ω values are used, with $A(\psi)$ as in (6.52), with $\Upsilon = 0.75$ and $A_0 = \frac{1-\Upsilon^2}{\Upsilon}$, so that $0 \leq A(\psi) \leq 4$. The colours denote correspondence, and the poles ($\theta = 0, \pi$) are indicated by the thick black line.

(6.45) and making the amplitude A a smooth, periodic function of the ψ coordinate:

$$A \to A(\psi) \equiv A(\psi + 2\pi).$$

$$f(\theta, \psi) \doteq \frac{1}{1 + A(\psi)} \left(\theta + A(\psi) \arccos \left(\frac{(1 + \Omega^2) \cos \theta - 2\Omega}{1 + \Omega^2 - 2\Omega \cos \theta} \right) \right). \quad (6.51)$$

One way to do this is to use the wrapped Cauchy distribution to obtain:

$$A(\psi) = A_0 \left[\frac{1 - \Upsilon^2}{1 + \Upsilon^2 - 2\Upsilon \cos\psi} - \frac{1 - \Upsilon^2}{(1 + \Upsilon)^2} \right], \tag{6.52}$$

where $0 < \Upsilon < 1$ is the width of the subsidiary Cauchy. We have chosen the formulation such that $A(\psi)$ has a minimum value of zero. An example of an asymmetric transformation is shown in Fig. 6.10, where A_0 was chosen so that the maximum value of $A(\psi)$ matched that of the *symmetric* transformation also shown in the figure.

Any of the symmetric theta transformations that have a single amplitude parameter A can obviously be generalized in this manner to produce a asymmetric theta transformation. Different choices of the periodic function $A(\psi)$ can also be used.

6.2.2.6 Shear Transformations

Finally, let us consider transformations of the ψ coordinate. This is equivalent to shearing and twisting the sphere about the axis defined by the point P. Consider a re-parameterisation of the form:

$$\psi \to \psi + g(\theta). \tag{6.53}$$

If $g(\theta)$ is a constant, we just have a rigid rotation of the unit sphere. In general, $g(\theta)$ just has to be some smooth function. One such choice is where the transformation $g(\theta)$ is given by a wrapped Cauchy distribution:

$$g(\theta) = B \frac{(1 - \Gamma^2)}{1 + \Gamma^2 - 2\Gamma \cos(\theta - \theta_0)}, \tag{6.54}$$

with amplitude parameter B, width parameter $0 < \Gamma < 1$, and centre position θ_0. This transformation is continuous with the identity at $B = 0$. In the limit $\Gamma \to 0$:

$$g(\theta) = B \left[1 + 2\Gamma \cos(\theta - \theta_0) \right] + O\left(\Gamma^2\right), \tag{6.55}$$

which hence approaches a rigid rotation. In the limit $\Gamma \to 1$:

$$g(\theta) = B \frac{1 - \Gamma}{1 - \cos(\theta - \theta_0)} + O\left((1 - \Gamma)^2\right). \tag{6.56}$$

We hence see that the effect becomes localized about $\theta \approx \theta_0$ in this limit, over some region where $|\theta - \theta_0| = O\left(\sqrt{1 - \Gamma}\right)$.

An example shear transformation is shown in Fig. 6.11.

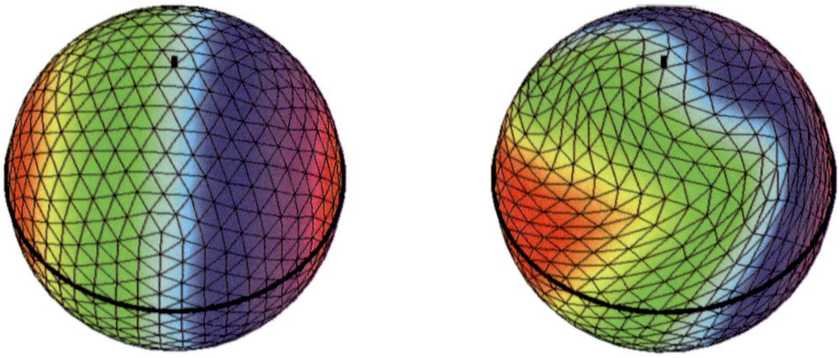

Fig. 6.11 **Left:** the original sphere, **Right:** the sphere after a shear transformation (6.54), with $\Gamma = 0.75$, $\theta_0 = \frac{\pi}{4}$, and B chosen so that $g(\theta_0) = \frac{\pi}{3}$. The colours denote correspondence, and the thick black lines indicate the line $\theta = \frac{\pi}{2}$, and the point $\theta = 0$.

6.2.3 Re-parameterisation of Other Topologies

To summarize, we have presented a variety of methods for re-parameterising open shapes with the topology of the disc or unit square, and closed shapes with spherical topology. Although this represents a significant proportion of shapes that we may wish to model, it is worth considering how the methods can be extended to other topologies.

One of the main difficulties in dealing with shapes with complex topologies is that of defining an initial parameterisation, or mapping onto the appropriate topological primitive. Another problem is that for some topologies, a single parameter space or a single chart does not suffice.

Consider first the case of shapes with more than one boundary (a few examples of which are shown in Fig. 6.12). An obvious approach is to extend the ideas presented in Sect. 6.1.1 for surfaces with a single boundary, and to map the shape surface onto some region of the plane by flattening it out, but without cutting or tearing the surface.

If we consider the examples in the figure, we see that some shapes (the tube, branched tube, and multiple branched tube) have the topology of the punctured disc.[7] These shapes can hence be flattened onto the plane, and then mapped onto the unit square. One boundary becomes the boundary of the square, and the other boundaries become the boundaries of holes (or punctures) within the square. However, the last case cannot be so flattened. If we compare this with the other shapes, we see that this complex case can be thought of as the branched tube, but with the addition of a tunnel through the common branch point. Topologically speaking, such a tunnel is equivalent to a handle.[8] This handle cannot be flattened. To parameterise this shape,

[7] Note that we are using *punctured disc* in the everyday sense, rather than in the strict mathematical sense of the disc with the point at the origin removed.

[8] The simplest example is a torus, which can be thought of as the usual ring doughnut shape, with a tunnel, or the topologically equivalent case of a sphere with a handle attached.

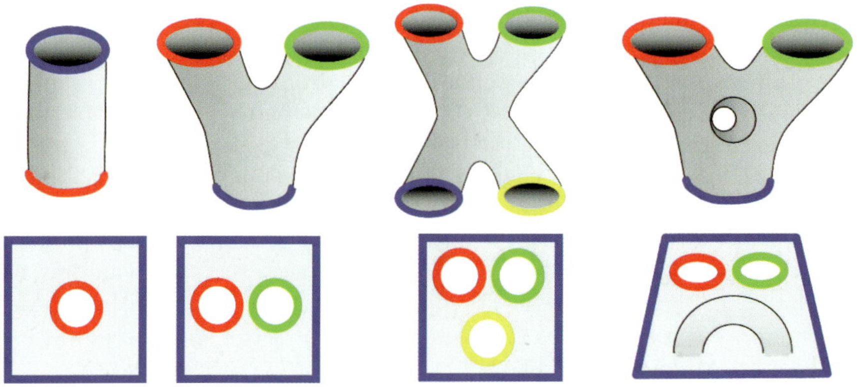

Fig. 6.12 Examples of shapes with boundaries. **Top:** The shapes (tube, branched tube, multiple branched tube, and a more complex case, a branched tube but with a tunnel driven through it). **Bottom:** The shapes flattened onto the plane. Note that the shape on the right gives a handle, which cannot be flattened. Thick coloured lines mark the boundaries on each shape and on the plane.

we would have to cut through the handle, producing two additional holes in the square, but now with the edges identified between the two holes, rather than the edges being true boundaries.

For shapes which we can flatten without cutting, we can produce an initial parameterisation by generalizing the approach given in Sect. 6.1.1. First, the nodes on each boundary are mapped to the corresponding boundaries in the punctured disc. The positions of all other nodes can be defined as weighted combinations of neighbouring nodes, and the weight values can be found by optimisation. A conformal parameterisation can then be established using a variational approach. Horkaew and Yang [95] used a similar approach to build a statistical shape model of the heart, although they mapped their surfaces to the topologically equivalent parameter space formed by conjoined, punctured spheres, rather than to the plane.

Surfaces without boundaries offer a similar range of increasingly complex topologies. The addition of a single handle to the sphere produces a shape topologically equivalent to a torus. We can also add multiple handles, the surface then being the surface of the resultant handlebody. For such shapes, once an initial parameterisation has been achieved, we can manipulate parameterisation by flattening regions of the surface onto the tangent plane in the same way as we did for the sphere in Sect. 6.2.2.2. In effect, this is equivalent to defining local coordinate patches by projection, but as in the case of the orthographic projection of the sphere (6.43), the size of the patch is limited by the nature of the topological primitive chosen, and the exact details of the projection. Each patch can then be re-parameterised using a local transformation, and a composition of such local transformations can be used to re-parameterise the surface. However, this suffers from the problem that it can generate *only* localized transformations.

An alternative approach is to take a lead from the computer graphics community, who have performed considerable research into mesh parameterisation of arbitrary topologies [162]. The most popular method is to cut the surface to form multiple charts and perform a planar parameterisation of each chart.[9] Given such an initial parameterisation, we can re-parameterise each planar chart using one of the formulations in Sect. 6.2.1. This approach also offers a somewhat limited transformation since each point can only be moved within the same chart. This limitation can be overcome by repeating the process several times and using a different set of 'cuts' to produce a different set of charts each time.

We can also extend the ideas used for constructing kernel-based re-parameterisations of closed surfaces (Sect. 6.2.2.2). The previous chapter described how a one-dimensional circle ($\mathbb{S}^1$) or a one-dimensional line ($\mathbb{R}^1$) could be re-parameterised. We can therefore re-parameterise any surface which is a product space formed by a combination of $\mathbb{S}^1$ and $\mathbb{R}^1$. For example, if we represent the cylinder or tube as the product space of $\mathbb{R}^1 \times \mathbb{S}^1$, we can re-parameterise either $\mathbb{S}^1$ or $\mathbb{R}^1$ using the representations described in Sects. 5.2 and 5.1, respectively. We can also view the tube as the unit square, but with periodic boundary conditions imposed on two opposing edges. Similarly, the torus can be viewed as a square with periodic boundary conditions on both pairs of edges.

So far, we have only considered the case of single-part shapes of fixed topology. Re-parameterisation of multi-part shapes of fixed topology is obviously a trivial extension of the single-part case, where each part is individually re-parameterised in the appropriate fashion. The case where the topology can change (for example, where two parts merge to become a single part, or where a single part changes its topology) is obviously much more difficult, and a complete discussion is beyond the scope of the current text.

6.3 Use in Optimisation

We conclude this chapter by considering how each of the homeomorphic transformations considered in this chapter can be incorporated into an optimisation algorithm – the precise details of the optimisation algorithm is the subject of the next chapter. We could consider optimising over all of the parameters that we have defined for our chosen set of parameterised homeomorphisms. However, as for the case of curves, we found that it is instead advisable to define some parameters as *auxiliary* parameters (which remain fixed), and others as free parameters, which are to be optimised over. This

[9] In differential geometry, such a collection of charts is called an *atlas*.

obviously has the advantage that it reduces the dimensionality of the search space for optimisation. However, this approach also makes sense if we consider the meaning of the various auxiliary parameters, inasmuch as they refer to the *position* and *scale* of localized transformations, or the *position* and *scale* of kernels. For a multiscale or coarse-to-fine optimisation strategy, it makes sense to define and fix the possible positions and scales *a priori*, and only search over the *amplitudes* of the localized deformations.

The parameters, and our split between auxiliary and free parameters is summarized in Table 6.1, along with any constraints on parameter values. We also initialize our transformations to the identity, and the values to achieve this are also listed.

Table 6.1 A summary of the representations of re-parameterisation described in this chapter.

Open Surfaces

Representation	Optimisation Parameters	Initial Parameter Values	Constraints	Auxiliary Variables
Recursive piecewise-linear (Sect. 6.2.1)	Node fractional distances, $\{\boldsymbol{\tau}^{\alpha\beta}\}$	$\{\boldsymbol{\tau}^{\alpha\beta}\}$ within allowed range.	$0 < \tau_{x,y}^{\alpha\beta} < 1$	
Clamped-plate spline (Sect. 6.2.1)	Node positions, $\{\mathbf{p}_k\}$	$\{\mathbf{p}_k\} = \{(0,0), \ldots, (0,0)\}$	$\|\|\mathbf{p}_k\|\| \leq \frac{e}{4}$	The width $\{w_k\}$ and centre point $\{\mathbf{a}_k\}$ of each local region.
Bounded polynomial (Sect. 6.2.1)	Node positions, $\{\mathbf{p}_k\}$	$\{\mathbf{p}_k\} = \{(0,0), \ldots, (0,0)\}$	$\|\|\mathbf{p}_k\|\| \leq \frac{3\sqrt{3}}{8}$	The width $\{w_k\}$ and centre point $\{\mathbf{a}_k\}$ of each local region.
Trigonometric (Sect. 6.2.1)	Node positions, $\{\mathbf{p}_k\}$	$\{\mathbf{p}_k\} = \{(0,0), \ldots, (0,0)\}$	$\|\|\mathbf{p}_k\|\| \leq \frac{2}{\pi}$	The width $\{w_k\}$ and centre point $\{\mathbf{a}_k\}$ of each local region.

Closed Surfaces

Representation	Optimisation Parameters	Initial Parameter Values	Constraints	Auxiliary Variables
Recursive piecewise-linear (Sect. 6.2.2.1)	Node fractional distances, $\{\boldsymbol{\tau}^{\alpha\beta\gamma}\}$	$\{\boldsymbol{\tau}^{\alpha\beta\gamma}\}$ within allowed range.	$0 < \tau_{1,2}^{\alpha\beta\gamma} < 1$	
Clamped-plate spline (Sect. 6.2.2.2)	Node positions, $\{\mathbf{p}_k\}$	$\{\mathbf{p}_k\} = \{(0,0), \ldots, (0,0)\}$	$\|\|\mathbf{p}_k\|\| \leq \frac{e}{4}$	The width $\{w_k\}$ and centre point $\{\mathbf{a}_k\}$ of each local region.
Bounded polynomial (Sect. 6.2.2.2)	Node positions, $\{\mathbf{p}_k\}$	$\{\mathbf{p}_k\} = \{(0,0), \ldots, (0,0)\}$	$\|\|\mathbf{p}_k\|\| \leq \frac{3\sqrt{3}}{8}$	The width $\{w_k\}$ and centre point $\{\mathbf{a}_k\}$ of each local region.
Kernel-based representations (Sect. 6.2.2.3)	Kernel magnitudes, $\{A_k\}$	$\{A_k\} = \{0, \ldots, 0\}$	$A_k \geq 0$	The width $\{w_k\}$ and position $\{\mathbf{a}_k\}$ of each kernel.

Chapter 7
Optimisation

In previous chapters, we considered objective functions and parametric methods of manipulating correspondence. It is now time to bring these together in the context of our framework for groupwise correspondence, and construct a suitable algorithm for locating the optimum value of the objective function.

Suppose that we have chosen an objective function $\mathcal{L}$ (Chap. 4) and a representation of re-parameterisation, that is controlled by a vector of parameters $\boldsymbol{\alpha}$ (Chaps. 5 and 6). Our goal is to find the values of these parameters for each shape, that minimises our chosen objective function:

$$\mathcal{L}_{\text{min}} = \min_{\{\boldsymbol{\alpha}^{(1)},\ldots,\boldsymbol{\alpha}^{(i)},\ldots\boldsymbol{\alpha}^{(n_S)}\}} \mathcal{L}\left(\mathbf{S}'_1,\ldots,\mathbf{S}'_i,\ldots\mathbf{S}'_{n_S}\right),$$
$$\mathbf{S}'_i(\mathbf{x}) \doteq \mathbf{S}_i(\mathbf{x}'), \quad \mathbf{x}' = \phi_i(\mathbf{x}) \doteq \phi(\mathbf{x};\boldsymbol{\alpha}^{(i)}), \tag{7.1}$$

where ϕ is the parametric re-parameterisation function[1] and $\boldsymbol{\alpha}^{(i)}$ are the set of parameters that control the re-parameterisation of shape i.

Finding a minimum of (7.1) is a difficult task since the number of parameters in the optimisation is large. If, for example, we have a training set of 50 shapes and we use 30 parameters to represent each re-parameterisation, then we have to find the optimum combination of 1500 parameter values. The problem is exacerbated by a non-linear objective function that contains many local minima.

Most standard optimisation algorithms cannot cope with an optimisation problem of this magnitude and invariably converge on a local optimum. There are, however, some specialized algorithms such as simulated annealing and genetic algorithm search that can handle such optimisation problems by using

[1] Note that for the purposes of implementation (as was previously indicated in Sect. 3.3.2), we are working with the alternative definition of re-parameterisation to that in (6.4). Our re-parameterisation function as defined here is actually the *inverse* of the re-parameterisation function according to the definition in (6.4). This difference is unimportant, since valid re-parameterisation functions are constrained to be invertible.

DOI: 10.1007/978-1-84800-138-1_7,

stochastic techniques to evade local minima. Genetic algorithm search was used in early work on model-building [101, 53] and demonstrated the feasibility of the optimisation approach. Genetic algorithm search is, however, an inefficient technique that leads to impractical run times, even for small training sets.

Producing an efficient minimisation routine was one of the most challenging aspects in the development of this optimisation approach to model-building. In recent years, many techniques were introduced that turned this optimisation problem into a tractable task – these are described in the first section of this chapter. The second section looks at how the optimisation can be tailored to deal with certain classes of objects, before the third section considers implementation issues. The chapter concludes by walking the reader through two examples of how the optimisation framework can be used in practice.

7.1 A Tractable Optimisation Approach

Many efficient 'off-the-shelf' optimisation algorithms exist, but they cannot cope with the optimisation problem in (7.1), since the number of parameters is high, leading to impractical computational demands and unreasonable memory requirements. We must, therefore, find a way of addressing this problem of dimensionality by breaking down the problem into a series of simpler ones. Note that any re-parameterisation can be decomposed as a composition of simpler re-parameterisations:

$$\phi = \phi^{(n_j)} \ldots \circ \phi^{(2)} \circ \phi^{(1)},$$

where each of the simpler re-parameterisation functions requires fewer parameters than the compound re-parameterisation function. We hence see that this decomposition can be used to construct an iterative optimisation scheme, where, at each iteration, we just optimise over one of these simpler re-parameterisation functions $\phi^{(j)}$. These are lower-dimensional optimisation problems, allowing the use of standard local optimisation algorithms, which are far more effective than genetic algorithm search, leading to a large reduction in convergence time [50]. Any suitable non-linear optimisation algorithm can be used for this purpose (see Chap. 10 in [139]); in practice we use the Nelder-Mead downhill simplex algorithm.

This iterative scheme works well for many classes of object, especially if the training shapes are simple curves (that is, one-dimensional shapes). However, many adaptations to the basic scheme have been proposed, leading to decreased convergence time and improved robustness – these are described in the remainder of this section.

7.1.1 Optimising One Example at a Time

Even if we employ the iterative approach above, the number of parameters to be optimised simultaneously can still prevent the local optimisation algorithm from converging reliably for relatively large training sets ($\gtrsim$ 100 examples). It is also not well suited to an iterative model-building scheme where examples are segmented and added one by one.

The number of variables optimised simultaneously can be reduced by optimising the parameterisation of only one example at a time. This is achieved by cycling through the training set, optimising the current re-parameterisation of each example before moving on to the next iteration. Note that we are still considering the *entire* training set (i.e., the model is built using the current parameterisations of *all* examples) but the parameterisation of each example is optimised independently. To remove any bias, the ordering of the training set is permutated at random before each iteration.

7.1.2 Stochastic Selection of Values for Auxiliary Parameters

Many of the representations of re-parameterisation described in Chaps. 5 and 6 have two sets of parameters: optimisation parameters and auxiliary parameters (see Tables 5.1 and 6.1). The optimisation parameters are manipulated to optimise $\mathcal{L}$, whereas the auxiliary parameters are chosen and fixed before optimisation.

The auxiliary parameter values can be chosen using some fixed scheme – in [49], for example, a kernel-based representation was used within a multi-resolution framework where re-parameterisation was initially defined by a small set of broad kernels and then refined by adding thinner kernels in between. However, only a limited number of kernel widths and positions were used, leading to a limited representation of re-parameterisation. Another disadvantage is that the scheme required an optimisation schedule (number of recursion levels, iterations for each level, etc.).

An alternative is to take a multiscale approach, where each auxiliary parameter value is chosen stochastically [55]. This approach adds more robustness to local minima and allows a richer set of re-parameterisation functions to be generated. It also avoids the need for an optimisation schedule: the only values to select are those that define the distribution from which auxiliary parameter values are sampled (a list of such distributions, that were found to be suitable for many classes of shape, in given in Table 7.1).

In order to use Table 7.1, we need a method of uniformly sampling the training surfaces. If we have followed the area-preserving parameterisation approach in the previous chapter (see Sect. 6.1.4), this can be achieved by uniformly sampling from the parameter space. A uniform sampling of the parameter space of curves and open surfaces is straightforward, but surfaces with spherical topology require a little more consideration – see Algorithm 7.1.

Table 7.1 The distributions from which values for auxiliary parameters are chosen. $\mathcal{N}(\mu, \sigma)$ is a Gaussian distribution with mean μ and standard deviation σ and $\mathcal{U}$ is a uniform distribution over the curve/surface. Note that we take the *modulus* of values drawn from a Gaussian, since the actual parameter is required to be positive.

Representation	**Auxiliary Parameter**	**Distribution**
CURVES		
Local re-parameterisations (Sect. 5.1.3)	kernel widths, $\{w_k\}$	$\mathcal{N}(0, 1/20)$
	centrepoints, $\{a_k\}$	$\mathcal{U}$
Kernel-based representations (Sects. 5.1.4.1 & 5.2.1)	kernel widths, $\{w_k\}$	$\mathcal{N}(0, 1/32)$
	centrepoints, $\{a_k\}$	$\mathcal{U}$
SURFACES		
Local re-parameterisations (Sects. 6.2.1 & 6.2.2.2)	kernel widths, $\{w_k\}$	$\mathcal{N}(0, 1/5)$
	centrepoints, $\{a_k\}$	$\mathcal{U}$
Kernel-based representations (Sect. 6.2.2.3)	kernel widths, $\{w_k\}$	$\mathcal{N}(0, 1/8)$
	centrepoints, $\{a_k\}$	$\mathcal{U}$

7.1.3 Gradient Descent Optimisation

Optimisation algorithms tend to perform better if they are supplied with information about the gradient of the objective function. The gradient of the objective function $\mathcal{L}$, with respect to the optimisation parameters $\boldsymbol{\alpha}^{(i)}$ of the ith example (that is, $\frac{\partial \mathcal{L}}{\partial \boldsymbol{\alpha}^{(i)}}$), can be estimated numerically using a finite difference scheme. However, this is computationally demanding since the entire model has to be re-built from scratch for the computation of each component of the gradient.

As we saw in Chap. 4 (specifically, Sect. 4.3.4, with detailed computations in Appendix B), a semi-analytic solution can be obtained by decomposing the gradient using the chain rule. For the case of infinite-dimensional shape representations, we obtain the expression:

Algorithm 7.1 : Generating Random Points on the Surface of a Sphere.

---

procedure $\mathbf{x} = random\_sphere\_points$

DESCRIPTION

- Generates a point $\mathbf{x}$, according to a random sampling of a uniform distribution over the sphere surface.

DECLARATIONS

- $\mathcal{R}(a, b)$ is a function that generates a number at random by sampling from a uniform distribution over the range $[a, b]$

SAMPLING

1. let $\mathbf{x} = (x, y, z)$ be a point on the sphere, whose coordinates are chosen as follows:

   1.1. let $z \leftarrow \mathcal{R}(-1, 1)$
   1.2. let $t \leftarrow \mathcal{R}(0, 2\pi)$
   1.3. let $r \leftarrow \sqrt{1 - z^2}$
   1.4. let $x \leftarrow r \cos t$
   1.5. let $y \leftarrow r \sin t$

return $\mathbf{x}$.

$$\boxed{\frac{\partial \mathcal{L}}{\partial \alpha_A^{(i)}} = \sum_{a=1}^{n_S-1} \frac{\partial \mathcal{L}}{\partial \lambda_a} \sum_{j=1}^{n_S} \sum_{k=1}^{n_S} \frac{\partial \lambda_a}{\partial \widetilde{D}_{jk}} \int \frac{\delta \widetilde{D}_{jk}}{\delta \mathbf{S}_i(\mathbf{x})} \frac{\delta \mathbf{S}_i(\mathbf{x})}{\delta \alpha_A^{(i)}} dA(\mathbf{x}),} \tag{7.2}$$

where $\alpha_A^{(i)}$ is the Ath optimisation parameter for the ith shape, $\tilde{\mathbf{D}}$ is the continuous form of the covariance matrix, and $\{\lambda_a\}$ are its eigenvalues. Practical details of how this expression can be evaluated in practice are given later in Sect. 7.3.2.

Optimisation can now be performed using a gradient descent algorithm such as steepest descent or conjugate gradients [139]. Although the latter is preferred in general optimisation problems, there is very little difference between their performance when used for model-building. In practice, steepest descent is used because of its simplicity: it only involves a series of one-dimensional line searches down the gradient.

7.1.4 Optimising Pose

The positions of corresponding points depend on the pose parameters of each example as well as the shape parameterisations. As we saw in Sect. 2.1.1, a shape can undergo a similarity pose transformation, consisting of: $\mathbf{t} \in \mathbb{R}^{dn_P}$ representing a translation in $\mathbb{R}^d$, a $dn_P \times dn_P$ rotation matrix $\mathbf{R}$, representing a rotation in $\mathbb{R}^d$, and a scaling, $s \in \mathbb{R}^+$:

$$\mathbf{x} \mapsto s\mathbf{R}(\mathbf{x} - \mathbf{t}). \tag{7.3}$$

These pose parameter values can be chosen using Procrustes analysis (see for example Algorithm 2.1), which is equivalent to minimising the squared distance between each shape and the mean. But as we have already seen in Sect. 4.1.1, the squared distance is not a good objective function. Better models can be built by optimising the pose parameters using one of the model-based objective functions described in Chap. 4. In practice, translation can be dealt with directly, by setting the centre of gravity of each re-parameterised shape to the origin. But both the scale factor and the rotation must be explicitly optimised. The techniques described above for optimisation of parameterisation (optimising one example at a time, and using gradient descent) also work well for pose optimisation.

It is important to note that the optimal pose parameters depend on the parameterisation of each shape, and must therefore be included in the iterative optimisation process. In practice, the optimisation of pose and the optimisation of parameterisation are decoupled, and performed sequentially, hence reducing the number of parameters that must be optimised concurrently.

Note that one trivial way of minimising the objective function is to simply reduce the size of all the training examples by a uniform scaling ($\mathbf{x}_i \mapsto s\mathbf{x}_i,\ s \leq 1$). To prevent this, we could just a constrained optimisation, so that $\sum_{i=1}^{n_S} s_i = n_S$; in other words, the mean scaling is fixed, and the mean shape is always roughly the same size. Unfortunately, this constraint cannot be applied if the pose parameters of only one shape are optimised concurrently. In practice, this can be addressed by limiting the scale factor to be within ±5% of its value before optimisation.

7.2 Tailoring Optimisation

The previous section described a general approach to solving the optimisation problem. We will now look at how the algorithm can be tailored to deal with certain classes of object.

7.2.1 Closed Curves and Surfaces

For closed curves and closed surfaces, the position of the parameterisation origin is a free parameter. If an obvious landmark is available on all training shapes, this can be used as the position of the origin. Alternatively, for each training example, we can move the origin on the curve or surface to find the position that minimises the sum of squared Euclidian distances between the moved points and the points on a fixed reference shape. This gives a good initial estimate of the position of the origin, but this can often be improved by considering the positions of the origins in the iterative optimisation process. If, however, the origin positions are included directly as parameters in the optimisation, they have a global effect that can disrupt any existing locally optimised correspondences. A better solution is to randomly place the point on the curve/surface using a uniform distribution. The position is then fixed *for that iteration*, giving every point on the curve/surface the freedom to move at some stage.

7.2.2 Open Surfaces

The approach to establishing an initial parameterisation of an open surface, described in Sect. 6.1.1, requires nodes on the surface boundary to be mapped to the boundary of the unit square. Choosing a suitable mapping is important because the parameterisation of non-boundary nodes are dependent on the position of the boundary nodes. Furthermore, correspondence on the edge of the surfaces is fixed during optimisation – a consistent boundary mapping must therefore be chosen across the training set before optimisation begins.

One approach is to position nodes so that the distance between neighbouring nodes on the square are proportional to the distance on the three-dimensional mesh, but this leads to an arbitrary correspondence between the boundaries of the training surfaces. Better results are obtained by optimising the correspondence between training boundaries. This is achieved by extracting the boundaries of each training surface and representing them as a set of closed curves. The correspondence can then be optimised using the optimisation method described above, using any model-based objective functions. After optimisation, the boundary nodes are re-positioned around the boundary according to the re-parameterisation found during optimisation. These re-positioned boundary nodes are used to obtain a solution for the parameterisation of non-boundary nodes, as described in Sect. 6.1.1.

7.2.3 Multi-part Objects

So far, we have only considered building models of structures where each training object consists of a single part. The extension to training objects with multiple sub-parts (where the corresponding sub-part has the same topology across the training set) is straightforward.

Optimisation is performed as in the general case, but with a separate re-parameterisation function for each sub-part of each training object. The points of all sub-parts of each training example are concatenated into a single shape vector and the objective function is evaluated as before – in this way the model represents the joint distribution of all sub-parts of all training shapes.

7.3 Implementation Issues

The evaluation of the objective function requires the calculation of an integral over each pair of training shapes, as we saw in Sect. 2.3. Although there are precise methods for evaluating the integral, they can take an impractical amount of time to compute. A quicker method is presented here that uses a discrete approximation with a fixed set of sample points. It is important that a suitable set of sample points is chosen to ensure that the integral is calculated precisely – two methods of achieving this are presented below.

We will also see that a discrete representation of re-parameterisation can contain singularities, despite the fact that the continuous re-parameterisation is non-singular. A fast method of detecting singularities is presented as well as ways of avoiding the problem in the first place.

7.3.1 Calculating the Covariance Matrix by Numerical Integration

In order to evaluate the objective function, we have to compute an integral over each pair of training shapes to calculate the normalized continuous version of the covariance matrix, $\widetilde{D}_{ij}$ (2.124). The integration ensures that each shape is represented sufficiently, and the true covariance of the data calculated. If the integral is not performed, the optimisation can 'cheat' by moving sample points away from 'difficult' areas on the shapes. As an extreme example, the sample points on each shape can be collapsed to a small region on each shape, which gives a small value of the objective function but the shapes are not represented properly. This behaviour only occurs in the later stages of optimisation when the models are almost optimal. In the early stages of the optimisation, it makes almost no difference whether the integral is computed or not.

There is no closed form solution to the integral in (2.124) so numerical quadrature methods [139] must be used to evaluate it. While any of the numerical methods described in Chap. 4 of [139] could be used as a basis for calculating the covariance matrix, they tend to take a long time to compute. A simple alternative approach is to use trapezoidal integration using a fixed set of abscissas. The integral in (2.124) can then be calculated using a simple finite approximation:

$$\begin{aligned}&\int (\mathbf{S}_i(\mathbf{x}) - \bar{\mathbf{S}}(\mathbf{x})) \cdot (\mathbf{S}_j(\mathbf{x}) - \bar{\mathbf{S}}(\mathbf{x}))\, dA(\mathbf{x}) \\ &\approx \sum_{k=1}^{n} (\mathbf{S}_i(\mathbf{x}_k) - \bar{\mathbf{S}}(\mathbf{x}_k)) \cdot (\mathbf{S}_j(\mathbf{x}_k) - \bar{\mathbf{S}}(\mathbf{x}_k))\, \Delta A(\mathbf{x}_k),\end{aligned} \tag{7.4}$$

where $\{\mathbf{x}_k\}$ are the parameters of some set of sample points. Remember that $dA(\mathbf{x})$ is the length/area measure on the mean shape. Hence for curves in two dimensions:

$$\Delta A(\mathbf{x}_k) = (d_{k,k-1} + d_{k,k+1}), \tag{7.5}$$

where $d_{k,k-1}$ is the Euclidian distance between the adjacent points k and $k-1$ as measured on the mean shape. In three dimensions, this becomes

$$\Delta A(\mathbf{x}_k) = \sum_{\alpha} A_{k\alpha}, \tag{7.6}$$

where $\{A_{k\alpha}\}$ is the area on the mean shape of all triangles that have point k as a vertex.

7.3.2 Numerical Estimation of the Gradient

In Sect. 7.1.3, we saw that the gradient of the objective function w.r.t. the Ath parameter of the ith training example ($\alpha_A^{(i)}$) is given by:

$$\frac{\partial \mathcal{L}}{\partial \alpha_A^{(i)}} = \sum_{a=1}^{n_S-1} \frac{\partial \mathcal{L}}{\partial \lambda_a} \sum_{j=1}^{n_S} \sum_{k=1}^{n_S} \frac{\partial \lambda_a}{\partial \widetilde{D}_{jk}} \int \frac{\delta \widetilde{D}_{jk}}{\delta \mathbf{S}_i(\mathbf{x})} \frac{\delta \mathbf{S}_i(\mathbf{x})}{\delta \alpha_A^{(i)}} dA(\mathbf{x}).$$

We saw in Sect. 4.3.4 that closed-form expressions are available for all terms except $\frac{\delta \mathbf{S}_i(\mathbf{x})}{\delta \alpha_A^{(i)}}$, to which we now turn our attention. An analytical solution was given in [94], but it was limited to a one-dimensional spline-based representation of re-parameterisation (described in Sect. 5.1.4.2). If we are willing to accept a possible small degree of numerical inaccuracy, then we can estimate the derivative using a simple finite difference scheme, thus allowing any representation of re-parameterisation to be used. From (7.1):

$$\mathbf{S}'_i(\mathbf{x}) \doteq \mathbf{S}_i(\mathbf{x}'), \quad \mathbf{x}' = \phi_i(\mathbf{x}) \doteq \phi(\mathbf{x}; \boldsymbol{\alpha}^{(i)}).$$

It is hence convenient to define a new shape function:

$$\mathbf{S}_i(\mathbf{x}; \boldsymbol{\alpha}^{(i)}) \doteq \mathbf{S}_i(\phi(\mathbf{x}; \boldsymbol{\alpha}^{(i)})).$$

The lowest-order finite difference approximation is then given by:

$$\frac{\delta \mathbf{S}_i(\mathbf{x})}{\delta \alpha_A^{(i)}} = \frac{\partial \mathbf{S}_i(\mathbf{x}; \boldsymbol{\alpha}^{(i)})}{\partial \alpha_A^{(i)}} \approx \frac{\mathbf{S}_i(\mathbf{x}; \alpha_A^{(i)} + \Delta \alpha_A^{(i)}) - \mathbf{S}_i(\mathbf{x}; \alpha_A^{(i)})}{\Delta \alpha_A^{(i)}},$$

where $\Delta \alpha_A^{(i)}$ is a small perturbation of the parameter $\alpha_A^{(i)}$. The exact details of how this can be implemented in practice are given in the example optimisation scheme at the end of this chapter (see Sect. 7.4.2, Algorithm 7.4).

The above expression for the gradient also contains an integral over the surface of the mean shape. We evaluate this using the same approach as that described in the previous section for calculating the covariance matrix, so that:

$$\int \frac{\delta \widetilde{D}_{jk}}{\delta \mathbf{S}_i(\mathbf{x})} \frac{\delta \mathbf{S}_i(\mathbf{x})}{\delta \alpha_A^{(i)}} dA(\mathbf{x}) \approx \sum_m \frac{\delta \widetilde{D}_{jk}}{\delta \mathbf{S}_i(\mathbf{x}_m)} \frac{\delta \mathbf{S}_i(\mathbf{x}_m)}{\delta \alpha_A^{(i)}} \Delta A(\mathbf{x}_m),$$

where $\{\mathbf{x}_m\}$ represent the parameters of some sample points. The values of the elements of area $\Delta A(\mathbf{x}_m)$ are calculated in the same way as for the covariance matrix integral (Sect. 7.3.1).

7.3.3 Sampling the Set of Shapes

In order to evaluate the covariance matrix using the method described above, we need to sample a number of points on each shape. These points must be chosen so that all shapes in the training set are sufficiently well represented. Two methods of achieving this are presented here. The first is based on digital half-toning techniques, and the second is a simpler scheme based on an adaptive uniform sampling.

The first method of re-sampling is one presented by Heimann et al. (see [88] and the references therein), which creates a set of surface points which has a more uniform distribution on the mean shape. The approach is illustrated in Fig. 7.1. The method is based on the work in [2]. The first stage consists of creating an *area distortion image*, formed by calculating the ratio $\text{Area}(\mathbf{v}(\mathbf{t}^\beta)/\text{Area}(\mathbf{x}(\mathbf{t}^\beta)$, where $\mathbf{v}(\mathbf{t}^\beta)$ is the set of positions of the vertices of triangle $\mathbf{t}^\beta$ on the shape, and $\mathbf{x}(\mathbf{t}^\beta)$ is the set of the positions of the vertices of the same triangle in parameter space. Area$(\cdot)$ is then just the area of the triangle in the appropriate space.

The mean of each distortion image is taken and digital half-toning techniques, employing efficient error diffusion methods, are used to create a binary image. As is shown in the figure, the black pixels are then used as our new sample vertices. The result is a sampling of the parameterisation that compensates for area distortion, thus producing a roughly uniform sampling of each shape's surface.

The use of this digital half-toning step can be avoided if we instead use the method described in Sect. 6.1.4 to produce initial parameterisations with minimal area distortion. Then a uniform sampling of the surface can be achieved by just uniformly sampling the parameter space. For simple open surfaces, the parameter space is just the unit square, whereas for closed surfaces with spherical topology, the corresponding parameter space is the unit sphere.

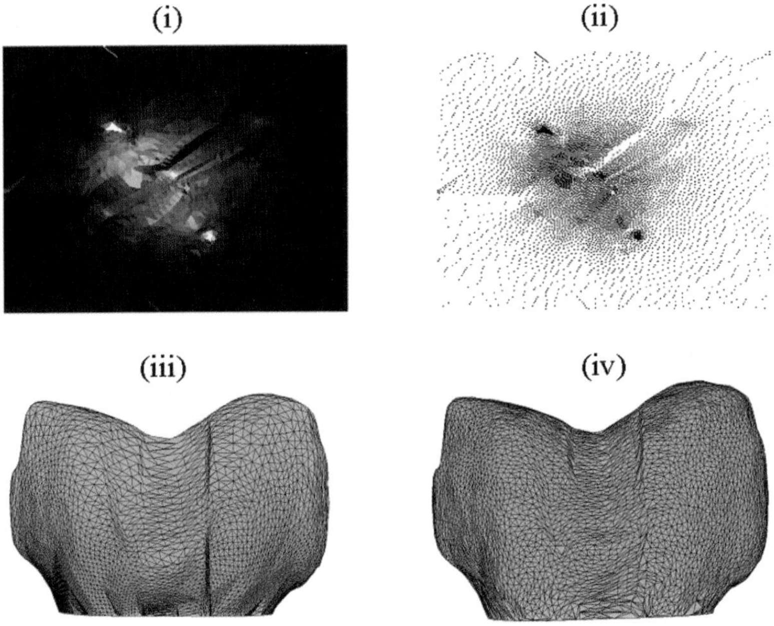

Fig. 7.1 Creating a more uniform surface sampling using digital half-toning. **(i)** An area distortion image (where white represents areas of high distortion) from the ratio of areas of triangles. **(ii)** Digital half-toning is used to dither the image: the image in (i) is sampled according to its intensity to produce a pre-defined number of black pixels, which are then used as new sample nodes. The two triangulated surfaces on the bottom row were created from the same shape, but using a different sampling technique. **(iii)** The shape sampled according to a uniform distribution over the parameter space and **(iv)** the same shape sampled using the nodes obtained from the half-toning technique in (ii). Note that the distribution of points over the surface is considerably more regular in (iv).

Uniformly sampling the unit square is trivial, but the unit sphere requires more careful consideration. It is not strictly possible to position an arbitrary number of equiareal points on the surface of a sphere. However, an approximation, that is sufficient for our purposes, can be achieved. As is shown in Fig. 7.2, this is achieved by subdivision of the faces of an icosahedron, to produce a larger number of finer triangles. The points are then projected onto the sphere to produce the final icosahedral sampling of the sphere.

Each parameterisation is manipulated during optimisation, so that we need to ensure that sufficient sampling is maintained across each training surface as the optimisation proceeds. Detecting a region where the sufficiency of the sampling may be degrading is simple, we just have to monitor the value of the area $\Delta A(\mathbf{x}_k)$ associated with the kth sample point. If this area begins to grow (for example, if it exceeds its initial value by some fixed proportion – 50% being a value typically used herein), we just add a new sample point to the parameterisation of each training example, placed at the centroid of its neighbours.

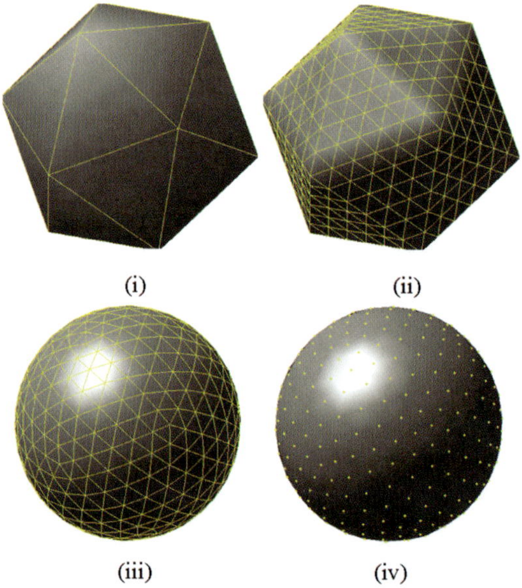

Fig. 7.2 An approximate uniform sampling of a sphere. A unit icosahedron is created (**i**), then each face is subdivided into smaller triangles (**ii**). The subdivided icosahedron is then projected onto the sphere (**iii**) and the nodes of the mesh are used as sample points (**iv**).

7.3.4 Detecting Singularities in the Re-parameterisations

The re-parameterisation functions for surfaces that we presented in Chap. 6 are diffeomorphic by construction. However, if we are using a triangulated mesh of sample points, the usual procedure is to consider just the motion of the nodes, then perform linear interpolation to find the motion of an arbitrary point. The new position of each edge of the triangulation is defined to be the *straight line* connecting the vertices at their new positions. There is then a possibility that the use of this discrete approximation can destroy the diffeomorphic property of the mapping. An illustration of such a case is shown in Fig. 7.3.

In practice, this situation occurs very rarely, and there is a straightforward method of detecting it. For each triangle in the mesh, we compute the normal vector before and after each re-parameterisation. A problematic transformation then corresponds to one where the direction of any normal vector deviates by more than 90° from its position before re-parameterisation. If such a transformation is detected, it can just be removed before continuing the optimisation.

Consider now Fig. 7.4, which illustrates how a shape is sampled according to its parameterisation. When we manipulate correspondence, we manipulate points on the parameter space (here given by the unit sphere). We then have a choice, we can either keep the sample points fixed on the sphere, and manipulate the nodes of the shape parameterisation, or keep the nodes of the shape

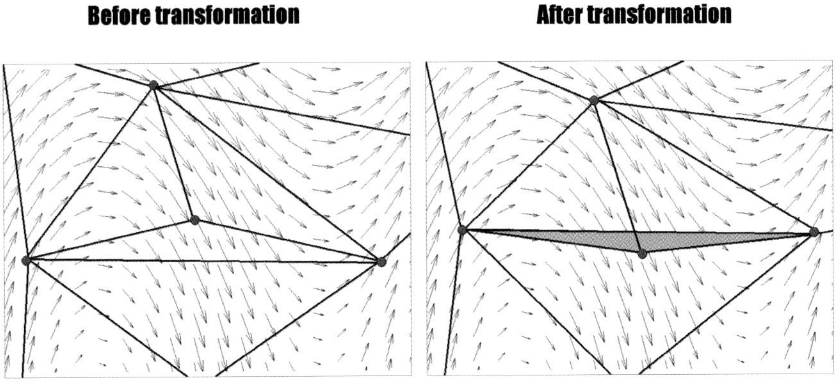

Fig. 7.3 An example of where a transformation may be homeomorphic in the continuous case, but not in the discrete case. The arrows represent a continuous transformation; the discrete representation is represented as a triangulation, drawn before (**left**) and after (**right**) applying the transformation. After the transformation has been applied, we can see that a triangle has folded, which implies that the discrete transformation is not a homeomorphism.

parameterisation fixed, and manipulate the sample points. If we consider the triangulations formed by each of these two sets of points, we see that, as in the example in the figure, the points of the shape parameterisation tend to give a more irregular mesh, which can contain many examples of long, thin triangles. And such triangles are more likely to fold (as in Fig. 7.3) when manipulated. Conversely, the sample points have a more regular triangulation, which is much less likely to experience the same sort of folding problems under manipulation. Hence choosing to manipulate the sample points instead of the parameterisation nodes greatly reduces the probability of obtaining singularities in the transformations.

7.4 Example Optimisation Routines

We conclude this chapter by giving an explicit illustration of how correspondence can be optimised in practice.

Two different training sets are used. The first is a set of hand outlines in two dimensions (see Fig. 7.5), and the second is a set of surfaces representing the distal end of the human femur (see Fig. 7.8).

We will also use various representations of re-parameterisation, as well as different objective functions and optimisation approaches. Operators, data types, and conventions used in the pseudocode are listed in the Glossary.

As previously (7.1), we use the alternative definition of re-parameterisation, so that:

$$\mathbf{S}'_i(\mathbf{x}) \doteq \mathbf{S}_i(\mathbf{x}'), \quad \mathbf{x}' = \phi_i(\mathbf{x})$$

$$\Rightarrow \ \mathbf{S}_i(\mathbf{x}) \sim \mathbf{S}_j(\mathbf{x}) \ \overset{\longmapsto}{\phi_i, \phi_j} \ \mathbf{S}'_i(\mathbf{x}) \sim \mathbf{S}'_j(\mathbf{x}), \ \mathbf{S}_i(\phi_i(\mathbf{x})) \sim \mathbf{S}_j(\phi_j(\mathbf{x})). \tag{7.7}$$

We begin with the hand training set, and example of shapes which are open curves.

7.4.1 Example 1: Open Curves

The raw data for the hand training set was obtained by taking a series of photographs of a person's hand, in a flat position, but with the fingers in various positions. Each hand outline was then segmented from the images, to produce the $n_S = 17$ training examples shown in Fig. 7.5.

For model-building, the following options were chosen:

Objective function: The minimum description length objective function (see (4.20) and (4.21)) will be used, with a fixed precision of $\Delta = 0.1$. For this particular training set, the optimisation is relatively simple in terms

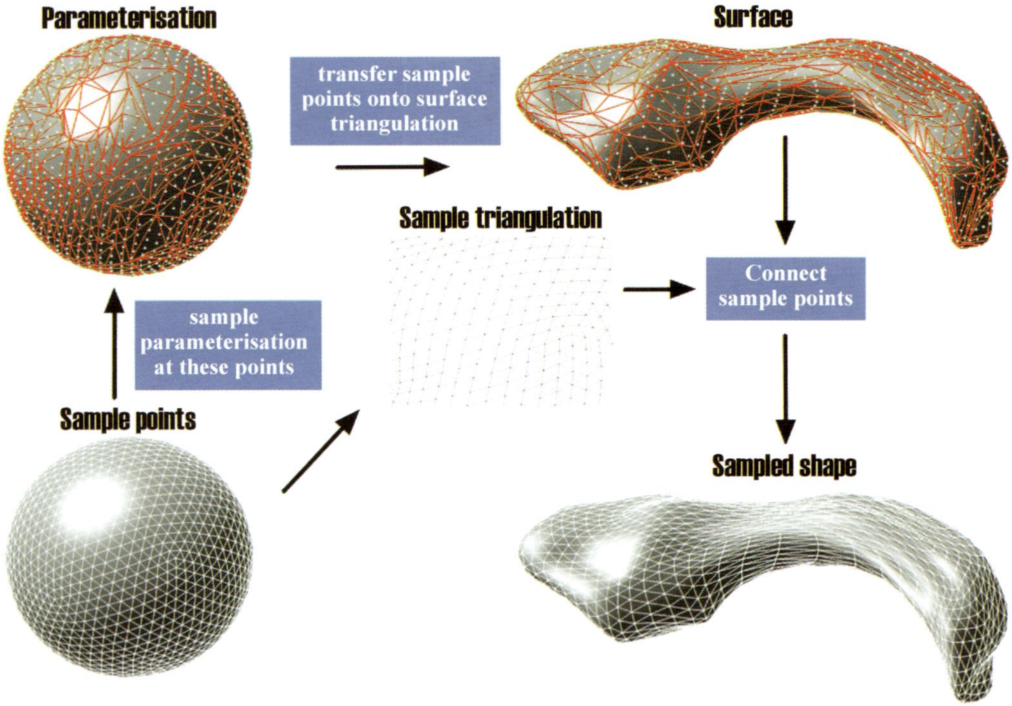

Fig. 7.4 How a shape is sampled. The nodes of a uniform subdivision of the sphere (white mesh, **bottom left**) are used to sample the parameterisation (green mesh, **top left**). The barycentric coordinates of each sample (white) point is then calculated with respect to the parameterisation and these coordinates are applied to the corresponding triangle on the surface mesh (**top right**). The projected sample points can then be connected using the same triangulation as the original parameterisation to create the sampled shape (**bottom right**). Note that for the purpose of illustration, only a small number of sample points were used.

of computational complexity, hence it is feasible to retain the full MDL objective function throughout the optimisation.

Representation of Re-parameterisation: For each shape at each iteration, we use a single Cauchy kernel (Sect. 5.1.4.1 with $n_k = 1$). The re-parameterisation is evaluated using (5.29), and is a function of the Cauchy's magnitude, position, and width.

Optimisation Approach: One shape parameterisation to be optimised at a time, using a simple one-dimensional line search optimisation, with stochastic selection of auxiliary parameter values. No gradient information to be used.

With these options, the pseudocode for the Algorithm is as given below.

Algorithm 7.2 : An Example of Optimising Correspondence on a Set of Curves.

---

procedure $\{\mathbf{shape\_vector}_i\} \leftarrow$ *optimisation_example_1* ↷
$(\{\mathbf{parameterisation}_i\}, \{\mathbf{Shape\_Points}_i\})$

VARIABLES

- $\{\mathbf{Shape\_Points}_i\}$ is a training set of n_S shape functions, where each shape is represented by an ordered set of points; the values of each shape function are defined at positions stored in the n_P-dimensional vector $\mathbf{parameterisation}_i$ and the coordinates for the points of each shape are stored in the rows of the $n_P \times 2$ matrix $\mathbf{Shape\_Points}_i$ (note that the number of points, n_P, may be different for each shape);
- the function returns a set of $2n$-dimensional shape vectors $\{\mathbf{shape\_vector}_i\}$, formed by sampling each training set according to its optimal parameterisation.

DECLARATIONS

- $b = linear\_interpolation(\mathbf{x}, \mathbf{y}, a)$
  the vector $\mathbf{y}$ holds values of a function, evaluated at the abscissas in vector $\mathbf{shape\_vector}$; linear interpolation is used to estimate the value, b, of the function at abscissa a;
- $\tilde{u} = cauchy(u, magnitude, position, width)$
  evaluates the re-parameterisation at abscissa u using a Cauchy with parameter values $magnitude$, $position$ and $width$, using (5.29).

INITIALIZATION

1. choose an index, ref, for a reference shape, whose pose and parameterisation do not change during optimisation;
2. for $i = 1 \ldots n_S$

   2.1. let $\mathbf{sample\_points}_i$ be a n-dimensional vector of points used to sample the ith shape ($n = 400$ is used here); each vector is initialized to a uniform sampling of the parameterisation:

$$\mathbf{sample\_points}_i \leftarrow (0, \frac{1}{n-1}, \frac{2}{n-1}, \ldots, 1);$$

   2.2. sample the ith shape function at the sample points and form a shape vector for each training example:
   for $j = 1 \ldots n$

% the x-coordinate component

$$\mathbf{shape\_vector}_i(j) \leftarrow linear\_interpolation \curvearrowright (\mathbf{parameterisation}_i, \mathbf{Shape\_Points}_i(\cdot, 1), \mathbf{sample\_points}_i(j)),$$

% the y-coordinate component

$$\mathbf{shape\_vector}_i(j+n) \leftarrow linear\_interpolation \curvearrowright (\mathbf{parameterisation}_i, \mathbf{Shape\_Points}_i(\cdot, 2), \mathbf{sample\_points}_i(j)),$$

3. align the set of shape vectors, $\{\mathbf{shape\_vector}_i\}$ using Algorithm 2.1, and also apply the pose parameters to the original shape points, $\{\mathbf{Shape\_Points}_i\}$.

OPTIMISATION

for $it = 1 \ldots n_{iterations}$, ($n_{iterations} = 6.5 \times 10^6$ here);

1. choose a shape i at random using a uniform distribution over the training set, with $i \neq ref$;
2. Optimise parameterisation:
   2.1. select values for the auxiliary parameters: choose values for the width $width$ and position $position$ of the Cauchy kernels by random sampling of the relevant distributions in Table 7.1;
   2.2. use a line-search algorithm to find the value of the parameter $magnitude$ in the range $[0, 0.01]$ that produces the minimum description length. For a given value of $magnitude$, the description length is calculated as follows:
      2.2.1. apply the re-parameterisation:
      for $j = 1 \ldots n$

$$\mathbf{temp\_sample}(j) \leftarrow \curvearrowright cauchy(\mathbf{sample\_points}_i(j); magnitude, position, width);$$

      2.2.2. sample the re-parameterised i^{th} shape to form a new shape vector:
      for $j = 1 \ldots n$

$$\mathbf{temp\_shape\_vector}(j) \leftarrow linear\_interpolation \curvearrowright (\mathbf{parameterisation}_i, \mathbf{Shape\_Points}_i(\cdot, 1), \mathbf{temp\_sample}(j)),$$

$$\mathbf{temp\_shape\_vector}(j+n) \leftarrow linear\_interpolation \curvearrowright (\mathbf{parameterisation}_i, \mathbf{Shape\_Points}_i(\cdot, 2), \mathbf{temp\_sample}(j)),$$

      2.2.3. concatenate **temp_shape_vector** with the other shape vectors, $(\{\mathbf{shape\_vector}_k; k = 1 \ldots n_S, k \neq i\})$, to give an updated training set;
      2.2.4. evaluate the description length of the updated training set with (4.20) & (4.21), using the approximation to the covariance matrix described in Sect. 7.3.1.
   2.3. apply the parameterisation found in optimisation:
      2.3.1. use the optimal value of $magnitude$ to re-parameterise the ith shape:
      for $j = 1 \ldots n$

$$\mathbf{sample\_points}_i(j) \leftarrow \curvearrowright cauchy(\mathbf{sample\_points}_i(j), magnitude, position, width),$$

      2.3.2. update the shape vector:
      for $j = 1 \ldots n$

$$\mathbf{shape\_vector}_i(j) \leftarrow linear\_interpolation \curvearrowright$$
$$(\mathbf{parameterisation}_i, \mathbf{Shape\_Points}_i(\cdot, 1), \mathbf{sample\_points}(j)),$$

$$\mathbf{shape\_vector}_i(j+n) \leftarrow linear\_interpolation \curvearrowright$$
$$(\mathbf{parameterisation}_i, \mathbf{Shape\_Points}_i(\cdot, 2), \mathbf{sample\_points}(j)),$$

2.3.3. set the centre of gravity of the updated shape to the origin, in preparation for the pose optimisation;

3. Optimise the scale factor:
 3.1. use the line search algorithm to find the value of $scale$ that leads to the minimum description length. The value of $scale$ is bounded to be in the range $\pm 5\%$ of the value found in the *initial* alignment (using Algorithm 2.1). For a given value of $scale$, the objective function is evaluated as follows:
  3.1.1. apply the scaling to the ith shape:

$$\mathbf{temp\_shape\_vector}_i \leftarrow scale \cdot \mathbf{shape\_vector}_i$$

  3.1.2. concatenate the scaled i^{th} shape with the other training shapes to give an updated training set:
  3.1.3. evaluate the description length of the updated training set using (4.20) & (4.21).
 3.2. once the optimal value of s is found, update the shape vector

$$\mathbf{shape\_vector}_i \leftarrow scale \cdot \mathbf{shape\_vector}_i$$

 and shape function:

$$\mathbf{Shape\_Points}_i \leftarrow scale \cdot \mathbf{Shape\_Points}_i;$$

4. optimise rotation:
 4.1. the rotation matrix, $\mathbf{R}$ is a function of a single parameter, θ that describes the angle of rotation in the plane; the optimal value is found over a range of $\pm 10^0$ in an identical manner to the scale factor, above.

return $\{\mathbf{shape\_vector}_i\}$.

The initial correspondence across the training set is shown in Fig. 7.5, along with some details of the model built using this initial correspondence. It can clearly be seen that even if we restrict the range of parameter values input to the model, choosing only values that vary by less than two standard deviations from the mean, this initial model still generates illegal shape examples. The model does show some movement of the hand, but with gross deformations of the fingers.

By contrast, consider the final optimised correspondence and the model built from it, as shown in Fig. 7.6. It is clear that the shapes generated by the model are plausible.[2] It is also clear that the model encompasses the full range of variation seen across the training set, without introducing any deformation of the hand or fingers.

---

[2] Apart from one case where the fingers cross, which is unlike any case provided in the training set. However, such possible intersections of parts in close proximity is a well-understood consequence of the linear assumptions we have made in the modelling, and does not indicate a problem with the found correspondence.

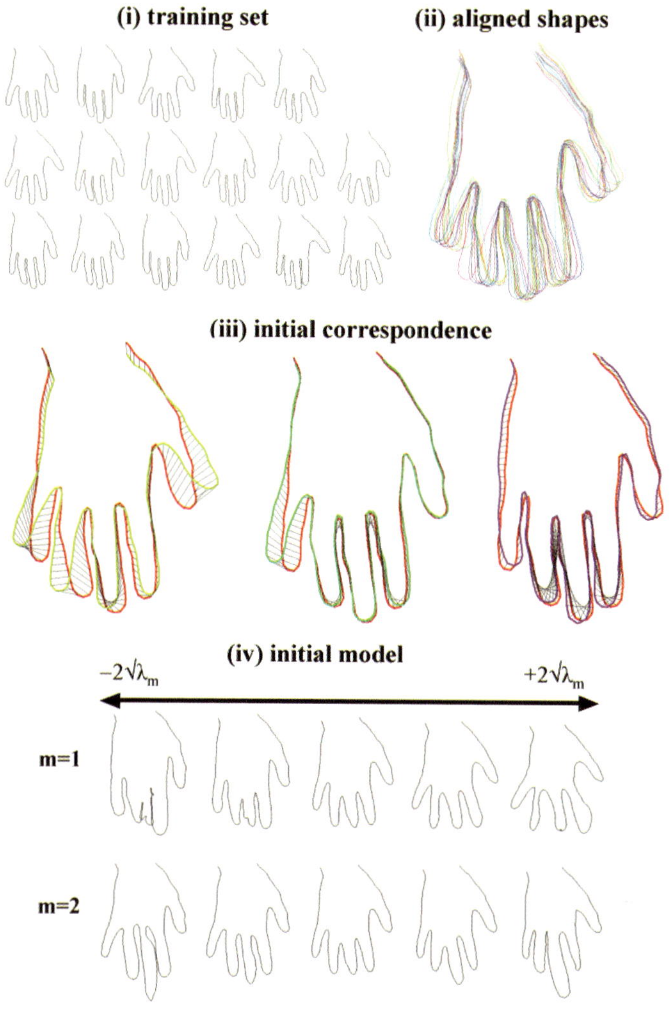

Fig. 7.5 The hand training set and initial model. **(i)** the training set; **(ii)** the initial alignment of the shapes; **(iii)** the initial correspondence shown between three examples (yellow, green, and magenta outlines) and the reference shape (red outline) – correspondence is shown by drawing lines between corresponding points on each curve. Note that correspondence is particularly poor around the ends of the middle three fingers. **(iv)** the model built from the initial correspondence, shown by varying the first two modes ($m = 1, 2$) by ± 2[standard deviations found over the training set] – it is clear that the model can produce illegal instances of the class of object.

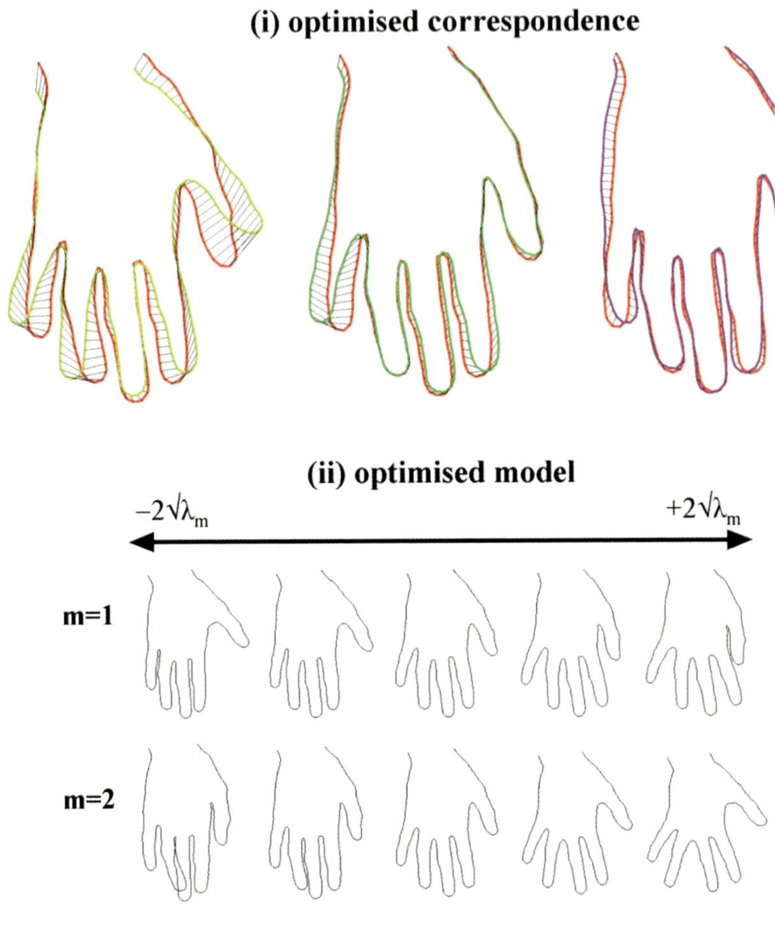

Fig. 7.6 The optimised correspondence and model. In **(i)**, the correspondence is shown between the same three examples (yellow, green, and magenta outlines) as shown in Fig. 7.5 and the reference shape (red outline) – it is clear that the correspondence is now correct at all points on the boundaries. The final model is shown in **(ii)** by varying the first two modes ($m = 1, 2$) by ± 2[standard deviations found over the training set] – the optimised model shows the sort of shape variation that one would expect to see, with no distortions.

7.4.2 Example 2: Open Surfaces

Our second example uses a training set of $n_S = 10$ segmentations of the human distal femur. The training set is shown in Fig. 7.8. Each training example is an open surface, represented by a triangulated mesh.

For the optimisation routine, we make the choices as listed below:

Objective function: We use the determinant of the covariance matrix (4.9), with regularization constant $\epsilon = 0.01$.
Representation of re-parameterisation: We use $n_k = 5$ clamped plate splines (Sect. 6.2.1) to transform each example: the re-parameterisation is evaluated using (6.31), and is a function of the displacement, position, and width of the spline.
Optimisation approach: A gradient descent optimisation (Sect. 7.1.3) was chosen, with all examples to be re-parameterised at the same time.

The parameter space for our open surfaces is the unit square. We thus need a method of transferring points from the parameter space onto the shape surface itself. As we saw in Sect. 6.1.3, given two surfaces with a common triangulation, we can transfer points between the two surfaces using barycentric coordinates. Pseudocode to achieve this is given in Algorithm 7.3. The pseudocode of an example groupwise optimisation is then given in Algorithm 7.4.

Algorithm 7.3 : Transferring a Point Between Triangulations.

---

procedure $\mathbf{y} \leftarrow transfer\_point(\mathbf{Points\_A}, \mathbf{Points\_B}, \mathbf{Triangulation}, \mathbf{x})$

DESCRIPTION

Transfers the point $\mathbf{x}$ from the surface defined by $(\mathbf{Points\_A}, \mathbf{Triangulation})$ to the surface $(\mathbf{Points\_B}, \mathbf{Triangulation})$ to form a new point $\mathbf{y}$, using barycentric areal coordinates.[3]

VARIABLES

- **Points_A** is a $n_P \times n_A$ matrix containing the coordinates of the nodes on the source triangulation; note that n_A can be 2 or 3;
- **Points_B** is a $n_P \times n_B$ matrix containing the coordinates of the nodes on the target triangulation; note that n_B can again be 2 or 3;
- **Triangulation** is a $n_t \times 3$ matrix containing the common triangulation that indexes **Points_A** and **Points_B** – see the glossary for details of the format;
- $\mathbf{x}$ is a n_A-dimensional vector that represents the coordinates of a point in the same space as the nodes of **Points_A**;
- the procedure returns $\mathbf{y}$, a n_B-dimensional vector that represents the coordinates of a point in the same space as the nodes of **Points_B**.

DECLARATIONS

---

[3] See Sect. 6.1.3 for further details.

- $area(\mathbf{a}, \mathbf{b}, \mathbf{c})$
  is a procedure that returns the area of a triangle with nodes $\mathbf{a}$, $\mathbf{b}$, and $\mathbf{c}$.

SAMPLING

1. find the barycentric coordinates of $\mathbf{y}$ with respect to the surface $(\mathbf{Points\_A}, \mathbf{Triangulation})$[4]
   for $j = 1 \ldots n_t$
   1.1. let $\mathbf{t} \leftarrow \mathbf{Triangulation}(j, \cdot)$;
   1.2. let $\{\mathbf{v}_k\}$ be the set of vertices indexed by triangle $\mathbf{t}$ (note that if $\mathbf{x}$ is three-dimensional then it needs to be projected onto the same plane as the triangle, $\mathbf{t}$);
   for $k = 1 \ldots 3$
   $$\mathbf{v}_k \leftarrow \mathbf{Points\_A}(\mathbf{t}(k), \cdot);$$
   1.3. let:
   $$\begin{aligned} d &\leftarrow area(\mathbf{v_1}, \mathbf{v_2}, \mathbf{v_3}), \\ a &\leftarrow \frac{area(\mathbf{x}, \mathbf{v_1}, \mathbf{v_2})}{d}, \\ b &\leftarrow \frac{area(\mathbf{x}, \mathbf{v_1}, \mathbf{v_3})}{d}, \\ c &\leftarrow \frac{area(\mathbf{x}, \mathbf{v_2}, \mathbf{v_3})}{d}; \end{aligned}$$
   1.4. if $((0 \le a \le 1)\ AND\ (0 \le b \le 1)\ AND\ (0 \le c \le 1)\ AND\ (a + b + c = 1))$
   then break from for loop (and save the value of $\mathbf{t}$) ;
2. apply the barycentric coordinates to the vertices of the corresponding triangle on the surface $(\mathbf{Points\_B}, \mathbf{Triangulation})$:
   for $k = 1 \ldots 3$,
   $$\mathbf{v}_k \leftarrow \mathbf{Points\_B}(\mathbf{t}(k), \cdot);$$
3. let $\mathbf{y} \leftarrow a\mathbf{v}_1 + b\mathbf{v}_2 + c\mathbf{v}_3$.

return $\mathbf{y}$.

---

[4] Note that there are many computational shortcuts that could be made here. For example, only searching triangles within a given distance of $\mathbf{x}$ or by remembering which triangle $\mathbf{x}$ lies inside from previous function calls (points typically move by only small amounts and therefore often stay within the same triangle between function calls). Some components of the barycentric coordinate calculation can also be precomputed.

Algorithm 7.4 : An Example of Optimising Correspondence on a Set of Surfaces.

procedure $\{\mathbf{shape\_vector}_i\} \leftarrow optimisation\_example\_2$
$(\{\mathbf{Shape\_Points}_i\}, \{\mathbf{Triangulation}_i\})$

VARIABLES

- $\mathbf{Shape\_Points}_i$ is a $n_P \times 3$ matrix whose rows contain the coordinates of nodes on the surface of the ith training example;
- $\mathbf{Triangulation}_i$ is a $n_t \times 3$ matrix that defines the connectivity of the nodes: each row defines a triangle by indexing the rows of $\mathbf{Shape\_Points}_i$ – see glossary for details of format;
- the function returns a set of $3n$-dimensional shape vectors $\{\mathbf{shape\_vector}_i\}$, formed by sampling each training set according to its optimal parameterisation.

DECLARATIONS

- $\mathbf{y} = transfer\_point(\mathbf{Points\_A}, \mathbf{Points\_B}, \mathbf{Triangulation}, \mathbf{x})$
  transfers the point $\mathbf{x}$ from triangulation $(\mathbf{Points\_A}, \mathbf{Triangulation})$ to triangulation $(\mathbf{Points\_B}, \mathbf{Triangulation})$ to form a new point $\mathbf{y}$ – pseudocode is given in Algorithm 7.3;
- $\tilde{\mathbf{u}} = cps\,(\mathbf{u}, \mathbf{displacement}, \mathbf{position}, width)$
  the point in the 2-dimensional vector $\mathbf{u}$ is transformed to a new position $\tilde{\mathbf{u}}$ by applying a clamped plate spline (cps) to a circular patch centred at a position described by a 2-dimensional vector $\mathbf{position}$ and of a width specified by the scalar $width$ – the transformation is formed by moving the origin of the patch to a new position determined by the 2-dimensional vector $\mathbf{displacement}$; note that the function only applies one cps at a time – see Sect. 5.1.3 for details;
- $\delta(i, j)$
  is the Kronecker delta; it returns 1 if $i = j$ and 0 otherwise.

INITIALIZATION

1. let ref be the index of a reference shape, whose pose and parameterisation do not change during optimisation;
2. parameterise each surface to produce a set of $n_P \times 2$ matrices $\{\mathbf{Parameterisation}_i\}$, whose rows contain the coordinates of nodes in parameter space (the unit square), corresponding to the surface nodes stored in $\{\mathbf{Shape\_Points}_i\}$:
   - 2.1. create an initial parameterisation of the reference shape using the method described in Sect. 6.1.1;
   - 2.2. for the reference shape only: use a set of five clamped plate splines to manipulate the parameterisation so as to minimise area distortion, as described in Sect. 6.1.4;
   - 2.3. parameterise the remaining surfaces:
     - 2.3.1. extract the boundary of each training surface (including the reference shape) and represent them as a set of two-dimensional curves;
     - 2.3.2. find the set of re-parameterisation functions that minimise the determinant-based objective function (4.9) – this can be achieved using Algorithm 7.2, above;
     - 2.3.3. map the nodes on the surface boundary to the boundary of the unit square such that the relative distance between neighbouring nodes are preserved;
     - 2.3.4. reposition the boundary nodes of all surface parameterisations (except for the reference shape) according to the re-parameterisation functions found during optimisation in step 2.3.2;
     - 2.3.5. solve for the positions of the internal, non-boundary nodes as described in Sect. 6.1.1;

2.4. improve the consistency of the parameterisations using the method described in Sect. 6.1.5 to match each shape to the reference shape;

3. initialize the sample points of each shape to be a $m \times m$ grid over the unit square ($m^2 = n = 1600$ was used here); the sample points of each shape are stored in a $n \times 2$ matrix $\mathbf{Sample\_Points}_i$:
for $i = 1 \dots n_S$

$$\mathbf{Sample\_Points}_i = \left((0,0)^T, \left(0, \frac{1}{m-1}\right)^T, \dots, \left(\frac{1}{m-1}, 0\right)^T, \curvearrowright \right.$$
$$\left. \left(\frac{1}{m-1}, \frac{1}{m-1}\right)^T, \dots, (1,1)^T \right)^T;$$

4. sample each surface according to its parameterisation to produce a set of $3n$-dimensional shape vectors:
for $i = 1 \dots n_S$

for $j = 1 \dots n$

$$\mathbf{v} \leftarrow transfer\_point(\mathbf{Parameterisation}_i, \curvearrowright$$
$$\mathbf{Shape\_Points}_i, \mathbf{Triangulation}_i, \mathbf{Sample\_Points}_i(j, \cdot)),$$
$$\mathbf{shape\_vector}_i(j) \leftarrow \mathbf{v}(1),$$
$$\mathbf{shape\_vector}_i(j+n) \leftarrow \mathbf{v}(2),$$
$$\mathbf{shape\_vector}_i(j+2n) \leftarrow \mathbf{v}(3);$$

5. align the set of shape vectors, $\{\mathbf{shape\_vector}_i\}$ using Algorithm 2.1, and also apply the pose parameters to the original shapes, $\{\mathbf{Shape\_Points}_i\}$.

OPTIMISATION

for $it = 1 \dots n_{iterations}$, (we used $n_{iterations} = 8000$);

1. optimise parameterisation:

1.1. let $\{\mathbf{widths}_i\}$ be a set of n_A-dimensional vectors that represent the widths of each clamped plate spline, let $\{\mathbf{Positions}_i\}$ represent a $n_A \times 2$ matrix whose rows contain the coordinates of the position of each clamped plate spline; the values of these parameters are generated by random sampling from the distributions in Table 7.1;

1.2. let $\{\mathbf{p}_i\}$ be a set of $2n_A$-dimensional vectors representing the coordinates of the control points of the splines – these are the parameters that we wish to optimise; note that the coordinates of the control point of each spline is 2-dimensional but these have been concatenated such that the coordinates of the Ath control point are given by the vector: $(\mathbf{p}_i(A), \mathbf{p}_i(A + n_A))$;

1.3. let $\{\mathbf{gradient}_i\}$ be a set of $2n$-dimensional vectors, which hold the values of the gradient of the objective function, $\mathcal{L}$, w.r.t. the vector of parameters, $\mathbf{p}_i$:

$$\mathbf{gradient}_i \leftarrow \left(\frac{\partial \mathcal{L}}{\partial p_1^{(i)}}, \dots \frac{\partial \mathcal{L}}{\partial p_A^{(i)}}, \dots \frac{\partial \mathcal{L}}{\partial p_{2n_A}^{(i)}} \right);$$

in Sect. 4.3.4, we saw that the gradient can be split into a product of simpler terms:

$$\frac{\partial \mathcal{L}}{\partial p_A^{(i)}} = \sum_{a=1}^{n_S - 1} \frac{\partial \mathcal{L}}{\partial \lambda_a} \sum_{j=1}^{n_S} \sum_{k=1}^{n_S} \frac{\partial \lambda_a}{\partial \widetilde{D}_{jk}} \int \frac{\delta \widetilde{D}_{jk}}{\delta \mathbf{S}_i(\mathbf{x})} \frac{\delta \mathbf{S}_i(\mathbf{x})}{\delta p_A^{(i)}} dA(\mathbf{x})$$

we also saw in Sect. 7.3.2 that the term involving the integral can be estimated using a simple finite sum:

$$\int \frac{\delta \widetilde{D}_{jk}}{\delta \mathbf{S}_i(\mathbf{x})} \frac{\delta \mathbf{S}_i(\mathbf{x})}{\delta p_A^{(i)}} dA(\mathbf{x}) \approx \sum_m \frac{\delta \widetilde{D}_{jk}}{\delta \mathbf{S}_i(\mathbf{x}_m)} \frac{\delta \mathbf{S}_i(\mathbf{x}_m)}{\delta p_A^{(i)}} \Delta A(\mathbf{x}_m)$$

where $\{\mathbf{x}_m\}$ represent the position of the sample points and $\Delta A(\mathbf{x}_m)$ represents the total area of all triangles connected to the mth sample point, calculated on the mean shape (see Sect. 7.3.2).
The gradient of the reference shape is set to zero:
for $A = 1 \ldots 2n_A$

$$\mathbf{gradient}_{ref}(A) \leftarrow 0.$$

For the other training examples, the components of each term are calculated as follows:

1.3.1. calculate the mean shape vector:

$$\mathbf{mean\_shape\_vector} \leftarrow \frac{1}{n_S} \sum_i \mathbf{shape\_vector}_i;$$

1.3.2. create a shape difference vector (2.133) for each shape by subtracting the mean shape vector:
for $i = 1 \ldots n_S$

$$\mathbf{centred\_shape\_vector}_i \leftarrow \mathbf{shape\_vector}_i - \mathbf{mean\_shape\_vector};$$

1.3.3. for each sample point: calculate the sum of the areas (calculated on the mean shape) of all triangles connected to that sample point and store the value in a n-dimensional vector **int_area** on the mean shape;

1.3.4. calculate a $n_S \times n_S$ covariance matrix, **Covariance**, using the approximation described in Sect. 7.3.1; each element is calculated as:

$$\mathbf{Covariance}(i,j) \leftarrow \sum_{k=1}^{n} \mathbf{int\_area}(k) [\curvearrowright$$
$$\mathbf{centred\_shape\_vector}_i(k) \cdot \mathbf{centred\_shape\_vector}_j(k) + \curvearrowright$$
$$\mathbf{centred\_shape\_vector}_i(k+n) \cdot \mathbf{centred\_shape\_vector}_j(k+n) + \curvearrowright$$
$$\mathbf{centred\_shape\_vector}_i(k+2n) \cdot \mathbf{centred\_shape\_vector}_j(k+2n)]$$

1.3.5. obtain the set of eigenvectors $\{\mathbf{eigenvector}_a\}$ and corresponding (ordered) eigenvalues $\{eigenvalue_a\}$ of **Covariance**;

1.3.6. use the eigenvalues to calculate

$$\frac{\partial \mathcal{L}}{\partial \lambda_a} = \frac{1}{eigenvalue_a + \epsilon};$$

1.3.7. normalize all eigenvectors to have unit length:
for $a = 1 \ldots n_m$

$$\mathbf{eigenvector}_a \leftarrow \frac{\mathbf{eigenvector}_a}{||\mathbf{eigenvector}_a||}$$

1.3.8. use the eigenvectors to calculate

$$\frac{\partial \lambda_a}{\partial \widetilde{D}_{jk}} = \mathbf{eigenvector}_a(j) \cdot \mathbf{eigenvector}_a(k);$$

1.3.9. calculate the components of $\frac{\delta \tilde{D}_{jk}}{\delta \mathbf{S}_i(\mathbf{x}_m)}$; the x coordinate component is given by:

$$\frac{1}{n_S \cdot \sum_m \mathbf{int\_area}(m)} \left[(n_S\delta(i,j) - 1)\mathbf{centred\_shape\_vector}_k(m) + \curvearrowright \right.$$
$$\left. (n_S\delta(i,k) - 1)\mathbf{centred\_shape\_vector}_j(m))\right];$$

the y and z coordinate components are obtained in a similar fashion by substituting m with $m+n$ and $m+2n$, respectively;

1.3.10. use a finite difference scheme to numerically approximate $\frac{\delta \mathbf{S}_i(\mathbf{x})}{\delta p_A^{(i)}}$:

- perturb the Ath control point of the ith shape by a small amount $\Delta = 10^{-5}$ parallel to the x-axis and use it to re-parameterise the ith shape and create a perturbed shape vector;
  for $j = 1 \ldots n$ `% do one sample point at a time`
  - re-parameterise by perturbing the x-coordinate of the clamped plate spline:

$$\mathbf{displacement} \leftarrow (\Delta, 0),$$
$$\tilde{\mathbf{u}} \leftarrow cps\left(\mathbf{Sample\_Points}_i(j,\cdot), \mathbf{displacement}, \curvearrowright \right.$$
$$\left. \mathbf{Positions}_i(A), \mathbf{widths}_i(A)\right),$$

  - sample the point and store in a new shape vector:

$$\tilde{\mathbf{v}} \leftarrow transfer\_point(\mathbf{Parameterisation}_i, \curvearrowright$$
$$\mathbf{Shape\_Points}_i, \mathbf{Triangulation}_i, \tilde{\mathbf{u}}),$$
$$\mathbf{perturbed\_shape\_vector}(j) \leftarrow \tilde{\mathbf{v}}(1),$$
$$\mathbf{perturbed\_shape\_vector}(j+n) \leftarrow \tilde{\mathbf{v}}(2),$$
$$\mathbf{perturbed\_shape\_vector}(j+2n) \leftarrow \tilde{\mathbf{v}}(3);$$

- estimate the derivative as a finite difference:

$$\frac{\delta \mathbf{S}_i(\mathbf{x})}{\delta p_A^{(i)}} = \frac{\mathbf{perturbed\_shape\_vector} - \mathbf{shape\_vector}_i}{\Delta};$$

- now perturb in a direction parallel to the y-axis:
  for $j = 1 \ldots n$

$$\mathbf{displacement} \leftarrow (0, \Delta),$$
$$\tilde{\mathbf{u}} \leftarrow cps\left(\mathbf{Sample\_Points}_i(j,\cdot), \mathbf{displacement}, \curvearrowright \right.$$
$$\left. \mathbf{Positions}_i(A), \mathbf{widths}_i(A)\right),$$
$$\tilde{\mathbf{v}} \leftarrow transfer\_point(\mathbf{Parameterisation}_i, \mathbf{Shape\_Points}_i, \curvearrowright$$
$$\mathbf{Triangulation}_i, \tilde{\mathbf{u}}),$$
$$\mathbf{perturbed\_shape\_vector}(j) \leftarrow \tilde{\mathbf{v}}(1),$$
$$\mathbf{perturbed\_shape\_vector}(j+n) \leftarrow \tilde{\mathbf{v}}(2),$$
$$\mathbf{perturbed\_shape\_vector}(j+2n) \leftarrow \tilde{\mathbf{v}}(3);$$

- estimate the derivative as a finite difference:

$$\frac{\delta \mathbf{S}_i(\mathbf{x})}{\delta p_{(A+n_A)}^{(i)}} = \frac{\mathbf{perturbed\_shape\_vector} - \mathbf{shape\_vector}_i}{\Delta};$$

1.4. stack the gradient vectors into a $n_S \times 2n_A$ matrix $\mathbf{L}$ such that: $\mathbf{L}(i, A)$ represents the partial derivative of the Ath element of $\mathbf{gradient}_i$;

1.5. use a line search algorithm to move along the gradient to find the optimal value of the objective function: using a scalar, μ, move along the gradient and calculate the objective function value $\mathcal{L}(\mu\mathbf{L})$; for a given value of μ, the objective function is evaluated as follows:

1.5.1. re-parameterise all sample points using one clamped plate spline at a time:
for $i = 1 : \ldots n_S$,
$\mathbf{New\_Sample\_Points}_i \leftarrow \mathbf{Sample\_Points}_i$
for $A = 1 \ldots n_A$
for $j = 1 \ldots n$

$$\begin{aligned}&\mathbf{displacement} \leftarrow \mu \cdot (\mathbf{L}(i, A), \mathbf{L}(i, A + n_A))\\&\mathbf{New\_Sample\_Points}_i(j, \cdot) \leftarrow cps \curvearrowright\\&\quad(\mathbf{New\_Sample\_Points}_i(j, \cdot), \mathbf{displacement}, \curvearrowright\\&\quad\mathbf{A}_i(A), \mathbf{widths}_i(A));\end{aligned}$$

1.5.2. sample the re-parameterised points on the shape surface:
for $i = 1 : \ldots n_S$,
for $j = 1 \ldots n$

$$\begin{aligned}&\mathbf{v} \leftarrow transfer\_point(\mathbf{Parameterisation}_i, \mathbf{Shape\_Points}_i, \curvearrowright\\&\quad\mathbf{Triangulation}_i, \mathbf{New\_Sample\_Points}_i(j, \cdot)),\\&\mathbf{new\_shape\_vector}_i(j) \leftarrow \mathbf{v}(1),\\&\mathbf{new\_shape\_vector}_i(j + n) \leftarrow \mathbf{v}(2),\\&\mathbf{new\_shape\_vector}_i(j + 2n) \leftarrow \mathbf{v}(3);\end{aligned}$$

1.5.3. use the determinant-based objective function to evaluate $\{\mathbf{new\_shape\_vector}_i\}$, using the approximation described in Sect. 7.3.1 to build the covariance matrix;

1.6. once the optimal value of μ is found, use it to update the sample points and shape vectors:

1.6.1. update the parameterisation by applying the clamped plate splines:
for $i = 1 \ldots n_S,\ i \neq ref$
for $A = 1 \ldots n_A$
for $j = 1 \ldots n$

$$\begin{aligned}&\mathbf{displacement} \leftarrow \mu \cdot (\mathbf{L}(i, A), \mathbf{L}(i, A + n_A))\\&\mathbf{Sample\_Points}_i(j, \cdot) \leftarrow cps\,(\mathbf{Sample\_Points}_i(j, \cdot), \curvearrowright\\&\quad\mathbf{displacement}, \mathbf{A}_i(A), \mathbf{widths}_i(A));\end{aligned}$$

1.6.2. use the new parameterisations to update each shape vector:
for $i = 1 : \ldots n_S,\ i \neq ref$
for $j = 1 \ldots n$

$$\begin{aligned}&\mathbf{v} \leftarrow transfer\_point(\mathbf{Parameterisation}_i, \mathbf{Shape\_Points}_i, \curvearrowright\\&\quad\mathbf{Triangulation}_i, \mathbf{Sample\_Points}_i(j, \cdot)),\\&\mathbf{shape\_vector}_i(j) \leftarrow \mathbf{v}(1),\\&\mathbf{shape\_vector}_i(j + n) \leftarrow \mathbf{v}(2),\\&\mathbf{shape\_vector}_i(j + 2n) \leftarrow \mathbf{v}(3);\end{aligned}$$

1.6.3. set the centre of gravity of each shape to the origin in preparation for pose optimisation;

2. optimise pose:

   2.1. let $\mathbf{m}_i$ be the pose parameters of shape i, formed by concatenating the rotation and scaling parameters;
   2.2. optimisation is performed in an identical manner to the parameterisation optimisation in the previous step, but using the pose parameters $\mathbf{m}_i$ instead of the re-parameterisation parameters $\mathbf{p}_i$;
   2.3. once the optimal values of $\{\mathbf{m}_i\}$ are found, separate out the rotation and scaling factor and use them to update each shape vector $\mathbf{shape\_vector}_i \leftarrow s_i \cdot \mathbf{R}_i \cdot \mathbf{shape\_vector}_i$, and each shape $\mathbf{Shape\_Points}_i \leftarrow s_i \cdot \mathbf{R}_i \cdot \mathbf{Shape\_Points}_i$; note that the rotation matrix must be reshaped appropriately before multiplication.

return $\{\mathbf{shape\_vector}_i\}$.

Our choice of objective function (the determinant of the covariance matrix), and the above algorithm, actually provides a very good estimate of the optimum of the full MDL objective function. If required, the found correspondence can be further refined, where the result of the above gradient-descent based implementation is used to initialize a further optimisation using the full MDL objective function. Note, however, that the gradient of the full MDL

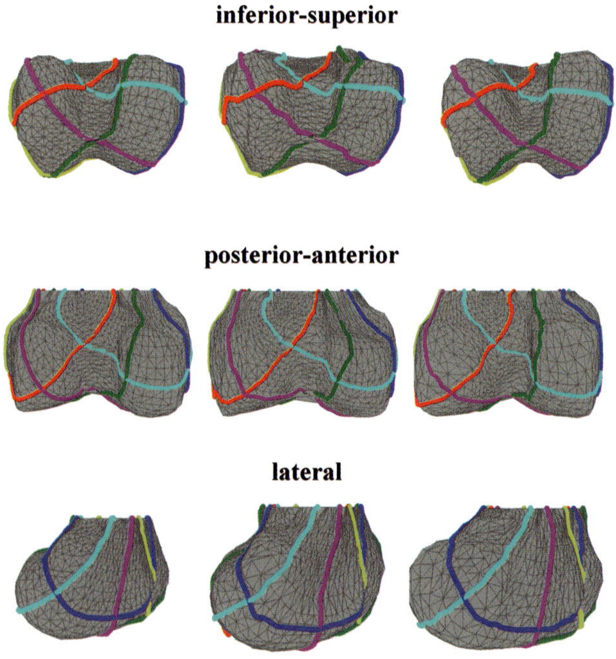

Fig. 7.7 A set of corresponding lines (colours), shown on three examples from the training set, from three different viewpoints.

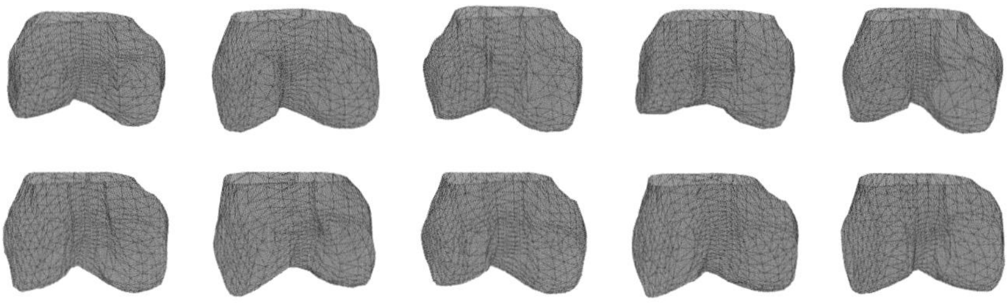

Fig. 7.8 The training set of $n_S = 10$ examples of the distal femur, in posterior-anterior view.

objective function cannot be estimated easily, so that rather than gradient-descent, a non-gradient method such as the Nelder-Mead simplex optimisation algorithm must be used instead.

The correspondence produced by Algorithm 7.4 is illustrated in Fig. 7.7, and the model built using this correspondence is shown in Fig. 7.9. Unlike the previous example, the visual assessment of these results is not straightforward, and ideally, specialist anatomical knowledge is required. And even with

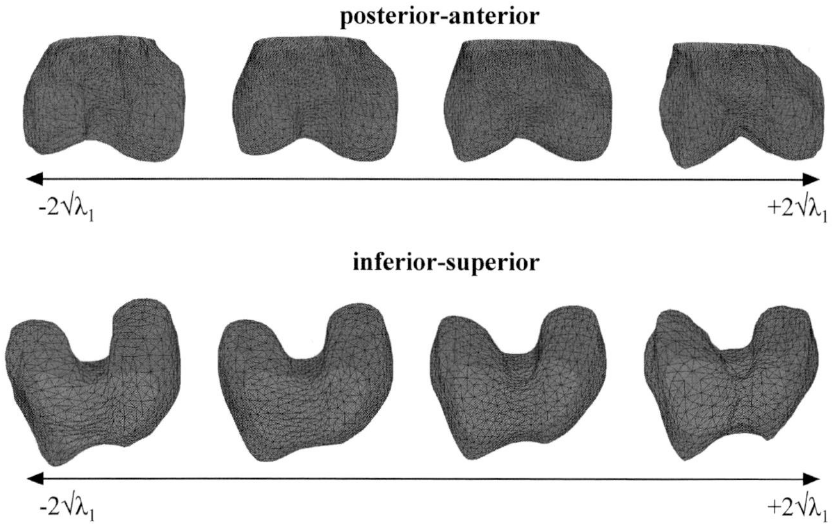

Fig. 7.9 The model built from the optimised correspondence. The first mode of variation of the model is shown by varying its value by ±2[standard deviations found over the training set]. If we compare this with Fig. 7.8 above, we can see that the model has captured variation that was evident in the training set.

such knowledge, the visual assessment is of limited utility. However, without the aid of such specialist knowledge, it is obvious that the correspondence shown in Fig. 7.7 is *reasonable*, with no obvious errors or inconsistencies. A comparison of the training set (Fig. 7.8) with the shapes produced by the model (Fig. 7.9), shows that the shapes produced by the model are also plausible. Detailed comparison of the modes of variation with the training set also indicates that the model modes do encompass the sort of variation that we can detect across the training set.

In summary, we have provided two detailed examples of the implementation of the optimisation framework for groupwise correspondence, and shown the results that are obtained on real datasets. But as is made clear by the last example, visual inspection is of somewhat limited utility when it comes to detailed assessment of results, or comparison of different approaches, except for the case where an approach has conspicuously failed. The question of detailed evaluation and comparison of different approaches to establishing groupwise correspondence is the subject of Chap. 9.

In the present chapter, we considered just the optimisation of correspondence for the parametric representation of re-parameterisation functions. There is a different approach to the representation of re-parameterisation, which also requires a different approach to optimisation, and this is the subject of the next chapter.

Chapter 8
Non-parametric Regularization

In this chapter, we return to the question of the representation of the re-parameterisation functions for shape surfaces.

In Chaps. 5 and 6, we considered how to build various parametric representations of homoeomorphic re-parameterisation functions for curves and surfaces respectively. Such parametric representations serve two purposes. First, they enable us to represent *continuous* re-parameterisation functions in terms of some small set of parameters, which is obviously advantageous from a computational point of view. The second advantage is one of regularization of the problem of finding the groupwise correspondence that minimises the objective function that we have chosen as defining the optimum groupwise correspondence. As mentioned previously (Sect. 6.2), such an optimisation problem is in general ill-posed in the Hadamard sense [86], and is not solvable without some form of regularization.

However, there are other approaches, both to representing our re-parameterisation functions, and to regularization of the optimisation problem, and it is such alternative approaches that we will consider in the present chapter.

8.1 Regularization

To recap, we construct parametric representations for each shape in the training set $\{S_i : i = 1, \dots n_S\}$ in terms of the shape functions $\{\mathbf{S}_i(\mathbf{x})\}$, where $\mathbf{x}$ lies in some parameter space X. The dense groupwise correspondence between the shapes is determined by the parameterisation, so that:

$$\mathbf{S}_i(\mathbf{x}) \sim \mathbf{S}_j(\mathbf{x}) \ \forall \ \mathbf{x} \in X.$$

DOI: 10.1007/978-1-84800-138-1_8,

The correspondence is manipulated by means of the homeomorphic re-parameterisation functions $\{\phi_i(\mathbf{x})\}$, which act on the parameter space and on the shape functions as follows:

$$\mathbf{x} \stackrel{\phi_i}{\longmapsto} \phi_i(\mathbf{x}) \;\&\; \mathbf{S}_i \stackrel{\phi_i}{\longmapsto} \mathbf{S}'_i, \text{ where } \mathbf{S}'_i(\phi_i(\mathbf{x})) \doteq \mathbf{S}_i(\mathbf{x}) \;\forall\; i = 1, \ldots n_S.$$

The optimum correspondence is defined as the minimum of the objective function $\mathcal{L}_S(\mathbf{S}_1, \mathbf{S}_2, \ldots \mathbf{S}_{n_S})$, which depends explicitly on the set of shape functions, and only implicitly on the re-parameterisation functions.

Posed in this form, we are trying to deduce, given a finitely sampled set of shapes or images, the position of the optimum re-parameterisation functions $\{\phi_i(\mathbf{x})\}$ in the *infinite-dimensional* space of all homeomorphic functions. And such problems are in general neither well-posed, nor well-conditioned.

From a computational and theoretical point of view, one obvious way to regularize the problem is to consider not the *infinite-dimensional* space of all possible re-parameterisation functions, but only a *finite-dimensional* sub-space of such functions. Such a sub-space can be constructed by considering re-parameterisation functions that can be described in terms of some finite set of parameters (such as those detailed in Tables 5.1 and 6.1). The search space for optimisation is then just the finite-dimensional space of function parameters.

This parametric method of representation and regularization can be thought of as *hard* regularization, in that the sub-space of allowed functions is held fixed, and no further distinction is made between functions other than that they are within or not within this sub-space. In coarse-to-fine approaches to optimization, the sub-space can be expanded as the algorithm proceeds. This finite-dimensional optimisation problem can then be solved by various methods, as detailed in Chap. 7.

These parametric approaches share the problem, well-known from image registration, that the required function can be hard to find, with the possibility of getting stuck in local minima. Or that the required function may not even be adequately represented within the space of parametric functions that we have constructed. Note that in this parametric approach, the parametric representation plays the dual rôle of both representation and regularizer.

We now decouple these two rôles of representation and regularization, and consider non-parametric methods of regularization.

8.1.1 Non-parametric Regularization

Rather than limiting the re-parameterisation to some parametric sub-space, we now formally allow the re-parameterisation function to be *any* homeomorphic function. The simplest method of non-parametric regularization is imposed by adding a term to the objective function thus:

$$\mathcal{L} = \mathcal{L}_S(\{\mathbf{S}_i\}) + \mathcal{L}_{\text{reg}}(\{\phi_i\}). \tag{8.1}$$

The total objective function now depends *implicitly* on the re-parameterisation functions $\{\phi_i\}$ via their effects on the shape-functions, and also *explicitly* via the regularization term $\mathcal{L}_{\text{reg}}(\{\phi_i\})$. The idea here is that more extreme re-parameterisation functions are allowed, provided the evidence from the data (i.e., the shape data via the shape-functions) warrants it. This can be described as *soft* regularization, with the previous approach of *hard* regularization being a special case with $\mathcal{L}_{\text{reg}}$ being zero on the sub-space of parametric functions, and infinite elsewhere. Typical choices for $\mathcal{L}_{\text{reg}}(\{\phi_i\})$ are based on the bending-energy[1] (e.g., see [151]) or curvature (e.g., see [68, 127]) of the re-parameterisation functions. Such regularization terms act to constrain the re-parameterisation to be smooth.

The first problem with this approach is that they tend to explicitly penalize large-amplitude deformations, which may be exactly what is required to solve the problem. The second problem is that the additional term shifts the position of the optimum.

Before proceeding further, it is convenient to introduce a mathematically-equivalent formulation of re-parameterisation. Rather than the re-parameterisation function $\phi(\mathbf{x})$, we instead introduce the displacement field $\mathbf{u}(\mathbf{y})$ where:

$$\mathbf{y} \overset{\phi}{\mapsto} \phi(\mathbf{y}) \doteq \mathbf{y} + \mathbf{u}(\phi(\mathbf{y})). \tag{8.2}$$

We hence see that the displacement field represents the inverse mapping to ϕ, where:

$$\mathbf{x} = \phi(\mathbf{y}) \;\Rightarrow\; \mathbf{y} = \mathbf{x} - \mathbf{u}(\mathbf{x}). \tag{8.3}$$

The regularization term can now be written as $\mathcal{L}_{\text{reg}} = \mathcal{L}_{\text{reg}}(\mathbf{u})$. One example of such a regularization is elastic registration [18]:

$$\mathcal{L}_{\text{elas}}(\mathbf{u}) = \int_X d\mathbf{x}\, \frac{l_1}{4}\left(\partial_\alpha u_\beta(\mathbf{x}) + \partial_\beta u_\alpha(\mathbf{x})\right)\left(\partial_\alpha u_\beta(\mathbf{x}) + \partial_\beta u_\alpha(\mathbf{x})\right) + \frac{l_2}{2}(\nabla\cdot\mathbf{u}(\mathbf{x}))^2,$$

where we have used our summation convention (see Glossary). The coefficients l_1, l_2 are the Lamé constants. A second example is that of diffusion registration [67]:

$$\mathcal{L}_{\text{diff}}(\mathbf{u}) = \frac{1}{2}\int_X d\mathbf{x}(\nabla u_\alpha(\mathbf{x})) \cdot (\nabla u_\alpha(\mathbf{x})). \tag{8.4}$$

[1] See Willmore energy in the Appendix.

Both these regularizers are based on the second-derivatives of the displacement field. Hence displacement fields which are linear functions of the coordinates entail no cost, and displacement fields which differ by a linear function of the coordinates have the same cost. The second derivatives mean that the regularizers tend to smooth the displacement field.

Rather than the objective function $\mathcal{L}$, we can also work in terms of the forces derived from this objective function. In general, the force at $\mathbf{x}$ is given by:

$$\mathbf{F}(\mathbf{x}) \doteq -\frac{\delta \mathcal{L}}{\delta \mathbf{u}(\mathbf{x})}, \tag{8.5}$$

where $\frac{\delta}{\delta \mathbf{u}(\mathbf{x})}$ is the *functional derivative.*[2] Hence the optimum of an objective function corresponds to zero net force. In more familiar terms, this is just the gradient of the objective function with respect to the coordinates in the search space. This is the gradient that would be used in gradient descent approaches to locating the minimum of $\mathcal{L}$. Since our search space here is the infinite-dimensional space of all homeomorphic displacement fields $\mathbf{u}(\mathbf{x})$, the more usual finite-dimensional gradient vector is here replaced by the infinite-dimensional functional derivative.

The first part of the objective function provides the driving force:

$$\mathbf{F}^{S}(\mathbf{x}) \doteq -\frac{\delta \mathcal{L}_S}{\delta \mathbf{u}(\mathbf{x})}. \tag{8.6}$$

Our regularizers provide a force in addition to the driving force. For elastic and diffusion regularization, the forces are:

$$\mathbf{F}^{\text{elastic}}(\mathbf{x}) = l_1 \nabla^2 \mathbf{u}(\mathbf{x}) + (l_1 + l_2)\nabla(\nabla \cdot \mathbf{u}(\mathbf{x})), \tag{8.7}$$

$$\mathbf{F}^{\text{diff}}(\mathbf{x}) = \nabla^2 \mathbf{u}(\mathbf{x}). \tag{8.8}$$

The Euler-Lagrange equations of the complete objective function(8.1) can then be written in the form:

$$\mathbf{F}^{S}(\mathbf{x}) + \mathbf{F}^{\text{reg}}(\mathbf{x}) = \mathbf{0}. \tag{8.9}$$

This is just a re-statement of the condition for a minimum of the total objective function. In general, this is a non-linear partial differential equation (PDE). For the case $\mathbf{F}^{\text{reg}} = \mathbf{F}^{\text{elastic}}$ (8.7), the regularizing force is the force term that appears in the Navier-Lamé equation describing the dynamics of homogenous, isotropic solids in the linear theory of elasticity. The body force applied to the solid (the driving force) is the force $\mathbf{F}^{S}(\mathbf{x})$. The diffusion regularizer (8.8) gives a PDE which describes a generalized diffusion problem, hence the name. It has the interesting property that the component $F_{\alpha}^{\text{diff}}(\mathbf{x})$

[2] For those unfamiliar with the calculus of variations, see the Appendix, Sect. A.2 for an example of the calculation of a functional derivative.

depends only on $u_\alpha(\mathbf{x})$. Interested readers should consult the appropriate chapters in Modersitzki [127] for more discussion and implementation details of both these forms of regularization within the context of image registration.

However, these forms of regularization share a problem, which is that in general, the regularizing force is non-zero for all non-zero displacements. Hence the optimum is shifted from the point $\mathbf{F}^S(\mathbf{x}) = \mathbf{0}$ to the point (8.9) $\mathbf{F}^S(\mathbf{x}) + \mathbf{F}^{\text{reg}}(\mathbf{x}) = \mathbf{0}$. Such regularization terms also tend to explicitly penalize large displacements, which may not be desirable if the required solution actually requires such a large displacement.

Nevertheless, the regularizing forces (8.7) and (8.8) do possess some desirable properties. As noted previously, since they only contain *second* derivatives of the displacement field, they do not penalize any deformation field which is a linear function of the coordinates plus a constant. That is, the value of the regularizing part of the objective function is invariant to affine transformations (rotations, reflections, directional or uniform scaling, and shear transformations, plus translation).

One way of solving these sorts of non-linear PDEs (8.9) is by introducing a *computational time* t. The displacement field is now made a function of time, so that $\mathbf{u} = \mathbf{u}(\mathbf{x}, t)$. The equation (8.9) is then solved by replacing the right-hand side, so that:

$$\mathbf{F}^S(\mathbf{x}, t) + \mathbf{F}^{\text{reg}}(\mathbf{x}, t) = \frac{\partial \mathbf{u}(\mathbf{x}, t)}{\partial t}.$$

The required solution is then obtained as the steady-state solution, where $\mathbf{u}(\mathbf{x}, t)$ becomes fixed.

However, the introduction of a time coordinate suggests other methods of regularization. Suppose we consider a time-dependant re-parameterisation function $\phi(\mathbf{y}; t)$. We can represent the action of this function by its action on a set of particles moving in $\mathbb{R}^n$. A particle that starts at a position $\mathbf{y}$ at time $t = 0$ then follows a trajectory, so that it is at a position $\phi(\mathbf{y}; t)$ at time t, with $\phi(\mathbf{y}; 0) \doteq \mathbf{y}$. In terms of the displacement field, this just means that we have zero displacement at $t = 0$.

When considering the motions of a set of such particles, there are two ways to describe the dynamics. The Lagrangian approach is to track a particular particle, which starts at a point $\mathbf{y}$ at time $t = 0$, say, and then has a position $\phi(\mathbf{y}; t)$ and velocity:

$$\dot{\phi}(\mathbf{y}; t) \doteq \left.\frac{\partial}{\partial t}\right|_{\mathbf{y}} \phi(\mathbf{y}; t),$$

at later times.

The alternative Eulerian approach is to consider instead a fixed point with respect to the coordinate frame, a point $\mathbf{x}$, say. At time t, the particle passing $\mathbf{x}$ obeys the relation:

$$\phi(\mathbf{y};t) = \mathbf{x} \;\Rightarrow\; \mathbf{y} = \phi^{-1}(\mathbf{x};t).$$

We now also have a time-dependant deformation field $\mathbf{u}(\mathbf{x},t)$, where as before:

$$\mathbf{x} = \phi(\mathbf{y};t), \;\Rightarrow\; \mathbf{y} \doteq \mathbf{x} - \mathbf{u}(\mathbf{x},t), \;\;\therefore\;\; \phi(\mathbf{y};t) = \mathbf{y} + \mathbf{u}(\phi(\mathbf{y};t),t).$$

The Eulerian velocity $\mathbf{v}(\mathbf{x},t)$ at a point $\mathbf{x}$ at time t is then defined as the velocity of the particles that are passing the point at that time. That is:

$$\mathbf{v}(\mathbf{x},t) \doteq \dot{\phi}(\mathbf{y};t) = \left.\frac{\partial}{\partial t}\right|_{\mathbf{y}} \phi(\mathbf{y};t) = \left.\frac{\partial}{\partial t}\right|_{\mathbf{y}} \left[\mathbf{y} + \mathbf{u}(\phi(\mathbf{y};t),t)\right],$$

$$\therefore \quad \boxed{\mathbf{v}(\mathbf{x},t) = \left.\frac{\partial}{\partial t}\right|_{\mathbf{x}} \mathbf{u}(\mathbf{x},t) + (\mathbf{v}(\mathbf{x},t)\cdot\nabla)\mathbf{u}(\mathbf{x},t).} \tag{8.10}$$

So we see that in the Lagrangian framework, the relevant variables are the particle trajectories $\phi(\mathbf{y};t)$ and the particle velocities $\dot{\phi}(\mathbf{y};t)$. Whereas in the Eulerian framework, they are the displacement field $\mathbf{u}(\mathbf{x},t)$ and the Eulerian velocity field $\mathbf{v}(\mathbf{x},t)$. Curvature regularization was based on the Lagrangian variable $\phi(\mathbf{x})$. Elastic or diffusion regularization uses a regularizing force based on the displacement field $\mathbf{u}(\mathbf{x},t)$. Instead, we will consider building a regularizer based on the other Eulerian variable as follows.

8.2 Fluid Regularization

Let us consider a regularizer based on the Eulerian velocity field $\mathbf{v}(\mathbf{x},t)$, where $\mathbf{x} \in \mathbb{R}^n$.

Within the context of image registration, such a regularizer, based on the concept of a viscous fluid was introduced by Christensen et al [22]. However, rather than considering the physical fluid model, we will instead derive the same form based on a few simple mathematical considerations.

Let us take the general case of a vector-valued regularizing force $\mathbf{F}^{\text{reg}}$ which is a functional of the Eulerian velocity field $\mathbf{v}(\mathbf{x},t)$. The simplest case would be to take:

$$\mathbf{F}^{\text{reg}}[\mathbf{v}] \propto \mathbf{v}(\mathbf{x},t).$$

However, this form explicitly penalizes uniform translations, where the velocity field is of the form $\mathbf{v}(\mathbf{x},t) = \mathbf{v}(t)$. This is usually not desirable in the context of image registration, where the variable location of objects within images means we must be free to perform affine or rigid-body alignment of the images before we beginning the non-rigid part of the registration. Allowing free uniform translations hence means that the regularizing force cannot depend on $\mathbf{v}(\mathbf{x},t)$ explicitly, but instead must depend only on its derivatives.

We hence consider the quantities:

$$M_{\alpha\beta}[\mathbf{v}] \doteq \partial_\alpha v_\beta(\mathbf{x}, t), \;\; \mathbf{x} \in \mathbb{R}^n, \;\; \alpha, \beta = 1, \dots n,$$

where $\{x_\alpha\}$, $\{\mathbf{v}_\alpha\}$ denote the Cartesian components of the vector position $\mathbf{x}$ and the vector velocity $\mathbf{v}$, and $\{\partial_\alpha \doteq \frac{\partial}{\partial x_\alpha}\}$ denotes the partial derivatives with respect to position. The $\{M_{\alpha\beta}\}$ are the components of the rate of strain tensor $\mathbf{M}$. We can rewrite any tensor in terms of a antisymmetric and a symmetric part thus:

$$A_{\alpha\beta}[\mathbf{v}] \doteq \frac{1}{2}\left(\partial_\alpha v_\beta(\mathbf{x}) - \partial_\beta v_\alpha(\mathbf{x})\right), \;\; S_{\alpha\beta}[\mathbf{v}] \doteq \frac{1}{2}\left(\partial_\alpha v_\beta(\mathbf{x}) + \partial_\beta v_\alpha(\mathbf{x})\right).$$

Let us now consider a rigid rotation of an image. For images in $\mathbb{R}^2$, with Cartesian components $\mathbf{x} = (x, y)$, a rigid rotation corresponds to the motion:

$$x(t) = r\cos\omega t, \;\; y(t) = r\sin\omega t,$$

where r is the radial coordinate $r^2 = x^2 + y^2$, and ω is the rate of rotation about the origin. This gives the associated velocity field:

$$v_x = \frac{dx(t)}{dt} = -\omega y, \;\; v_y = \frac{dy(t)}{dt} = \omega x.$$

For this particular velocity field, we have:

$$A_{xx} = A_{yy} = 0, \;\; A_{xy} = \omega, \;\; S_{xy} = S_{xx} = S_{yy} = 0.$$

Allowing the freedom to rotate the image hence means that the regularizer should depend on only the *symmetric* part of the rate of strain tensor. Similar results hold in higher numbers of dimensions.

The simplest vector-valued function we can construct from $\{S_{\alpha\beta}[\mathbf{v}]\}$ is:

$$F^{\text{reg}}_\alpha[\mathbf{v}] = a\partial_\alpha S_{\beta\beta} + b\partial_\beta S_{\alpha\beta},$$

where (a, b) are some coefficients. Re-writing this in terms of the velocity field, we obtain the form:

$$\begin{aligned} F^{\text{reg}}_\alpha &= a\partial_\alpha(\partial_\beta v_\beta) + \frac{b}{2}\partial_\beta(\partial_\alpha v_\beta + \partial_\alpha v_\beta), \\ &= \left(a + \frac{b}{2}\right)\partial_\alpha(\partial_\beta v_\beta) + \frac{b}{2}\partial_\beta\partial_\beta v_\alpha. \\ \therefore \;\; \mathbf{F}^{\text{reg}} &= \frac{b}{2}\nabla^2\mathbf{v}(\mathbf{x}, t) + \left(a + \frac{b}{2}\right)\nabla(\nabla \cdot \mathbf{v}(\mathbf{x}, t)). \end{aligned} \tag{8.11}$$

Rewriting the coefficients, we obtain the final form:

$$\boxed{\mathbf{F}^{\text{visc}} = \mu\nabla^2\mathbf{v}(\mathbf{x},t) + (\lambda+\mu)\,\nabla(\nabla\cdot\mathbf{v}(\mathbf{x},t)).} \tag{8.12}$$

This is the expression used by Christensen et al. [22], and represents the viscous forces in a physical fluid, where μ is the shear viscosity, and λ is the second viscosity coefficient (related to the bulk viscosity). For physical fluids, shear viscosity acts to resist non-uniform velocity gradients, whereas bulk viscosity acts to resist non-uniform compression/rarefaction. In fluid dynamics, the existence of viscosity reflects the physical effect that internal forces act within a moving fluid, which tend to resist non-uniform velocities existing within the fluid. It is analogous to the physical effect of friction for solid objects, but can also be related to the resistance of an elastic solid to deformation. In the particle picture we introduced earlier, elasticity reflects the fact that we can model a solid as a collection of particles connected by springs. A solid hence possesses a definite undeformed shape, where the forces between the particles balance, and any deformation from this is resisted. Whereas for fluids, they have no definite shape. The resistance of a fluid is resistance to flow, not resistance to deformation. Hence when we include the freedom to perform affine transformations, we have a similar mathematical form for both elastic regularization and fluid regularization, but using the deformation field $\mathbf{u}(\mathbf{x},t)$ for solids, whereas for fluids, the flow field $\mathbf{v}(\mathbf{x},t)$ is the relevant variable.

As regards the dynamics, we use the force-balance equation we had before (8.9):

$$\mathbf{F}^S(\mathbf{x},t) + \mathbf{F}^{\text{reg}}(\mathbf{x},t) = \mathbf{F}^S[\mathbf{u}] + \mathbf{F}^{\text{visc}}[\mathbf{v}] = \mathbf{0}. \tag{8.13}$$

Note that unlike the case of elastic registration, there is no need to modify the right-hand side in order to solve the PDE. The dynamics is obtained directly from the above equation. For image or shape correspondence, the driving force $\mathbf{F}^S(\mathbf{x})$ depends only on the displacement field $\mathbf{u}(\mathbf{x},t)$, and not on the velocity field $\mathbf{v}(\mathbf{x},t)$. Hence with the displacement field given at some time t, we can determine the driving force at this time. Balancing this driving force against the viscous forces determines the velocity field (8.12) at the time t. The velocity field determines the time-dependance of the displacement field via (8.10), hence determines the displacement field at an incremental future time $t+\Delta t$. The values of the displacement and velocity fields can hence be integrated numerically until we reach the static limit, where $\mathbf{v}(\mathbf{x},t)\to\mathbf{0}$. In this limit, flow vanishes, hence the viscous forces vanish, giving a static-limit solution satisfying:

$$\mathbf{F}^S(\mathbf{x}) = \mathbf{0}.$$

Hence we can conclude that unlike the elastic or diffusion regularization schemes, based on $\mathbf{F}^{\text{reg}}[\mathbf{u}]$, the fluid regularizer based on $\mathbf{F}^{\text{reg}}[\mathbf{v}]$ *does not* shift the position of the minimum.

If we view the above equations in the context of the physics of actual fluids, we have a driving force $\mathbf{F}^S(\mathbf{x})$ acting on the fluid. But unlike real fluids, the only other force acting is the viscosity, we have no pressure terms and no gravitational forces. The force balance equation also means that the fluids have to be considered as massless and inertialess. Christensen et al. [22] derived this regularization scheme based on the physics of a visco-elastic fluid. However, given the above peculiar nature of the fluid, the current authors found that this physically-based picture of a fluid often caused confusion, since we found that our physical intuition usually fails when it comes to trying to imagine the behaviour of massless and inertialess fluids. Hence our reason for giving this simple mathematical justification for such a regularization term.

As for the elastic case (8.7), the viscous force (8.12) depends only on the second derivatives of the flow field $\mathbf{v}(\mathbf{x}, t)$. It hence gives zero resistance to flows fields which are linear functions of the coordinates, and these represent instantaneous and infinitesimal affine transformations.

Within the context of image registration, the displacement fields, velocity fields and the forces lie in the space $\mathbf{x} \in \mathbb{R}^n = X$, which is just the space of the image itself. Calculating derivatives and applying viscous forces directly in the space of the image is then a sensible procedure. But for the case of shapes, $X = \mathbb{R}^n$ is just the parameter space for the shape. Hence calculating derivatives, viscous forces and therefore the regularizer itself directly in X depends on the details of the parameterisation for each shape. In particular, the regularizer will not be invariant to a *global* re-parameterisation of all the shapes, where:

$$\phi_i(\mathbf{x}) \mapsto \psi(\phi_i(\mathbf{x})),$$

and $\psi(\mathbf{x})$ is the *global* re-parameterisation function. Such a global re-parameterisation makes no difference to the groupwise correspondence between shapes, hence a principled regularizer should also be invariant to such a global re-parameterisation.

To make the fluid regularizer independent of the parameterisation, we need to consider the manipulation of correspondence, hence the movement of fluid particles, directly on the surfaces of the shapes themselves. This means that we need to calculate the viscous forces for fluids flowing on the actual shape surfaces, and the mathematical tools for this are the subject of the next section.

8.3 The Shape Manifold

The formal mathematics of calculating derivatives on the curved surface of a shape is just the usual mathematics of Riemannian differential geometry. All that we have to do is write down the form of the Riemannian metric in the coordinates on parameter space, and replace all the ordinary derivatives with covariant derivatives. However, for the benefit of readers who may have only a passing acquaintance with such topics, or none at all, we instead present a simple derivation of the required calculation, which requires little more than an understanding of the chain rule for partial derivatives. For those readers who have a more intimate acquaintance with differential geometry, we note that we will not employ any of the usual machinery of contravariant and covariant vectors, and just restrict ourselves to the usual definitions of vectors and matrices.

Suppose we have a single shape. We will consider the abstract shape manifold $\mathcal{S}$, which has some definite topology (e.g., a sphere). An *actual* shape $\mathbf{S}$ is what we obtain when we embed this shape manifold into ordinary Euclidean space $\mathbb{R}^m$, where $\mathbf{S} \subset \mathbb{R}^m$ is the entire shape. For the same shape manifold, we can obtain a variety of actual shapes, depending on the exact details of the embedding.

We will present a simple example, to help explain this notion of embedding. Consider a simple shape manifold with the topology of the open line. This line can then be embedded into $\mathbb{R}^2$, to create a variety of physical shapes in the plane. But we can also embed the line into $\mathbb{R}^3$, to create shapes which cannot be flattened onto the plane.

The distinction between intrinsic topology and the topology of the embedding is important. Suppose we have a physical object in $\mathbb{R}^3$ with the topology of the open line (i.e., no self-intersections). Such an object could be though of as a thin flexible wire. Such a wire cannot self intersect, although it can be formed into a variety of shapes in $\mathbb{R}^3$ by bending the wire. The intrinsic topology of the wire is always maintained. We now consider photographing a selection of such objects, and studying the shapes formed by the image of the wire in the photographs. It is obviously possible to obtain shapes in $\mathbb{R}^2$ where the object appears to self-intersect. To enable us to appropriately model the shapes we obtain from these *images* of our original object, we have to recognize that this apparent change of topology is just an artifact of the imaging process. Rather than using the topology of the images of the shape, which is a result of the way we have obtained the images, we would instead retain the intrinsic topology of the object as the topology of our shape.

To summarize, the abstract shape manifold knows about the intrinsic topology of the shape, and this topology is maintained when we move to the actual shapes $\mathbf{S} \subset \mathbb{R}^m$.

When we embed the shape, we also gain a definition of distance on the shape manifold. Suppose we have two points on the shape manifold, with a continuous path between them. This path on the shape manifold then becomes a path in $\mathbb{R}^m$, between the points on the surface[3] of the actual shape, lying wholly within the shape surface. The length of this path is then

[3] Note than similar considerations hold for n-dimensional shape manifolds embedded in $\mathbb{R}^m$, $m \geq n+1$. However, for the sake of clarity, we will from now on just refer to shape surfaces.

defined as the length of the path in $\mathbb{R}^m$, calculated using the usual definition of distance in Euclidean spaces.

We then parameterise our actual shape $\mathbf{S}$ as before (2.117):

$$X \stackrel{\mathfrak{X}}{\longmapsto} S, \ \mathbf{x} \stackrel{\mathfrak{X}}{\longmapsto} \mathbf{S}(\mathbf{x}), \tag{8.14}$$

where X is the parameter space, and $\mathbf{S}(\mathbf{x}) \in \mathbb{R}^m$ is the shape function that describes our actual shape. Since this mapping $\mathfrak{X}$ is taken to be one-to-one and onto (that is, every point in the parameter space maps to a point on the actual shape, and every point on the shape has a unique parameter value), we will employ a slight abuse of notation, and use $\mathfrak{X}$ to denote both the mapping from X to $\mathbf{S}$, and the mapping from $\mathbf{S}$ to X. For a n-dimensional shape manifold, the parameter space is usually taken to be some subset of $\mathbb{R}^n$, that is, $X \subset \mathbb{R}^n$. A single point in parameter space is then given by $\mathbf{x}$, with components $\{x_\mu : \mu = 1, \dots n\}$.

We can now identify our parameter space with the shape manifold itself, since they share the same intrinsic topology. As $\mathfrak{X}$ maps us from parameter space to the shape function $\mathbf{S}(\mathbf{x})$, so embedding takes us from the shape manifold to the actual shape. A particular parameterisation then defines a chart on the shape manifold, in that it assigns a unique parameter value $\mathbf{x}$ to every point in the shape manifold. The actual shape $\mathbf{S}$ can then be considered as a vector-valued function on the shape manifold $\mathbf{S}(\mathbf{x}) \in \mathbb{R}^m$, where we will use $\{S_A(\mathbf{x}) : A = 1, \dots m\}$ to denote the m Cartesian components of $\mathbf{S}(\mathbf{x})$.

This vector-valued function on the shape manifold then defines what we mean by distances between points.

As an example, consider the entire surface of the earth. We can take as our parameters the usual latitude and longitude. The entire parameter space can be represented as the unit square in $\mathbb{R}^2$. The top edge is identified as a single point (the north pole), and similarly for the bottom edge, which maps to the south pole. The side edges are identified using periodic boundary conditions, to give the correct spherical topology. The shape function then gives the shape of the actual surface of the earth. So, we would then have a square map of the entire earth, where the latitude and longitude coordinates have a simple relation to position on the spherical earth. The final component of the actual shape is the elevation of points on the surface of the earth, a scalar function on our map, such as that usually represented graphically by contour lines. To calculate distances between points on the surface of the earth, we obviously cannot use the distances on the unit square in parameter space, but have to incorporate the knowledge about the actual shape. Similarly, if we wish to compute the slopes of surfaces (i.e., the derivatives of height data) on the real earth, we again have to use the shape information to do this.

The relation between infinitesimal distances in parameter space, and distances on the shape surface can be described in terms of the induced Riemannian metric $\mathbf{g}$ as follows.

8.3.1 The Induced Metric

Suppose we have two infinitesimally close points in parameter space with values $\mathbf{x}$ and $\mathbf{x} + \Delta\mathbf{x}$. The corresponding points on the shape surface are separated by an infinitesimal distance Δd in $\mathbb{R}^m$. The way that Δd varies with $\Delta\mathbf{x}$ and $\mathbf{x}$ is described by the matrix-valued metric function $\mathbf{g}(\mathbf{x})$, defined by:

$$(\Delta d)^2 \doteq g_{\alpha\beta}(\mathbf{x})\Delta x_\alpha \Delta x_\beta. \tag{8.15}$$

The metric $\mathbf{g}(\mathbf{x})$ is hence an $n \times n$ symmetric matrix at each point $\mathbf{x}$. The metric is computed from the shape function according to the following Theorem:

Theorem 8.1. Induced Metrics.
For a shape with shape function $\mathbf{S}(\mathbf{x}) \in \mathbb{R}^m$, where $\mathbf{x} \in \mathbb{R}^n$ is the parameter value, the induced Riemannian metric $\mathbf{g}(\mathbf{x})$ in parameter space coordinates is given by:

$$\boxed{g_{\alpha\beta}(\mathbf{x}) = (\partial_\alpha \mathbf{S}(\mathbf{x})) \cdot (\partial_\beta \mathbf{S}(\mathbf{x})), \;\; \alpha, \beta = 1, \ldots n,} \tag{8.16}$$

where $\cdot$ is the Euclidean dot product in $\mathbb{R}^m$.

Proof. *Consider the points $\mathbf{x}$ and $\mathbf{x} + \Delta\mathbf{x}$ in parameter space, where $\|\Delta\mathbf{x}\|$ is infinitesimal. The corresponding points on the shape are $\mathbf{S}(\mathbf{x} + \Delta\mathbf{x})$ and $\mathbf{S}(\mathbf{x})$. Using a Taylor expansion, we have:*

$$\mathbf{S}(\mathbf{x} + \Delta\mathbf{x}) = \mathbf{S}(\mathbf{x}) + \Delta x_\alpha \partial_\alpha \mathbf{S}(\mathbf{x}) + O\left(\|\Delta\mathbf{x}\|^2\right).$$
$$\therefore \;\; \Delta\mathbf{S}(\mathbf{x}) \doteq \mathbf{S}(\mathbf{x} + \Delta\mathbf{x}) - \mathbf{S}(\mathbf{x}) \approx \Delta x_\alpha \partial_\alpha \mathbf{S}(\mathbf{x}).$$

Using the definition of distances in $\mathbb{R}^m$ in terms of the Euclidean dot product $\cdot$ in $\mathbb{R}^m$:

$$\begin{aligned}
(\Delta d)^2 &= (\Delta\mathbf{S}(\mathbf{x})) \cdot (\Delta\mathbf{S}(\mathbf{x})), \\
&= \Delta x_\alpha \Delta x_\beta \left(\partial_\alpha \mathbf{S}(\mathbf{x})\right) \cdot \left(\partial_\beta \mathbf{S}(\mathbf{x})\right), \\
\textit{From (8.15):}\;\; (\Delta d)^2 &\doteq g_{\alpha\beta}(\mathbf{x})\Delta x_\alpha \Delta x_\beta, \\
\therefore \;\; g_{\alpha\beta}(\mathbf{x}) &= (\partial_\alpha \mathbf{S}(\mathbf{x})) \cdot (\partial_\beta \mathbf{S}(\mathbf{x})).
\end{aligned}$$

□

Note that the metric is calculated purely in terms of derivatives in parameter space, of functions defined on that parameter space. Now that we can relate distances on the shape to distances in parameter space, we are partway to being able to calculate derivatives on the shape. The other thing needed is the ability to relate directions on the shape to directions in parameter space.

8.3.2 Tangent Space

Let P be an arbitrary point on the shape, with parameter value $\mathbf{x}_P$, and position $\mathbf{S}_P \doteq \mathbf{S}(\mathbf{x}_P)$. We construct the space T_P, which is the tangent space to the shape at P.

For example, for a shape surface in $\mathbb{R}^3$, the tangent space is just the tangent plane at a point. For a one-dimensional shape in $\mathbb{R}^2$, the tangent space is the line tangent to the shape at a point. And for a one-dimensional shape in $\mathbb{R}^3$, the tangent space is the tangent line.

We then define Cartesian coordinates in the tangent space. For an n-dimensional shape, there will be n such coordinates. For the αth such coordinate axis in the tangent space, we hence have a unit vector in $\mathbf{e}^{(\alpha)} \in \mathrm{T}_P \subset \mathbb{R}^m$, where:

$$\mathbf{e}^{(\alpha)} \cdot \mathbf{e}^{(\beta)} \doteq \delta_{\alpha\beta}, \;\; \alpha, \beta = 1, \ldots n, \;\; \mathbf{e}^{\alpha} = \{e_A^{(\alpha)}\}, \;\; A = 1, \ldots m. \tag{8.17}$$

We then also have the additional unit vectors $\mathbf{e}^{(n+1)}, \ldots \mathbf{e}^{(m)}$, which are orthogonal to the tangent space. The complete set of vectors form an orthonormal basis for $\mathbb{R}^m$.

A general point on the shape, $\mathbf{S}(\mathbf{x})$ say, can hence be written in terms of this basis at P thus:

$$\begin{aligned} S^{(\alpha)}(\mathbf{x}) &\doteq (\mathbf{S}(\mathbf{x}) - \mathbf{S}_P) \cdot \mathbf{e}^{(\alpha)}, \;\; \alpha = 1, \ldots n, \\ S^{(n+p)}(\mathbf{x}) &\doteq (\mathbf{S}(\mathbf{x}) - \mathbf{S}_P) \cdot \mathbf{e}^{(n+p)}, \;\; p = 1, \ldots m-n, \\ \mathbf{S}(\mathbf{x}) &= \mathbf{S}_P + \sum_{\alpha=1}^{n} S^{(\alpha)}(\mathbf{x})\mathbf{e}^{(\alpha)} + \sum_{p=1}^{m-n} S^{(n+p)}(\mathbf{x})\mathbf{e}^{(n+p)}. \end{aligned} \tag{8.18}$$

What we have done here is create a new Cartesian coordinate system in $\mathbb{R}^m$, with the origin at the point P, and the axes of the system chosen so that the first n such axes lie in the tangent space T_P, whereas the remaining axes are perpendicular to this tangent space. The reason for creating such a coordinate system is that the coordinate derivatives on the shape surface are particularly simple if we chose tangent-space coordinates.

Within some neighbourhood about P, there is a one-to-one mapping between points in the tangent space, and points on the shape. Hence within this Monge patch, the tangent space coordinates give a local parameterisation of the shape. Let $\boldsymbol{\tau}$ denote a point in the tangent space, with coordinates:

$$\tau_\alpha \doteq \boldsymbol{\tau} \cdot \mathbf{e}^{(\alpha)}. \tag{8.19}$$

The corresponding point on the shape is the point $\mathbf{S}(\mathbf{x})$ such that:

$$S^{(\alpha)}(\mathbf{x}) = \tau_\alpha.$$

Since we are within the Monge patch, this has a single solution. It hence relates the coordinate $\mathbf{x}$ in parameter space to the coordinate in the tangent space $\boldsymbol{\tau} = \boldsymbol{\tau}(\mathbf{x})$. The tangent-space coordinate derivatives on the shape at P are related to the coordinate derivatives in parameter space via:

$$d_\alpha \doteq \left.\frac{\partial}{\partial t_\alpha}\right|_P, \quad \partial_\alpha \doteq \left.\frac{\partial}{\partial x_\alpha}\right|_P = \left.\frac{\partial \tau_\beta(\mathbf{x})}{\partial x_\alpha}\right|_P d_\beta. \tag{8.20}$$

The Jacobian of the transformation between $\mathbf{x}$ and $\boldsymbol{\tau}$ at P is $\mathbf{J}_P$, where:

$$(J_P)_{\alpha\beta} \doteq \left.\frac{\partial \tau_\beta(\mathbf{x})}{\partial x_\alpha}\right|_P, \quad \partial_\alpha = (J_P)_{\alpha\beta} d_\beta, \quad \boxed{d_\beta = \left(J_P^{-1}\right)_{\beta\alpha} \partial_\alpha.} \tag{8.21}$$

The Jacobian matrix is hence the relation we seek between the coordinate derivatives on the shape at P (d_α), and the coordinate derivatives in the parameter space (∂_α).

The Jacobian is related to the metric previously defined by the following Theorem:

Theorem 8.2. The Jacobian Matrix.
Given the definition of the Jacobian $\mathbf{J}_P$ above (8.21), this Jacobian is related to the induced Riemannian metric $\mathbf{g}(\mathbf{x})$ via:

$$\boxed{\mathrm{g}(\mathbf{x}_P) = (\mathbf{J}_P)(\mathbf{J}_P^T).} \tag{8.22}$$

The Jacobian relates derivatives with respect to the tangent space coordinates to derivatives in parameter space as stated previously (8.21). But the Jacobian also relates vectors on the shape to vectors in the parameter space. If $\tilde{\mathbf{v}}$ is a vector in the shape at P, then it is related to the corresponding vector $\mathbf{v}$ in the parameter space via:

$$\boxed{\mathbf{v} = (\mathbf{J}_P^T)^{-1}\tilde{\mathbf{v}}.} \tag{8.23}$$

Proof. *To compute the Jacobian, we take the expression for $\mathbf{S}(\mathbf{x})$ (8.18), and substitute the expression for $S^{(\alpha)}(\mathbf{x})$:*

$$\mathbf{S}(\mathbf{x}) = \mathbf{S}_P + \sum_{\alpha=1}^{n} \tau_\alpha \mathbf{e}^{(\alpha)} + \sum_{p=1}^{m-n} S^{(n+p)}(\mathbf{x})\mathbf{e}^{(n+p)}. \tag{8.24}$$

Since we are within a Monge patch, the other components $\{S^{(n+p)}(\mathbf{x})\}$ can be written as a function of $\boldsymbol{\tau}$. And since we have taken these local parameters $\boldsymbol{\tau}$

to be defined via the tangent space *at P, these functions are purely quadratic to leading order. Hence, taking the derivative d_α at P, we obtain:*

$$d_\alpha \mathbf{S}(\mathbf{x})|_P = \mathbf{e}^{(\alpha)} = \left(J_P^{-1}\right)_{\alpha\mu} \partial_\mu \mathbf{S}(\mathbf{x})|_P \,.$$

Using the orthonormality of $\{\mathbf{e}^{(\alpha)}\}$ (8.17) and the definition of the metric (8.16) then gives:

$$\delta_{\alpha\beta} = \left(J_P^{-1}\right)_{\alpha\mu} \left(J_P^{-1}\right)_{\beta\nu} (\partial_\mu \mathbf{S}(\mathbf{x})) \cdot (\partial_\nu \mathbf{S}(\mathbf{x})) \;\Rightarrow\; \boxed{\mathrm{g}(\mathbf{x}_P) = (\mathbf{J}_P)(\mathbf{J}_P^T).}$$

Now consider a vector $\tilde{\mathbf{v}}$ in the shape surface at P. It hence lies purely in the tangent space at P:

$$\tilde{\mathbf{v}} = \sum_{\alpha=1}^{n} \tilde{v}^{(\alpha)} \mathbf{e}^{(\alpha)}.$$

This vector is specified by a direction *and a* magnitude. *To relate this vector to one in the parameter space, we have to relate directions at P in the tangent space to directions at $\mathbf{x}_P$ in the parameter space. Let us consider points $\mathbf{x}_P$ and $\mathbf{x}_P + \Delta\mathbf{x}$ in parameter space as before. Using (8.20) and (8.21), and remembering that $\boldsymbol{\tau}(\mathbf{x}_P) = \mathbf{0}$ (i.e., the origin of tangent space coordinates), then gives to leading order:*

$$\tau_\beta(\mathbf{x}_P + \Delta\mathbf{x}) \doteq \Delta\tau_\beta = (J_P)_{\alpha\beta}\Delta x_\alpha \;\Rightarrow\; \Delta x_\alpha = \left(J_P^{-1}\right)_{\beta\alpha} \Delta\tau_\beta. \qquad (8.25)$$

This is then the relation we require between a direction at P in the shape, described by $\{\Delta\tau_\beta\}$, and a direction at $\mathbf{x}_P$ in the parameter space, described by $\{\Delta x_\alpha\}$.[4] *Hence for the vector $\tilde{\mathbf{v}}$ in the shape at P, the corresponding vector in the parameter space at $\mathbf{x}_P$ obeys:*

$$\mathbf{v} \propto (\mathbf{J}_P^T)^{-1}\tilde{\mathbf{v}}, \;\; v_\alpha \propto \left(J_P^{-1}\right)_{\beta\alpha} \tilde{v}_\beta.$$

The unknown coefficient of proportionality reflects the fact that we have the correct direction for $\mathbf{v}$, but are not yet sure of its required magnitude. Remember that distances in parameter space are calculated using the induced metric $\mathbf{g}(\mathbf{x})$. Hence the length of the vector $\mathbf{v}$ is computed as:

$$\|\mathbf{v}\|^2 \doteq v_\alpha v_\beta \mathrm{g}_{\alpha\beta}(\mathbf{x}_P).$$

Hence $\|(\mathbf{J}_P^T)^{-1}\tilde{\mathbf{v}}\|$ is given by:

$$\|(\mathbf{J}_P^T)^{-1}\tilde{\mathbf{v}}\|^2 = \left(J_P^{-1}\right)_{\mu\alpha} \tilde{v}_\mu \left(J_P^{-1}\right)_{\nu\beta} \tilde{v}_\nu \mathrm{g}_{\alpha\beta}(\mathbf{x}_P) = \tilde{v}_\mu\tilde{v}_\mu = \|\tilde{\mathbf{v}}\|^2.$$

Therefore the relation between $\tilde{v}$ and $\mathbf{v}$ is given by:

---

[4] This has to be calculated using *infinitesimal* coordinate displacements to represent direction, since in general, a straight line in parameter space *does not* map to a straight line in tangent space.

$$\mathbf{v} = (\mathbf{J}_P^T)^{-1}\tilde{\mathbf{v}},$$

which correctly relates both direction and magnitude. □

We can hence relate derivatives in terms of the tangent space coordinates at P to derivatives on parameter space, and vectors on the shape at P to vectors in the parameter space using the Jacobian at P, $\mathbf{J}_P$.

We now consider extending this to all points $\mathbf{x}$. At each point of the shape, we define the tangent-space frame vectors $\{\mathbf{e}^{(\alpha)}(\mathbf{x}) \in \mathbb{R}^m\}$, which lie in the tangent space at the point $\mathbf{S}(\mathbf{x})$. Provided that this tangent-space frame varies smoothly with position, we then have a Jacobian $\mathbf{J}(\mathbf{x})$, which is a smooth, matrix-valued function of position.

The metric $\mathbf{g}(\mathbf{x})$ is totally determined by the shape via (8.16). Note that the Jacobian $\mathbf{J}(\mathbf{x})$, although related to the metric $\mathbf{g}(\mathbf{x})$ via (8.22) is not totally determined. This is because we are free to rotate the frame vectors in the tangent space. In fact, we can rotate the frame vectors by differing amounts at neighbouring points. This means that the Jacobian function is undetermined up to a local gauge transformation $\mathbf{R}(\mathbf{x})$, where $\mathbf{R}(\mathbf{x})$ is the matrix representation of a rotation in $\mathbb{R}^n$ centred at the point $\mathbf{S}(\mathbf{x})$. This rotation varies smoothly with position, and we have the gauge invariance of the metric:

$$\begin{aligned} &\mathbf{J}(\mathbf{x}) \mapsto \mathbf{J}(\mathbf{x})\mathbf{R}(\mathbf{x}) \\ \Rightarrow\ &\mathbf{g}(\mathbf{x}) = \mathbf{J}(\mathbf{x})\mathbf{J}^T(\mathbf{x}) \mapsto \mathbf{J}(\mathbf{x})\mathbf{R}(\mathbf{x})\mathbf{R}^T(\mathbf{x})\mathbf{J}^T(\mathbf{x}) \equiv \mathbf{J}(\mathbf{x})\mathbf{J}^T(\mathbf{x}) = \mathbf{g}(\mathbf{x}), \end{aligned}$$

since the inverse of a rotation $\mathbf{R}(\mathbf{x})$ is just the transpose $\mathbf{R}^T(\mathbf{x})$.

8.3.3 Covariant Derivatives

Let us now recall the aim of these calculations. We want to describe the dynamics of fluids moving on the surface of a shape, under the action of viscous forces acting within the fluid. The fields that describe the velocities of the particles obviously lie in the shape surface, since the particles do. The viscous forces that act on the particles are constructed in terms of derivatives of the velocities, and since the particles remain on the surface of the shape, the viscous forces that we compute must also be restricted to lie wholly within the surface. We hence need to use an appropriately defined derivative when computing these forces.

If we consider just the shape *surface*, without the interior or exterior of the shape, this shape surface can then be described as an n-dimensional manifold, the shape manifold. One global coordinate system for this manifold is just the parameter space for the shape $\mathbf{x} \in X$. In this parameter space, distances are defined by the Riemannian metric induced on the manifold $\mathbf{g}(\mathbf{x})$ (8.15, 8.16). The metric can be considered as a matrix-valued function of the coordinate $\mathbf{x}$.

This shape manifold is not flat, and this property of non-flatness does not just refer to the fact that the surface viewed from the perspective of $\mathbb{R}^m$ is curved, but that the shape manifold with distances defined using the induced metric is *intrinsically* curved. The appropriate derivatives within a curved space are the *covariant* derivatives of Riemannian differential geometry [4].

For the benefit of those unfamiliar with differential geometry, we will attempt to give a brief explanation as to why considering a problem defined on the surface of a shape means that we have to introduce a new type of derivative.

We start by considering the simplest case, the derivative of a scalar function $f(\mathbf{x})$ on the surface of the shape. As previously, at each point $\mathbf{S}(\mathbf{x}) \in \mathbb{R}^m$ on the shape, we will define a set of orthonormal frame vectors $\{\mathbf{e}^{(\alpha)}(\mathbf{x}) \in \mathbb{R}^m : \alpha = 1, \dots n\}$, which lie wholly in the tangent space at that point. We also constrain our tangent frame so that for each tangent-space direction α, $\mathbf{e}^{(\alpha)}(\mathbf{x})$ is a smooth function of the parameter value $\mathbf{x}$ as we move across the physical surface.

We will consider computing a derivative in the direction $\mathbf{e}^{(\alpha)}(\mathbf{x})$. The derivative is defined thus:

$$D_\alpha f \doteq \lim_{\|\Delta\mathbf{x}\| \to 0} \left(\frac{f(\mathbf{x} + \Delta\mathbf{x}) - f(\mathbf{x})}{\|\mathbf{S}(\mathbf{x} + \Delta\mathbf{x}) - \mathbf{S}(\mathbf{x})\|} \right), \tag{8.26}$$

where $\Delta\mathbf{x}$ is such that:

$$\begin{aligned} \mathbf{S}(\mathbf{x} + \Delta\mathbf{x}) - \mathbf{S}(\mathbf{x}) &= \Delta\tau_\alpha \mathbf{e}^{(\alpha)}(\mathbf{x}) + O\left((\Delta\tau_\alpha)^2\right), \ \ \Delta\tau_\alpha \geq 0 \\ \|\mathbf{S}(\mathbf{x} + \Delta\mathbf{x}) - \mathbf{S}(\mathbf{x})\| &= \Delta\tau_\alpha + O\left((\Delta\tau_\alpha)^2\right). \end{aligned}$$

That is, to leading order the two points $\mathbf{S}(\mathbf{x} + \Delta\mathbf{x})$ and $\mathbf{S}(\mathbf{x})$ at which we are taking the values of the function are separated in the direction $\mathbf{e}^{(\alpha)}(\mathbf{x})$. By the definition of the tangent space, we already know that the separation of these two points in the directions perpendicular to the tangent space is quadratic in terms of the tangent-plane coordinate $\Delta\tau_\alpha$ (see Theorem 8.2).

Using the results in (8.21) and (8.25), it is simple to calculate the variation of f:

$$\begin{aligned} f(\mathbf{x} + \Delta\mathbf{x}) - f(\mathbf{x}) &= \Delta x_\beta \partial_\beta f(\mathbf{x}) + O\left(\|\Delta\mathbf{x}\|^2\right) \\ &= \Delta\tau_\mu J^{-1}_{\mu\beta}(\mathbf{x}) \partial_\beta f(\mathbf{x}) + O\left(\|\Delta\boldsymbol{\tau}\|^2\right) \\ &= \Delta\tau_\mu d_\mu f(\mathbf{x}) + O\left(\|\Delta\boldsymbol{\tau}\|^2\right) \\ &= \Delta\tau_\alpha d_\alpha f(\mathbf{x}) + O\left(\|\Delta\boldsymbol{\tau}\|^2\right) \ \text{(no sum on } \alpha). \end{aligned}$$

The *covariant* derivative of a scalar as defined in (8.26) is then:

$$D_\alpha f \doteq \lim_{\|\Delta\mathbf{x}\|\to 0}\left(\frac{f(\mathbf{x}+\Delta\mathbf{x})-f(\mathbf{x})}{\|\mathbf{S}(\mathbf{x}+\Delta\mathbf{x})-\mathbf{S}(\mathbf{x})\|}\right)=\lim_{\Delta\tau_\alpha\to 0}\left(\frac{\Delta\tau_\alpha d_\alpha f(\mathbf{x})}{\Delta\tau_\alpha}\right)=d_\alpha f(\mathbf{x}). \tag{8.27}$$

Hence we see that in tangent-plane coordinates, the covariant derivative of a scalar at the origin of the coordinate system is just the partial derivative with respect to the tangent-space coordinates.

Let us now consider calculating the derivative on the shape of a vector field lying wholly within the surface of the shape. The vector field $\tilde{\mathbf{v}}(\mathbf{x})$, when viewed on the physical shape, is a vector field in $\mathbb{R}^m$. The fact that it is everywhere in the surface means that $\tilde{\mathbf{v}}(\mathbf{x})$ is always tangential to the surface at $\mathbf{x}$. We have tangent frame vectors $\mathbf{e}^{(\alpha)}(\mathbf{x})$ at $\mathbf{x}$, hence $\tilde{\mathbf{v}}(\mathbf{x})$ can everywhere be written in the form:

$$\tilde{\mathbf{v}}(\mathbf{x})=\sum_{\alpha=1}^{n}\left(\tilde{\mathbf{v}}(\mathbf{x})\cdot\mathbf{e}^{(\alpha)}(\mathbf{x})\right)\mathbf{e}^{(\alpha)}(\mathbf{x})\doteq\sum_{\alpha=1}^{n}\tilde{v}_\alpha(\mathbf{x})\mathbf{e}^{(\alpha)}(\mathbf{x}),$$

where $\cdot$ is the Euclidean dot product in $\mathbb{R}^m$. The set $\{\tilde{v}_\alpha(\mathbf{x})\}$ are then the *components* of the vector field $\tilde{\mathbf{v}}(\mathbf{x})$ with respect to the tangent frame $\{\mathbf{e}^{(\alpha)}(\mathbf{x})\}$.

A naïve (and incorrect) approach to calculating a derivative of this vector field would be to take (8.26), but replace $f(\mathbf{x})$ by each of the components $\tilde{v}_\beta(\mathbf{x})$ in turn. But a vector field is more than just a set of scalar fields which are its components. Obviously, different coordinate systems give different components, yet the vector itself, as a physically relevant entity, is unchanged. The relation between the components $\{\tilde{v}_\alpha(\mathbf{x})\}$ and the vector $\tilde{\mathbf{v}}(\mathbf{x})$ is defined by our frame vectors. The frame vectors themselves vary with position, so that what we mean by the direction α, hence what exactly we mean by the αth component of a vector, depends on where we are on the shape.

In general, $\mathbf{e}^{(\alpha)}(\mathbf{x}+\Delta\mathbf{x})$ and $\mathbf{e}^{(\alpha)}(\mathbf{x})$ are not the same, since they lie in *different* tangent spaces. These tangent spaces are only the same if the shape is flat at $\mathbf{x}$, hence it can be seen that their non-equivalence relates to curvature of the shape surface. And even if the two tangent spaces are the same, the directions can still rotate.

A simple example is provided by the surface of a sphere. At every point on the sphere, we can choose frame vectors $\{\mathbf{e}^{(\alpha)}\}$ as the unit vectors tangential to the surface pointing to the north and to the east, respectively.[5] If we consider any two points on the sphere, we see that these frame vectors are in general not the same. Similarly if we consider a curve in the plane, the

[5] Note that we cannot define such vectors everywhere on the sphere, since what we mean by north and east is not defined at the poles themselves. Another way to view this is as a consequence of the Hairy-Ball Theorem, which states that for any continuous tangent vector field on the sphere, there must be at least one point where the value of the field is zero.

tangent line at any point on the curve is in general not parallel to the tangent line at some other point.

We hence see that there are two possible sources for the variation of a component $\tilde{v}_\alpha$. The first is that the value of the vector field itself just varies with position, the sort of variation that is normally associated with the usual concept of partial coordinate derivatives. However, there is a secondary source of variation, in that the vector field may not vary, but what we mean by the αth direction is position dependent.

We hence need a way of dealing with this issue of the position-dependence of directions and hence components. The usual approach is to define a mechanism whereby we take the vector $\tilde{\mathbf{v}}(\mathbf{x} + \Delta\mathbf{x})$, and compute its new value when it is transported to the point $\mathbf{x}$. This means that we can then take the difference of the two vectors *at the same point*, an operation which has a definite meaning independent of the coordinate system. This transportation is referred to in the literature as the operation of *parallel transport*, and the precise details of this process are described in terms of the *connection*. For Riemannian manifolds, there is a unique connection that can be defined entirely in terms of the Riemannian metric, called the *Levi-Civita connection*, which is described in terms of the set of *Christoffel symbols*. The derivative constructed using this connection is the *covariant* derivative of Riemannian geometry. The importance of covariant derivatives of vectors or tensors is that they transform correctly as a tensor under a change of coordinate system, a property that the partial coordinate derivatives themselves do not possess.

Let us consider the computation of covariant derivatives in a little more detail. We will start by defining the coordinate system for our calculations. We take as our local parameterisation of the shape the tangent-space coordinates $\boldsymbol{\tau}$ at P, rather than parameter space coordinates $\mathbf{x}$, since results are simpler in this particular coordinate system. As in (8.24), within some Monge patch based at P, the shape function can be expanded in terms of the frame vectors at P thus:

$$\mathbf{S}(\mathbf{x}) = \mathbf{S}_P + \sum_{\alpha=1}^{n} \tau_\alpha \mathbf{e}^{(\alpha)}(\mathbf{x}_P) + \frac{1}{2} \sum_{p=1}^{m-n} \left(C^{(p)}_{\alpha\beta} \tau_\alpha \tau_\beta + O\left(\tau^3\right) \right) \mathbf{e}^{(n+p)}(\mathbf{x}_P), \tag{8.28}$$

where we have now made the quadratic dependence explicit, in terms of the set of coefficients $\{C^{(p)}_{\alpha\beta} : p = 1, \ldots m-n, \alpha, \beta = 1, \ldots n\}$.

In (8.16), we gave the definition of the metric $\mathbf{g}(\mathbf{x})$ in *parameter* space coordinates. We can hence obtain the metric $\mathbf{g}(\boldsymbol{\tau})$ in *tangent-space* coordinates by simply replacing the parameter-space coordinate derivatives ∂_α in (8.16) by tangent-space coordinate derivatives d_α thus:

$$g_{\alpha\beta}(\boldsymbol{\tau}) \doteq (d_\alpha \mathbf{S}(\mathbf{x})) \cdot (d_\beta \mathbf{S}(\mathbf{x})). \tag{8.29}$$

Previously, when we were using an orthonormal frame basis, the vectors $\{\mathbf{e}^{(\alpha)}(\mathbf{x})\}$ defined the direction(s) α at each point. We discussed how the need for a covariant derivative rests on the fact that what we mean by the direction α changes with position. A similar argument applies to our current coordinate basis, as follows.

Consider the set of vectors $\{d_\alpha \mathbf{S}(\mathbf{x})\}$. They obviously indicate the direction of the αth coordinate axis at the point $\mathbf{S}(\mathbf{x})$ on the shape. Since they are formed from derivatives of the shape function, they obviously lie within the tangent space at $\mathbf{S}(\mathbf{x})$. And as long as $\mathbf{S}(\mathbf{x})$ lies within the Monge patch at P, these vectors span the tangent space. The vectors $\{\mathbf{c}^{(\alpha)}(\mathbf{x}) \doteq d_\alpha \mathbf{S}(\mathbf{x})\}$ hence form a basis for the tangent space at $\mathbf{S}(\mathbf{x})$ – the coordinate basis vectors. But unlike the orthonormal frame vectors $\{\mathbf{e}^{(\alpha)}(\mathbf{x})\}$ that we considered earlier, the basis formed is not in general an orthogonal basis.[6]

In our current coordinate system, which is based on the tangent-space at P, it is the coordinate basis vectors $\{\mathbf{c}^{(\alpha)}(\mathbf{x})\}$ that define what we mean by direction α. The metric (8.29) can obviously be written in terms of these basis vectors thus:

$$\mathrm{g}_{\alpha\beta}(\boldsymbol{\tau}) \equiv \mathbf{c}^{(\alpha)}(\mathbf{x}) \cdot \mathbf{c}^{(\beta)}(\mathbf{x}).$$

The way that the direction α changes as we move on the shape can be encoded in terms of the set of quantities:

$$\mathbf{c}^{(\mu)}(\mathbf{x}) \cdot d_\alpha \mathbf{c}^{(\beta)}(\mathbf{x}) \equiv (d_\mu \mathbf{S}(\mathbf{x})) \cdot (d_\alpha d_\beta \mathbf{S}(\mathbf{x})). \tag{8.30}$$

In most texts on differential geometry, the covariant derivative is described in terms of the Christoffel symbols,[7] which are related to the metric [4] thus:

$$\begin{aligned}
\Gamma_{\alpha\beta\mu}(\boldsymbol{\tau}) &\doteq \frac{1}{2}\left(d_\alpha \mathrm{g}_{\beta\mu}(\boldsymbol{\tau}) + d_\beta \mathrm{g}_{\alpha\mu}(\boldsymbol{\tau}) - d_\mu \mathrm{g}_{\alpha\beta}(\boldsymbol{\tau})\right), \\
&= (d_\mu \mathbf{S}(\mathbf{x})) \cdot (d_\alpha d_\beta \mathbf{S}(\mathbf{x})), \\
&= \sum_{p=1}^{m-n} C^{(p)}_{\alpha\beta} C^{(p)}_{\mu\eta} \tau_\eta + O\left(\tau^2\right).
\end{aligned} \tag{8.31}$$

---

[6] The distinction made here between the frame vectors $\{\mathbf{e}^{(\alpha)}(\mathbf{x})\}$ and the coordinate basis vectors $\{\mathbf{c}^{(\alpha)}(\mathbf{x})\}$ can be understood as follows. In the coordinate basis, we chose Cartesian coordinates in the tangent space at P. These can then be mapped back onto the shape as long as we remain within the Monge patch. Although the coordinate directions are everywhere orthonormal in the tangent space at P, the same is not true when we map the coordinate directions back onto the physical surface of the shape, hence into the tangent space at some other point. Therefore the coordinate basis vectors are not in general orthogonal at any point other than P. Whereas for the frame vectors, we define a *separate* tangent space *at each point* of the surface, and a set of orthonormal basis vectors within each such tangent space.

[7] What is given here is a Christoffel symbol of the *first* kind, whereas in applications such as general relativity [126], what is usually referred to as the Christoffel symbol is actually the Christoffel symbol of the *second* kind $\Gamma^{\mu}_{\alpha\beta}$. They are related via multiplication by $\mathbf{g}^{-1}(\boldsymbol{\tau})$ in our matrix notation.

We hence see that the quantities we used to describe the variation in direction of the basis vectors (8.30) are just the Christoffel symbols. This hence demonstrates the link between the usual definition of the Christoffel symbols in terms of the metric (8.31), and our earlier, rather heuristic discussion in terms of the variation of directions.

There are two important things to note about the result (8.31) for the Christoffel symbols. The first is that at P (the point $\boldsymbol{\tau} = \mathbf{0}$) the Christoffel symbols vanish. This is the simplification that occurs by our use of tangent-space coordinates at P. The second is that although the symbols themselves vanish at P, in general their derivatives at P will not.

Covariant derivatives of vectors and tensors [4] involve various combinations of the Christoffel symbols and their derivatives. In particular, second derivatives of a vector field involve first derivatives of the Christoffel symbols. The viscous force we are trying to calculate involves second-derivatives of the velocity field. Even though the Christoffel symbols are zero at P, their derivatives are not, and hence give a contribution to the viscous force at P.

We could proceed using the full machinery of covariant derivatives.[8] This would mean calculating the set of second derivatives of the shape at each point. And without going into further detail as regards the exact definitions of covariant derivatives, it can already be seen that the calculation is becoming rather unwieldy.

We will instead make a simplifying assumption. In practice, our fields and forces are only defined at a finite set of sample points on the surface of the shape. If we then treat the surface as piecewise-flat about these sample points, then the quadratic terms in (8.28) vanish, as do the corresponding terms in (8.31). We can hence use just the ordinary partial derivatives with respect to the tangent-frame coordinates $\{d_\alpha\}$, rather than the full covariant derivatives $\{D_\alpha\}$. In effect, this approximation retains the information encoded in the Jacobian at each point, which gives a linear approximation to the relation between parameter space coordinates and coordinates on the surface. But it ignores the higher-order terms in the Christoffel symbols, related to the curvature of the shape.

The viscous force in tangent-frame coordinates in this piecewise linear approximation is given by:

$$\tilde{F}_\alpha^{\text{visc}} = \mu d_\beta d_\beta \tilde{v}_\alpha(\mathbf{x}, t) + (\lambda + \mu)\, d_\alpha(d_\beta \tilde{v}_\beta(\mathbf{x}, t)),$$

where $\tilde{\mathbf{F}}^{\text{visc}}$ is the viscous force, and $\tilde{\mathbf{v}}$ the Eulerian velocity in tangent-space coordinates. These are related to the vectors in the parameter-space via (8.23):

$$\mathbf{v}(\mathbf{x}, t) = (\mathbf{J}^T(\mathbf{x}))^{-1}\tilde{\mathbf{v}}(\mathbf{x}, t) \;\Rightarrow\; \tilde{\mathbf{v}}(\mathbf{x}, t) = \mathbf{J}^T(\mathbf{x})\mathbf{v}(\mathbf{x}, t).$$

[8] Readers interested in further explicit details of covariant derivatives and frame fields on surfaces should consult the excellent book by Koenderink [100], which takes a heuristic rather than a formal approach to the subject.

We can also relate derivatives via (8.21)

$$d_\beta = \left(\mathbf{J}^{-1}(\mathbf{x})\right)_{\beta\alpha} \partial_\alpha.$$

Note that our piecewise-linear assumption means that we must ignore the derivatives of the Jacobian, and hence we can pass the Jacobian through derivatives without generating extra terms. The above expression for the viscous force in tangent-frame components is related to the viscous force in parameter space components thus:

$$\tilde{F}^{\text{visc}}_\alpha(\mathbf{x}, t) = J_{\eta\alpha}(\mathbf{x}) F^{\text{visc}}_\eta(\mathbf{x}).$$

After some algebra, we obtain the final result for the viscous force in parameter space:

$$\boxed{F^{\text{visc}}_\eta(\mathbf{x}, t) = \mu g^{-1}_{\mu\nu}(\mathbf{x})\partial_\mu\partial_\nu v_\eta(\mathbf{x}, t) + (\lambda + \mu) g^{-1}_{\eta\nu}(\mathbf{x})\partial_\nu\partial_\mu v_\mu(\mathbf{x}, t).} \quad (8.32)$$

Note that the Jacobian does not explicitly appear in this expression, since products of Jacobians either combine to give the identity matrix, or appear in a combination that is equal to the inverse of the metric given in (8.16). We hence do not have to worry about the local gauge degree of freedom possessed by the Jacobian, as we would expect for a physically meaningful expression.

This final expression gives the relation between the derivatives of the Eulerian velocity field, defined in the parameter space, and the viscous force defined in the parameter space. The effect of the shape appears via the inverse of the metric matrix $\mathbf{g}^{-1}(\mathbf{x})$. As noted previously, the metric is calculated (8.16) entirely in terms of the shape function on the parameter space and derivatives of it within the parameter space. We hence have a position-dependant combination of the second derivatives of the velocity field. Although this combination depends on position, the matrix $\mathbf{g}^{-1}(\mathbf{x})$ has only to be calculated once at each sample point, since in the Eulerian framework, the sample point $\mathbf{x}$ remains fixed on the shape as the fluid, hence the deformation field, evolves.

From the point of view of implementation, the problem is the construction of a suitable set of sample points in parameter space, at which the values of the shape function, the displacement field, the Eulerian velocity field, and the viscous and driving forces are to be defined or calculated. The computation of the metric and of the viscous forces involve spatial derivatives on parameter space, hence ideally we require a set of sample points where we can calculate the numerical derivatives efficiently. A computationally efficient sampling for shapes with spherical topology is provided by the use of the concept of shape images, as described in the next section.

8.4 Shape Images

Let us consider the case of closed shape surfaces in three dimensions, with spherical topology. We showed previously (Sect. 6.1.2) how to construct an initial parameterisation for these shapes using the unit sphere, so that $X = \mathbb{S}^2$. The set of training shapes is typically represented by a set of triangulated meshes (Sect. 6.1.2). Hence working with such a parameterisation involves interpolation on a general triangulated mesh, which is computationally intensive.

To implement fluid regularization, we could just generate sample points as densely as required on the unit sphere. Although sampling directly on the unit sphere means that it is relatively easy to ensure that the topology is maintained, the computation of derivatives of the various fields involves another computationally intensive step if we perform it on the sphere.

The computational cost of any interpolation or differentiation would be considerable reduced if our parameter space X was instead some region of $\mathbb{R}^2$. We could then generate a set of sample points on parameter space which were part of a *regular* grid. And such regular grids allow much more straightforward and faster methods of interpolation and numerical differentiation.

For the case of shapes with spherical topology, it is well-known that the sphere cannot be covered by a single chart, so that any such single chart will have at least one point where derivatives cannot be defined.[9]

A naïve approach would be to map the surface of the sphere onto the square, where the Cartesian coordinates on the square are just the usual polar angles on the sphere, (θ, ψ), with $0 \leq \theta \leq \pi$, $0 \leq \psi < 2\pi$. However, this has the problem that the upper and lower edges of the square are mapped to single points on the sphere (the poles $\theta = 0$ and $\theta = \pi$.) Hence we have two problematic points in our chart. Better charts for the sphere can be constructed using the work of Praun and Hoppe [138], as we will explain.

We start by considering shapes which have the topology of the sphere, but are simpler, in the sense that they are formed of flat faces – polyhedra. There are nine regular polyhedra in $\mathbb{R}^3$, all of which have the property that they can be contained within a sphere (a *circumsphere*), with each vertex lying on that sphere. However, of these nine, only five are convex, and these are the Platonic solids of classical geometry (tetrahedron, cube, octahedron, dodecahedron and icosahedron). Each Platonic solid has a polyhedral net [58], which is an arrangement of edge-joined polygons in the plane, from which the final solid can be constructed by folding along these joins (see Fig. 8.1). Hence these polyhedral nets give a mapping from the Platonic solid to a region of the plane. It is obvious that there exist one-to-one mappings between each Platonic solid and the *smallest* possible circumsphere. The Platonic solids

[9] A sphere requires at least two charts for each chart to be differentiable everywhere. In differential geometry, such a collection of charts is called an *atlas*.

hence provide possible methods for mapping the unit sphere onto an inscribed Platonic solid, hence onto a region of the plane.[10]

The final step is then to deform the polyhedral net, so that it lies within some more convenient region of the plane. This process is simplest [138] for the tetrahedron and the octahedron, as illustrated in Fig. 8.1, both of which can be mapped onto the unit square. The case of the cube is slightly more complicated, and in [138], Praun and Hoppe retained all six faces of the cube, hence had an atlas of six square charts, appropriately connected. They also retained the rectangular mapping for the tetrahedron, since this then retained the important property that sampling regularly on each of the faces of the mapping gave regular sampling on the faces of the polyhedron.

Similar square charts can be constructed for shape surfaces with topologies other than spherical. For example, a shape with cylindrical topology can be mapped to the unit square, with periodic boundary conditions in one direction. Whereas a shape with the topology of the torus would require periodic boundary conditions in both directions.

Geometry images were originally introduced by Gu et al. [84], who considered the case of mapping arbitrary surfaces to regularly-sampled square domains. They used the re-sampled, pixellated square grid to store information about the *geometry* of the original shape, in terms of the Cartesian coordinates of the surface, but they also stored information about the surface normals and colour of the surfaces. Davies et al. [56, 181] used a simplified version, storing just the values of the shape function, in the context of shape modelling, and called them *shape images*.

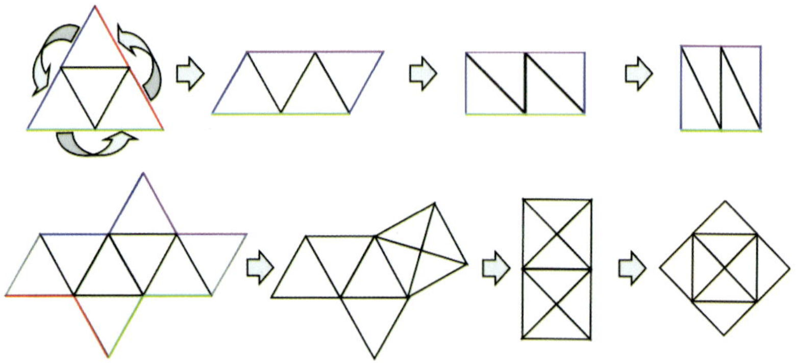

Fig. 8.1 Top: Polyhedral net for the tetrahedron, cut, re-stitched, and stretched to fit into the square. The coloured lines indicate where edges should be joined, and the black lines indicate where the net should be folded to create the final shape. **Bottom:** The net for the octahedron, cut, re-stitching, and distorted, and the final mapping to a square. Note that the coloured lines are sometimes omitted for reasons of clarity.

[10] A similar approach was utilized by Buckminster Fuller in 1946, when he used an unfolded cuboctahedron (a non-regular polyhedron) or a slightly modified regular icosahedron to create his Dymaxion^{TM}map of the earth [19, 20, 81].

We now introduce some notation. We map the original shape to the surface of the sphere as described previously in Sect. 6.1.2. We then place the octahedron inside the sphere, and map points from the surface of the sphere onto the octahedron using a gnomic projection. That is, we just take the radius that passes through the given point on the sphere, and where this line cuts the polyhedron is the point that it maps to. Hence the edges of the polyhedron map to arcs of great circles on the sphere, and great circles on the sphere map to straight lines within a face. Cutting and unfolding the octahedron as shown in Fig. 8.1, then maps the points from the sphere to the unit square.

Our parameter space X is now the unit square. We create a regular grid of points (the *pixel* positions), with Cartesian coordinates (i, j). We will concatenate the positions of all the pixel positions, collecting them into a variable which we will denote by $\mathbf{X}$, the entire pixel grid. The shape function can now be interpolated from the original triangulated mesh to the regular grid, to give the vector-valued shape image $\mathbf{S}(i, j)$.

If we are going to apply fluid regularization on the surface of the shape, we first need to compute the metric at each point (8.16). We hence need to compute the set of spatial derivatives:

$$\frac{\partial}{\partial i}\mathbf{S}(i,j),\ \frac{\partial}{\partial j}\mathbf{S}(i,j).$$

For quantities defined on a regular grid, spatial derivatives are computing using a simple finite-difference approximation. If Δ is the grid spacing, then to lowest order:

$$\partial_i f(i+\frac{\Delta}{2},j) = \frac{f(i+\Delta,j)-f(i,j)}{\Delta},$$
$$\partial_i^2 f(i,j) = \frac{\partial_i f(i+\frac{\Delta}{2},j)-\partial_i f(i-\frac{\Delta}{2},j)}{\Delta},$$

and so on. If $\mathbf{S} \doteq \{\mathbf{S}(i,j) : (i,j) \in \mathbf{X}\}$, the collected set of shape values across the entire grid, then the set of all metric values across the grid can be written symbolically in the form:

$$\mathbf{g} = (\mathbf{SG})^T\mathbf{GS},$$

where $\mathbf{G}$ is some *static* matrix, since the derivatives required to construct $\partial_i\mathbf{S}$ are just a simple linear operation when taken as finite-differences. The exact details of $\mathbf{G}$ obviously depend on how we have concatenated the set of values, hence are implementation-dependent.

For fluid regularization, we now have the displacement field defined at each point on the pixel grid $\mathbf{X}$, and the corresponding Eulerian velocity field which is also defined at each point of the grid. We will collect the entire set of such values into the variables:

Algorithm 8.1 : Fluid Regularization.

Pre-compute the metric $\mathbf{g}$ and hence the derivative matrix $\mathbf{D}[\mathbf{g},\lambda,\mu]$.
Initialize variables with $\mathbf{U} = \mathbf{0}$, $\mathbf{V} = \mathbf{0}$.
Repeat:

- Given $\mathbf{U(t)}$, find $\mathbf{F^S(t)}$
- Compute the velocity field $\mathbf{V(t)}$ using:
  $\mathbf{F}^{\text{visc}}\mathbf{(t)} = -\mathbf{F^S(t)}$, $\mathbf{F}^{\text{visc}}\mathbf{(t)} = \mathbf{D}[\mathbf{g},\lambda,\mu]\mathbf{V(t)}$
- Update the displacement field $\mathbf{U(t+\Delta t)}$:
  $\mathbf{U(t+\Delta t)} = \mathbf{U(t)} + \mathbf{\Delta t}\left[\mathbf{V(t)} - (\mathbf{V(t)}\cdot\nabla)\mathbf{U(t)}\right]$
- Check if re-gridding is required

Until convergence.

$$\mathbf{U}(t) \doteq \{\mathbf{u}(\mathbf{x},t) : \mathbf{x} \in \mathbf{X}\},\;\; \mathbf{V}(t) \doteq \{\mathbf{v}(\mathbf{x},t) : \mathbf{x} \in \mathbf{X}\}. \tag{8.33}$$

The viscous force is computed from the spatial derivatives of the velocity field (8.32), hence can be written in the general form:

$$\mathbf{F}^{\text{visc}}(t) = \mathbf{D}[\mathbf{g},\lambda,\mu]\mathbf{V}(t), \tag{8.34}$$

where $\mathbf{D}$ is a matrix representing the finite-difference operations required to construct the set of second derivatives in (8.32). It depends on the values of the viscosities, and on the metric $\mathbf{g}$ we have previously calculated. The point is note is that like $\mathbf{G}$ defined above, this matrix is also static, hence can be pre-computed.

The final relation we need is that between the displacement field $\mathbf{U}(t)$ and the velocity field $\mathbf{V}(t)$. From (8.10):

$$\mathbf{v}(\mathbf{x},t) = \left.\frac{\partial}{\partial t}\right|_{\mathbf{x}} \mathbf{u}(\mathbf{x},t) + (\mathbf{v}(\mathbf{x},t)\cdot\nabla)\mathbf{u}(\mathbf{x},t).$$

If we let $\boldsymbol{\nabla}$ denote the operation of spatial differentiation taken across the entire grid, then the above can be written in the form:

$$\mathbf{U}(t+\Delta t) = \mathbf{U}(t) + \Delta t\left[\mathbf{V}(t) - (\mathbf{V}(t)\cdot\boldsymbol{\nabla})\mathbf{U}(t)\right],$$

where Δt is now the temporal step-size. This is the required update rule for the displacement field. The basic fluid regularization algorithm is then as given in Algorithm 8.1.

We see that the sequence of operations within the loop is as follows. Given the value of the displacement field at a time t, this determines the *driving* force $\mathbf{F}^S(t)$ (8.6), since as already stated, the driving force depends only on the shape correspondence (hence $\mathbf{U}(t)$), and not on the velocity. The exact details of the driving force obviously depends on the objective function chosen, and the details of the computation will be considered in the next

Algorithm 8.2 : Solving for $\mathbf{V}(\mathbf{t})$.

Pre-compute the derivative matrix $\mathbf{D}[\mathbf{g}, \lambda, \mu]$.
Pre-compute the LU decomposition: $\mathbf{D} = \mathbf{PLUQ}$, where $\mathbf{L}$ and $\mathbf{U}$ are lower and upper triangular matrices, respectively, and $\mathbf{P}$ and $\mathbf{Q}$ are permutation matrices.
Within the main loop of Algorithm 8.1:

- Solve $\mathbf{F}^{\text{visc}}(\mathbf{t}) = \mathbf{PLW}$ for $\mathbf{W}$
- Solve $\mathbf{W} = \mathbf{UQV}(\mathbf{t})$ for $\mathbf{V}(\mathbf{t})$

section. The balance of forces (8.9) means that knowing $\mathbf{F}^S(t)$, we then know $\mathbf{F}^{\text{visc}}(t)$. The next step is non-trivial, in that we have to deduce, given the viscous forces, what the flow field is that yields those forces. We hence have to solve the matrix equation (8.34) $\mathbf{F}^{\text{visc}}(t) = \mathbf{D}[\mathbf{g}, \lambda, \mu]\mathbf{V}(t)$ for $\mathbf{V}(t)$. It is important to note that since $\mathbf{D}$ involves second derivatives, it is a sparse matrix. As noted previously, it is static. Christensen et al. [22] solved this equation using an iterative method, However, since we are only considering two-dimensional shapes, hence two-dimensional shape images, it is quicker to use a pre-computed LU decomposition, as shown in Algorithm 8.2.

The velocity field is then used to update the displacement field. Note that the temporal step-length Δt only enters at this stage. It is hence convenient to determine it here, where, for example, Δt can be chosen so that the additional maximum displacement is equal to some pre-determined value. In effect, this determines the time-scale of our problem. Note that scaling both viscosities λ, μ just scales the velocity field, hence is equivalent to a temporal re-scaling. We then have only one physically-relevant parameter, the ratio of the viscosities, rather than the two the viscosities might initially have suggested. The final step allows for re-gridding if the displacement field is close to becoming non-homeomorphic (i.e., folding).

We have now provided a brief sketch of the fluid regularization algorithm. In the following sections, we first consider several issues that arise when we begin to consider the practicalities of implementation, followed by explicit examples of the regularizer in action.

8.5 Implementation Issues

In this section, we deal with several issues that arise whilst implementing the fluid regularization algorithm using shape images. The first is a purely practical one, involving memory limitations. The second issue is how we avoid singular transformations.

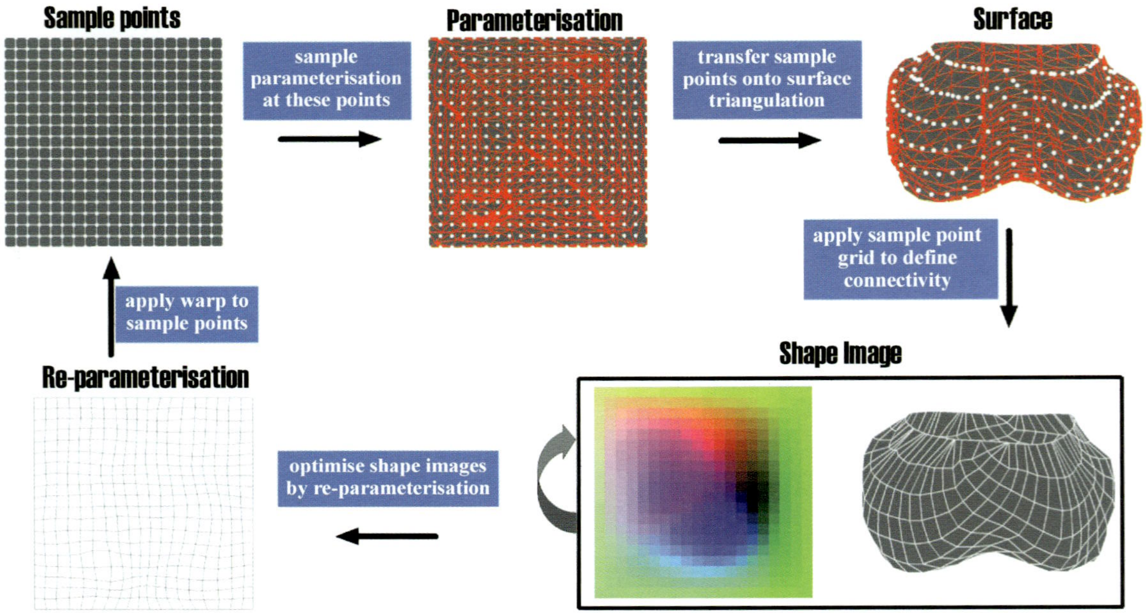

Fig. 8.2 An iterative scheme for updating shape images – for reasons of clarity, only one of the training examples is shown. Starting at the top left, a set of sample points (white points) are created by means of a regular square grid. These sample points are then projected onto the parameterisation (green mesh) and transferred to the surface of the shape using barycentric coordinates. Applying the connectivity of the sample points (in the top left figure), a regularly connected grid is then constructed on the surface. A shape image is then constructed from this. For the purposes of illustration, the shape image has been rendered using the values of the Cartesian coordinates (x, y, z) as the channels of an RGB image. The shape image (along with the shape images of the other members of the training set) is optimised by re-parameterisation. The optimal re-parameterisation is then applied to the sample points. We then proceed to the next iteration, using the updated sample point positions.

8.5.1 Iterative Updating of Shape Images

For some computer systems, memory limitations can mean that it is impractical to store an entire set of high-resolution shape images. An alternative scheme, that works well in practice, uses an iterative approach, where shape images are re-sampled as required.

At each iteration, we start by creating a lower resolution shape image for each training example, which is sampled *according to the current parameterisation*. Correspondence across the set of shape images is then optimised by re-parameterisation. These optimised re-parameterisation functions are then used to update each shape parameterisation, which then enables a new set of shape images to be created by sampling, ready for the start of the next iteration.

In practice, as was discussed previously in Sect. 7.3.4, we chose to manipulate the set of sample points rather than the parameterisations themselves, since this decreases the chance of creating a non-homeomorphic transformation. We consider first the case of open surfaces, since the issue of boundary conditions on the shape image is simpler in this case than for the case of closed surfaces (we will deal with the case of closed surfaces in the next section).

We give here the outline of an efficient method, which is also illustrated diagrammatically in Fig. 8.2. Complete pseudocode for this case is given later in Algorithm 8.4. The method is:

1. Create a regular grid on the unit square.
2. For each training example:
   - Set the initial positions of the sample points to be the gridpoints created in step 1.
3. For each training example:
   - Construct a shape image by using the current sample points to sample the training surface according to its parameterisation.
4. Optimise the set of shape images. This produces a set of re-parameterisation functions, one for each shape.
5. For each training example:
   - Use the re-parameterisation functions found during optimisation to re-parameterise the positions of the sample points.
6. Go back to step 3.

For open surfaces, the task of creating a shape image is simplified, since we just have to ensure that the boundary of the shape image is mapped to the boundary of the shape. However, for closed surfaces, the issue of the placement of the boundary of the shape image has to be dealt with explicitly, as we will see in the next section.

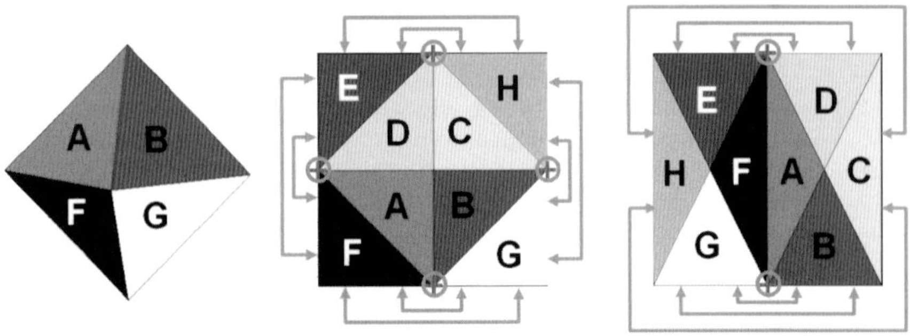

Fig. 8.3 Cutting and unfolding an octahedron and flattening it onto the plane. The $\oplus$ sign denotes singular points, where a fold line meets the edge of the image.

8.5.2 Dealing with Shapes with Spherical Topology

As was discussed previously in Sect. 8.4, in order to create a shape image from a surface with spherical topology, we have to cut open the parameter space of the sphere, and flatten and map this to the unit square. There are several constructions which we can use to perform this task, but in practice, we use the octahedron scheme shown in Figs. 8.1 and 8.3.

Whichever way the octahedron is cut and unfolded, there will always be singular points on the shape image (e.g., see Fig. 8.3). These arise at the points where a fold line meets the edge of the image. As a result, there will always be some points on the shape image that cannot be re-parameterised, and must be fixed throughout optimisation. This hence limits the possible choices of correspondence.

If we use the iterative scheme described in the previous section, we rebuild the shape image at each iteration. We can therefore ensure that the positions of these fixed points change at some point, by cutting the octahedron differently at each iteration. This then means all parts of the parameterisation can be manipulated, even if they cannot all be manipulated at each iteration.

One easy way of varying the octahedral mapping from the sphere to the plane is to rotate the parameterisation (i.e., the sphere) with respect to the octahedron before mapping it to the plane.

An even simpler scheme that also works well in practice is to simply change the sign of the z-coordinates of the sphere (i.e., the parameterisation) on each iteration. We thus alternate between two different positions of the singular points. The basic idea is illustrated in the following algorithm:

1. For each training example:
   - Change the sign of the z-coordinates of the parameterisation.
2. For each training example:

- Construct a shape image by sampling the training surface according to its current parameterisation.

3. Optimise the set of shape images, thus producing a set of re-parameterisation functions, one for each shape.
4. For each training example:
   - Apply the re-parameterisation function found during optimisation to the parameterisation.
5. Go back to step 1.

For the reasons given in the case of open surfaces above, we chose to manipulate the sample points, rather than the parameterisation. This is a little more complicated for closed surfaces, but there is an efficient method, which relies on the use of two sets of sample points for each training example. As before, we here give a brief outline of the method, a diagrammatic explanation in Fig. 8.4, and detailed pseudocode later in this chapter (see Algorithm 8.6):

1. Create a regular grid on the unit square.
2. For each training example:
   - Project the gridpoints onto the unit sphere via the flattened octahedron – call these sample points A.
   - Make a copy of sample points A and reverse the sign of the z-coordinate – these are sample points B.
3. For each training example:
   - If the iteration number is odd, use sample points A, otherwise use sample points B.
   - Construct a shape image using the sample points to sample the training surface according to its parameterisation.
4. Optimise the set of shape images, producing a set of re-parameterisation functions, one for each shape.
5. For each training example:
   - Project the re-parameterisation onto the sphere (reversing the sign of the z-coordinates if sample points B were used), using the flattened octahedron.
   - Use the re-parameterisation functions found during optimisation to re-parameterise the positions of sample points A *and* sample points B.
6. Go back to step 3.

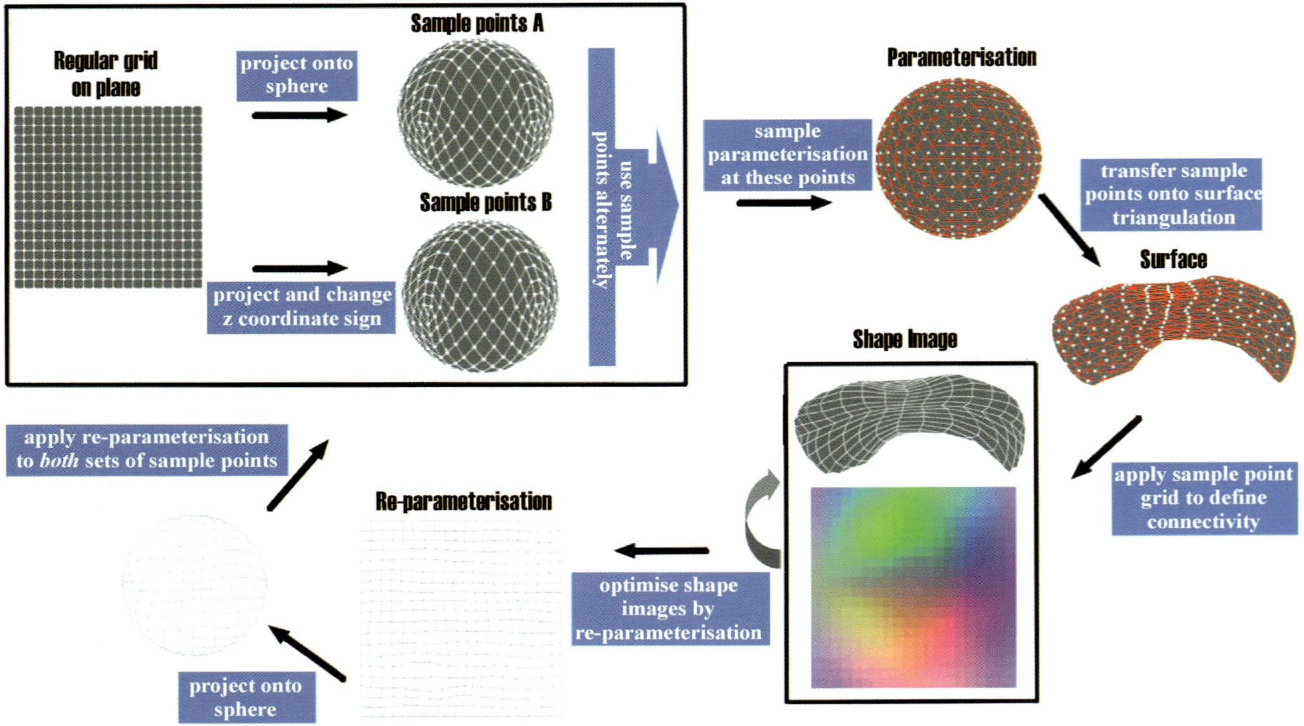

Fig. 8.4 An iterative scheme for updating shape images of surfaces with spherical topology. For reasons of clarity, only one of the training examples is shown. Please refer to the text for a complete explanation.

8.5.3 Avoiding Singularities by Re-gridding

We saw in Fig. 7.3 that although a continuous transformation may be homeomorphic, a finitely sampled version of it, which is then interpolated using a piecewise flat function, may not be. A similar problem can occur for the fluid-based transformations, since we are using a finitely sampled approximation to the actual continuous fluid evolution equations.

Since the fluid-based transformations are defined on a regular grid, it is straightforward to calculate the Jacobian matrix of the transformation at each point. Monitoring the value of the determinant of this Jacobian provides a convenient method of predicting and avoiding singularities (that is, folding) in the transformation [22].

A singularity occurs if the determinant of the Jacobian is zero or less at any point. We can therefore avoid such singularities if we only allow transformations whose Jacobian is greater than some suitable pre-defined value at each point. But if we follow this approach, we can only produce a limited set of transformations, which is insufficient for establishing correspondence in many practical applications. Remember that one advantage of the fluid approach is that it explicitly allows large deformations to be considered, and it is typically such large deformations of the grid which can lead to problems.

An alternative approach to the one of placing constraints on the Jacobian, is to instead use a regridding method. Rather than trying to consider a single deformation field, which may become singular when the deformation becomes large, we instead decompose the transformation into a series of smaller, non-singular ones [22]. Regridding is performed if a transformation approaches singularity at any point. The process involves saving the current transformation and applying it to the shape image, thus creating a new propagated shape image. The working transformation is reset to the identity and optimisation is restarted using this new shape image. Once optimisation is complete, the set of transformations are then composed into the required single transformation.

8.6 Example Implementation of Non-parametric Regularization

We will now describe an efficient algorithm for optimising correspondence between a training set of *shape images* using a fluid-based regularizer. The algorithm can be used by itself if the shape images are of sufficient resolution, but as we have stated above this is impractical on many computer systems. The alternative approach, and the one that we follow here, is the iterative method described in Sect. 8.5.1.

The following options were used in this implementation:

- **Objective function**: The determinant-based objective function ((4.9), Sect. 4.2.1), with regularization constant $\Delta = 0.001$.
- **Number of iterations:** $n_{iterations}$ was set to 25.[11]
- **Fluid Equations:** A finite difference scheme, employing LU decomposition was used to solve the fluid equations.
- **Pose:** pose transformations were not optimised.

If the shape images represent open surfaces (with a single boundary edge), then boundary conditions must be set on the shape image so that points do not move off the edge. For closed surfaces, using the octahedron mapping shown in Fig. 8.3, only the four singular points of the shape image need to be fixed. For the rest of the boundary of the shape image, boundary conditions may be imposed as shown on the figure, which then allows points to move appropriately across the boundary. Fixing only four points allows us to produce more flexible transformations than fixing the entire boundary, but it requires a more complicated implementation. In what follows, we make the simpler choice to fix the entire boundary. This then allows the algorithm to be used within an iterative scheme for either open or closed surfaces, as described in Sects. 8.7.1 and 8.7.2, respectively.

Algorithm 8.3 : Optimisation Using Fluid Regularization.

---

procedure $\{\textbf{Reparameterisation}_i\} \leftarrow optimise\_fluid$
$(\{\textbf{Shape\_Image}_i\}, \textbf{Grid\_Points})$

DESCRIPTION

Uses fluid regularization to find the set of re-parameterisation functions that minimises a group-wise objective function.

VARIABLES

- $\textbf{Shape\_Image}_i$ is a $n \times 3$ shape image representing the ith training shape; the pixel coordinates are stored in the rows of a $n \times 2$ matrix **Grid_Points**;
- $\{\textbf{Reparameterisation}_i\}$ is a set of $n \times 2$ matrices that represent the re-parameterisation functions that can be applied to the corresponding shape images to produce the optimal value of the objective function.

DECLARATIONS

- $\mathbf{b} = bilinear\_interpolation(\mathbf{X}, \mathbf{Y}; \mathbf{a})$
  the $n \times 2$ matrix $\mathbf{X}$, holds values of a (vector-valued) function, evaluated at points whose coordinates are stored in the $n \times 2$ matrix $\mathbf{Y}$; the function returns $\mathbf{b}$, the value of the function at point $\mathbf{a}$, which is estimated by bilinear interpolation;
- $a_{max} = max(\mathbf{a})$
  returns the element of the vector $\mathbf{a}$ with maximum value.

---

[11] Note that this number would be considerably higher if the non-iterative approach was used.

Initialization

1. let ref be in the index of the reference shape;
2. initialize the regrid counter to zero for all shapes:
   for $i = 1 \ldots n_S$
   $$regrid\_counter_i \leftarrow 0;$$
3. make a copy of the original shape images:
   for $i = 1 \ldots n_S$
   $$\mathbf{Original\_Shape\_Image}_i \leftarrow \mathbf{Shape\_Image}_i;$$
4. concatenate the coordinates of the $m \times m$ regular grid **Grid_Points** into a $2n$-dimensional vector (hence $m^2 = n$):[12]
   $$\mathbf{grid\_points} \leftarrow \left(\mathbf{Grid\_Points}(\cdot, 1)^T, \mathbf{Grid\_Points}(\cdot, 2)^T\right),$$
   where the components are ordered as:
   $$(x(1,1), x(2,1), \ldots, x(1,2), x(2,2), \ldots, x(m-1,m), x(m,m), \curvearrowright$$
   $$y(1,1), y(2,1), \ldots, y(1,2), y(2,2), \ldots, y(m-1,m), y(m,m));$$
5. let $\mathbf{displacement}_i$ be a $2n$-dimensional vector representing the displacement field of the ith shape – the vector is arranged so that each component of **displacement** corresponds to a component of **grid_points**; each element of each vector is initialized to zero:
   for $i = 1 \ldots n_S$
   $$\mathbf{displacement}_i \leftarrow (0, 0, \ldots, 0);$$
6. create a $3n$-dimensional shape vector for each training example;
   for $i = 1 \ldots n_S$

   for $j = 1 \ldots n$

   $$\mathbf{sample\_point} \leftarrow \curvearrowright$$
   $$\begin{pmatrix} \mathbf{grid\_points}(j) - \mathbf{displacement}_i(j), \\ \mathbf{grid\_points}(j+n) - \mathbf{displacement}_i(j+n) \end{pmatrix},$$
   $$\mathbf{v} \leftarrow bilinear\_interpolate\left(\mathbf{Grid\_Points}, \mathbf{Shape\_Image}_i, \mathbf{sample\_point}\right),$$
   $$\mathbf{shape\_vector}_i(j) \leftarrow \mathbf{v}(1),$$
   $$\mathbf{shape\_vector}_i(j+n) \leftarrow \mathbf{v}(2),$$
   $$\mathbf{shape\_vector}_i(j+2n) \leftarrow \mathbf{v}(3);$$
7. let **free_node** be a n-dimensional boolean-valued vector with value $true$ if the node is allowed to move and, $false$ if its position remains fixed – in this implementation, the free nodes correspond to all nodes not at the boundary of the unit square;
8. precompute the $2n \times 2n$ (sparse) matrix, $\mathbf{D}$, which holds the finite difference operations required to construct the set of second derivatives[13]:

   8.1. initialize by setting all elements of $\mathbf{D}$ to zero;

---

[12] For clarity, we have assumed that the number of pixels in the shape image was the same as the number of sample points, but only simple modifications are needed to remove this assumption.

[13] Note that we are here making the approximation that all fluids flow on the surface of the mean shape, rather than on each individual shape. There is hence only *one* matrix involved in computing the second derivatives, rather than one such for each shape.

8.2. let:

$$a \leftarrow \frac{2\mu + \lambda}{h^2}, \quad b \leftarrow \frac{\mu}{h^2}, \quad c \leftarrow \frac{\mu + \lambda}{4h^2}, \quad d \leftarrow \frac{\lambda + \mu}{h^2}, \quad e \leftarrow \frac{\lambda + 3\mu}{4h^2}$$

where the grid-spacing $h = 1/m$; fluid viscosity values of $\mu = 1$ and $\lambda = 1$ were used here;

8.3. create a $3n$-dimensional vector to store the shape function and set it to the mean shape:

$$\textbf{shape\_function} \leftarrow \frac{1}{n_S} \sum_i \textbf{shape\_vector}_i;$$

8.4. populate the matrix using a finite difference scheme:
for $k = 1 \dots n$
if ($free\_node(k)$)
– calculate the gradient of the shape function in the i and j directions and store them in three-dimensional vectors:

$$\textbf{grad\_i} \leftarrow \curvearrowright \begin{pmatrix} \textbf{shape\_function}(k) - \textbf{shape\_function}(k+1), \\ \textbf{shape\_function}(k+n) - \textbf{shape\_function}(k+1+n), \\ \textbf{shape\_function}(k+2n) - \textbf{shape\_function}(k+1+2n) \end{pmatrix},$$

$$\textbf{grad\_j} \leftarrow \curvearrowright \begin{pmatrix} \textbf{shape\_function}(k) - \textbf{shape\_function}(k+m), \\ \textbf{shape\_function}(k+n) - \textbf{shape\_function}(k+m+n), \\ \textbf{shape\_function}(k+2n) - \textbf{shape\_function}(k+m+2n) \end{pmatrix};$$

– calculate 2×2 the matrix **Metric** – this corresponds to g in (8.32):

$$\textbf{Metric} \leftarrow \frac{1}{h^3} \begin{pmatrix} \textbf{grad\_i} \bullet \textbf{grad\_i}, & \textbf{grad\_i} \bullet \textbf{grad\_j} \\ \textbf{grad\_j} \bullet \textbf{grad\_i}, & \textbf{grad\_j} \bullet \textbf{grad\_j} \end{pmatrix},$$

where • represents the dot product operator;
– calculate the inverse of the metric matrix:

$$\textbf{Inverse\_Metric} \leftarrow \textbf{Metric}^{-1};$$

– fill in the finite difference operations for the i component – note that the comments after the % sign denotes the element of $\mathbf{v}$ to which the entry corresponds:

let:
$\tilde{a} \leftarrow a \cdot \textbf{Inverse\_Metric}(1,1)$, $\tilde{b} \leftarrow b \cdot \textbf{Inverse\_Metric}(2,2)$,
$\tilde{c} \leftarrow c \cdot \textbf{Inverse\_Metric}(1,1)$, $\tilde{d} \leftarrow d \cdot \textbf{Inverse\_Metric}(1,2)$,
$\tilde{e} \leftarrow e \cdot \textbf{Inverse\_Metric}(1,2)$;

$$
\begin{array}{lll}
\mathbf{D}(k,k) & \leftarrow -2\tilde{a} - 2\tilde{b} & \%\ v_i(i,j) \\
\mathbf{D}(k,k-1) & \leftarrow \tilde{a} & \%\ v_i(i+1,j) \\
\mathbf{D}(k,k+1) & \leftarrow \tilde{a} & \%\ v_i(i+1,j) \\
\mathbf{D}(k,k-m) & \leftarrow \tilde{b} & \%\ v_i(i,j-1) \\
\mathbf{D}(k,k+m) & \leftarrow \tilde{b} & \%\ v_i(i,j+1) \\
\mathbf{D}(k,k+1+m) & \leftarrow \tilde{e} & \%\ v_i(i+1,j+1) \\
\mathbf{D}(k,k-1+m) & \leftarrow -\tilde{e} & \%\ v_i(i-1,j+1) \\
\mathbf{D}(k,k+1-m) & \leftarrow -\tilde{e} & \%\ v_i(i+1,j-1) \\
\mathbf{D}(k,k-1-m) & \leftarrow \tilde{e} & \%\ v_i(i-1,j-1) \\
\mathbf{D}(k,k+n) & \leftarrow -2\tilde{d} & \%\ v_j(i,j) \\
\mathbf{D}(k,k+m+n) & \leftarrow \tilde{d} & \%\ v_j(i,j+1) \\
\mathbf{D}(k,k-m+n) & \leftarrow \tilde{d} & \%\ v_j(i,j-1) \\
\mathbf{D}(k,k+1+m+n) & \leftarrow \tilde{c} & \%\ v_j(i+1,j+1) \\
\mathbf{D}(k,k-1+m+n) & \leftarrow -\tilde{c} & \%\ v_j(i-1,j+1) \\
\mathbf{D}(k,k+1-m+n) & \leftarrow -\tilde{c} & \%\ v_j(i+1,j-1) \\
\mathbf{D}(k,k-1-m+n) & \leftarrow \tilde{c} & \%\ v_j(i-1,j-1)
\end{array}
$$

– now do the same for the j component:

let:
$\tilde{a} \leftarrow a \cdot \textbf{Inverse\_Metric}(2,2)$, $\tilde{b} \leftarrow b \cdot \textbf{Inverse\_Metric}(1,1)$,
$\tilde{c} \leftarrow c \cdot \textbf{Inverse\_Metric}(2,2)$, $\tilde{d} \leftarrow d \cdot \textbf{Inverse\_Metric}(2,1)$,
$\tilde{e} \leftarrow e \cdot \textbf{Inverse\_Metric}(2,1)$;

$$
\begin{array}{lll}
\mathbf{D}(k+n,k+n) & \leftarrow -2\tilde{a} - 2\tilde{b} & \%\ v_j(i,j) \\
\mathbf{D}(k+n,k-m+n) & \leftarrow \tilde{a} & \%\ v_j(i,j-1) \\
\mathbf{D}(k+n,k+m+n) & \leftarrow \tilde{a} & \%\ v_j(i,j+1) \\
\mathbf{D}(k+n,k-1+n) & \leftarrow \tilde{b} & \%\ v_j(i+1,j) \\
\mathbf{D}(k+n,k+1+n) & \leftarrow \tilde{b} & \%\ v_j(i+1,j) \\
\mathbf{D}(k+n,k+1+m+n) & \leftarrow \tilde{e} & \%\ v_j(i+1,j+1) \\
\mathbf{D}(k+n,k-1+m+n) & \leftarrow -\tilde{e} & \%\ v_j(i-1,j+1) \\
\mathbf{D}(k+n,k+1-m+n) & \leftarrow -\tilde{e} & \%\ v_j(i+1,j-1) \\
\mathbf{D}(k+n,k-1-m+n) & \leftarrow \tilde{e} & \%\ v_j(i-1,j-1) \\
\mathbf{D}(k+n,k) & \leftarrow -2\tilde{d} & \%\ v_i(i,j) \\
\mathbf{D}(k+n,k+m) & \leftarrow \tilde{d} & \%\ v_i(i,j+1) \\
\mathbf{D}(k+n,k-m) & \leftarrow \tilde{d} & \%\ v_i(i,j-1) \\
\mathbf{D}(k+n,k+1+m) & \leftarrow \tilde{c} & \%\ v_i(i+1,j+1) \\
\mathbf{D}(k+n,k-1+m) & \leftarrow -\tilde{c} & \%\ v_i(i-1,j+1) \\
\mathbf{D}(k+n,k+1-m) & \leftarrow -\tilde{c} & \%\ v_i(i+1,j-1) \\
\mathbf{D}(k+n,k-1-m) & \leftarrow \tilde{c} & \%\ v_i(i-1,j-1)
\end{array}
$$

8.5. fill in the values for the fixed nodes:
for $k = 1 \dots n$
if $NOT(\textbf{free\_node}(k))$
$\mathbf{D}(k,k) \leftarrow 1$, $\mathbf{D}(k+n,k+n) \leftarrow 1$,
for $j = 1 \dots 2n$
$\mathbf{D}(k,j) \leftarrow 0$, $\mathbf{D}(j,k) \leftarrow 0$,
$\mathbf{D}(k+n,j) \leftarrow 0$, $\mathbf{D}(j,k+n) \leftarrow 0$;

8.6. perform an LU decomposition of $\mathbf{D}$ to get a lower triangular matrix $\mathbf{L}$ and an upper triangular matrix $\mathbf{U}$.

OPTIMISATION

1. for $it = 1 \ldots n_{iterations}$

   1.1. select a shape number, i, at random using a uniform distribution over the training set with $i \neq ref$;

   1.2. let $\mathbf{gradient}_i$ be a $2n$-dimensional vector that holds the value of the gradient of the objective function, $\mathcal{L}$, w.r.t. the displacement of the sample points of the ith shape – for convenience, we will hold the position of the sample points in a $2n$-dimensional vector, $\mathbf{p}$:

$$\mathbf{gradient}_i \leftarrow \left(\frac{\partial \mathcal{L}}{\partial p_1^{(i)}}, \ldots \frac{\partial \mathcal{L}}{\partial p_A^{(i)}}, \ldots \frac{\partial \mathcal{L}}{\partial p_{2n}^{(i)}} \right),$$

   where $p_A^{(i)} = \mathbf{grid\_points}(A) - \mathbf{displacement}_i(A)$.
   In Sect. 4.3.4, we saw that the gradient can be split into a product of simpler terms:

$$\frac{\partial \mathcal{L}}{\partial p_A^{(i)}} = \sum_{a=1}^{n_S-1} \frac{\partial \mathcal{L}}{\partial \lambda_a} \sum_{j=1}^{n_S} \sum_{k=1}^{n_S} \frac{\partial \lambda_a}{\partial \widetilde{D}_{jk}} \int \frac{\delta \widetilde{D}_{jk}}{\delta \mathbf{S}_i(\mathbf{x})} \frac{\delta \mathbf{S}_i(\mathbf{x})}{\delta p_A^{(i)}} dA(\mathbf{x})$$

   we also saw in Sect. 7.3.2 that the term involving the integral can be estimated using a simple finite sum:

$$\int \frac{\delta \widetilde{D}_{jk}}{\delta \mathbf{S}_i(\mathbf{x})} \frac{\delta \mathbf{S}_i(\mathbf{x})}{\delta p_A^{(i)}} dA(\mathbf{x}) \approx \sum_m \frac{\delta \widetilde{D}_{jk}}{\delta \mathbf{S}_i(\mathbf{x}_m)} \frac{\delta \mathbf{S}_i(\mathbf{x}_m)}{\delta p_A^{(i)}} \Delta A(\mathbf{x}_m)$$

   where $\{\mathbf{x}_m\}$ represent the position of the sample points and $\Delta A(\mathbf{x}_m)$ represents the total area of all triangles connected to the mth sample point, calculated on the mean shape (see Sect. 7.3.2).

   - set the value of the gradient of the reference shape to zero:
     for $j = 1 : 2n$
       $\mathbf{gradient}_{ref}(j) \leftarrow 0$,
   - set the value at all fixed nodes to be zero:
     for $i = 1 : n_S,\ i \neq ref$
       for $j = 1 : 2n$
         if $NOT(\mathbf{free\_node}(j))$
           $\mathbf{gradient}_i(j) \leftarrow 0$;

   1.3. The components of each other term are calculated as follows:

   1.3.1. calculate the mean shape vector:

$$\mathbf{mean\_shape\_vector} \leftarrow \frac{1}{n_S} \sum_i \mathbf{shape\_vector}_i;$$

   1.3.2. create a shape difference vector for each shape by subtracting the mean shape vector:
   for $i = 1 \ldots n_S$

$$\mathbf{centred\_shape\_vector}_i \leftarrow \mathbf{shape\_vector}_i - \mathbf{mean\_shape\_vector};$$

   1.3.3. for each sample point: calculate the sum of the areas (on the mean shape) of all triangles[14] connected to that sample point and store it in a n-dimensional vector $\mathbf{int\_area}$;

---

[14] In this implementation, a triangulation of the sample points was created by performing a Delaunay triangulation of **Grid_Points**. One could also define connectivity using quadrangles, but it is usually easier to work with triangles.

1.3.4. calculate a $n_S \times n_S$ covariance matrix, **Covariance**, using the approximation described in Sect. 7.3.1; each element is calculated as:

$$\mathbf{Covariance}(i,j) \leftarrow \sum_{k=1}^{n} \mathbf{int\_area}(k) \, [\curvearrowright$$
$$\mathbf{centred\_shape\_vector}_i(k) \cdot \mathbf{centred\_shape\_vector}_j(k) + \curvearrowright$$
$$\mathbf{centred\_shape\_vector}_i(k+n) \cdot \mathbf{centred\_shape\_vector}_j(k+n) + \curvearrowright$$
$$\mathbf{centred\_shape\_vector}_i(k+2n) \cdot \mathbf{centred\_shape\_vector}_j(k+2n)]$$

1.3.5. obtain the $n_S - 1$ set of eigenvectors $\{\mathbf{eigenvector}_a\}$ and the corresponding (ordered) eigenvalues $\{eigenvalue_a\}$ of **Covariance**;

1.3.6. use the eigenvalues to calculate

$$\frac{\partial \mathcal{L}}{\partial \lambda_a} = \frac{1}{eigenvalue_a + \epsilon};$$

1.3.7. normalize all eigenvector to have unit length:
for $a = 1 \ldots n_S - 1$

$$\mathbf{eigenvector}_a \leftarrow \frac{\mathbf{eigenvector}_a}{||\mathbf{eigenvector}_a||}$$

1.3.8. use the eigenvectors to calculate

$$\frac{\partial \lambda_a}{\partial \widetilde{D}_{jk}} = \mathbf{eigenvector}_a(j) \cdot \mathbf{eigenvector}_a(k);$$

1.3.9. calculate the components of $\frac{\delta \widetilde{D}_{jk}}{\delta \mathbf{S}_i(\mathbf{x}_m)}$; the x coordinate component is given by:

$$\frac{1}{n_S \cdot \sum_m \mathbf{int\_area}(m)} \left[(n_S \delta(i,j) - 1) \mathbf{centred\_shape\_vector}_k(m) \curvearrowright \right.$$
$$\left. + \, (n_S \delta(i,k) - 1) \mathbf{centred\_shape\_vector}_j(m)) \right];$$

the y and z coordinate components are obtained in a similar fashion by substituting m with $m+n$ and $m+2n$, respectively;

1.3.10. use a finite difference scheme to numerically approximate the x-coordinate components of $\frac{\delta \mathbf{S}_i(\mathbf{x})}{\delta p_A^{(i)}}$ for each free grid point:

- perturb the Ath point of the ith shape by a small amount $\Delta = 10^{-5}$ parallel to the x-axis:
  - re-parameterise by perturbing the x-coordinate of the sample point:

$$\mathbf{sample\_point} \leftarrow \begin{pmatrix} \mathbf{grid\_point}(A) - \Delta \\ \mathbf{grid\_point}(A+n) \end{pmatrix}^T$$

  - sample the the perturbed point on the shape:

$$\mathbf{v} \leftarrow bilinear\_interpolation \curvearrowright$$
$$(\mathbf{Grid\_Points}, \mathbf{Shape\_Image}_i, \mathbf{sample\_point}),$$

  - create a copy of the current shape vector for the ith shape and replace the Ath point with the perturbed point:

$$\begin{aligned}&\mathbf{perturbed\_shape\_vector} \leftarrow \mathbf{shape\_vector}_i\\&\mathbf{perturbed\_shape\_vector}(A) \leftarrow \mathbf{v}(1),\\&\mathbf{perturbed\_shape\_vector}(A+n) \leftarrow \mathbf{v}(2),\\&\mathbf{perturbed\_shape\_vector}(A+2n) \leftarrow \mathbf{v}(3);\end{aligned}$$

– estimate the derivative as a finite difference:

$$\frac{\delta \mathbf{S}_i(\mathbf{x})}{\delta p_A^{(i)}} = \frac{\mathbf{perturbed\_shape\_vector} - \mathbf{shape\_vector}_i}{\Delta},$$

1.3.11. now do the same to the y-coordinate components of $\frac{\delta \mathbf{S}_i(\mathbf{x})}{\delta p_A^{(i)}}$ for each free grid point:

- perturb the Ath point of the ith shape by a small amount $\Delta = 10^{-5}$ parallel to the y-axis:
  – re-parameterise by perturbing the y-coordinate of the sample point:

$$\mathbf{sample\_point} \leftarrow \begin{pmatrix}\mathbf{grid\_point}(A)\\ \mathbf{grid\_point}(A+n) - \Delta\end{pmatrix}^T$$

  – sample the the perturbed point on the shape:

$$\begin{aligned}\mathbf{v} \leftarrow &\ \mathit{bilinear\_interpolation} \curvearrowright\\ &(\mathbf{Grid\_Points}, \mathbf{Shape\_Image}_i, \mathbf{sample\_point}),\end{aligned}$$

  – create a copy of the current shape vector for the ith shape and replace the Ath point with the perturbed point:

$$\begin{aligned}&\mathbf{perturbed\_shape\_vector} \leftarrow \mathbf{shape\_vector}_i\\&\mathbf{perturbed\_shape\_vector}(A) \leftarrow \mathbf{v}(1),\\&\mathbf{perturbed\_shape\_vector}(A+n) \leftarrow \mathbf{v}(2),\\&\mathbf{perturbed\_shape\_vector}(A+2n) \leftarrow \mathbf{v}(3);\end{aligned}$$

  – estimate the derivative as a finite difference:

$$\frac{\delta \mathbf{S}_i(\mathbf{x})}{\delta p_{A+n}^{(i)}} = \frac{\mathbf{perturbed\_shape\_vector} - \mathbf{shape\_vector}_i}{\Delta},$$

2. for $i = 1 \ldots n_S, i \neq ref$; % for each shape

2.1. solve the linear PDE

$$-\mathbf{gradient}_i = \mathbf{D}\,\mathbf{velocity}_i$$

for the velocity field of the ith shape, $\mathbf{velocity}_i$; this is achieved by solving

$$\mathbf{La} = -\mathbf{gradient}_i$$

for $\mathbf{a}$ by forward substitution, then solving

$$\mathbf{U}\,\mathbf{velocity}_i = \mathbf{a}$$

for $\mathbf{velocity}_i$ by back substitution (where $\mathbf{L}$ and $\mathbf{U}$ are the pre-computed matrices produced by the LU decomposition computed in the Initialization);

2.2. calculate the n-dimensional vectors corresponding to the gradient of the displacement field of the ith shape, $\mathbf{displacement}_i$, along the x and y axis:
for $j = 1 \ldots n$
  if $\mathbf{free\_node}(j)$

$$\mathbf{duxdx}(j) \leftarrow \frac{\mathbf{displacement}(j+m) - \mathbf{displacement}(j-m)}{2h},$$
$$\mathbf{duxdy}(j) \leftarrow \frac{\mathbf{displacement}(j+1) - \mathbf{displacement}(j-1)}{2h},$$
$$\mathbf{duydx}(j) \leftarrow \frac{\mathbf{displacement}(j+m+n) - \mathbf{displacement}(j-m+n)}{2h},$$
$$\mathbf{duydy}(j) \leftarrow \frac{\mathbf{displacement}(j+1+n) - \mathbf{displacement}(j-1+n)}{2h},$$

  else

$$\mathbf{duxdx}(j) \leftarrow 0,$$
$$\mathbf{duxdy}(j) \leftarrow 0,$$
$$\mathbf{duydx}(j) \leftarrow 0,$$
$$\mathbf{duydy}(j) \leftarrow 0,$$

2.3. calculate the $n \times 2$ regularization matrix $\mathbf{R}$:
for $j = 1 \ldots n$

$$\mathbf{R}(j,1) \leftarrow \mathbf{velocity}(j) - (\mathbf{velocity}(j)\ \mathbf{duxdx}(j) + \curvearrowright$$
$$\mathbf{velocity}(j+n)\ \mathbf{duxdy}(j),)$$
$$\mathbf{R}(j,2) \leftarrow \mathbf{velocity}(j+n) - (\mathbf{velocity}(j)\ \mathbf{duydx}(j) + \curvearrowright$$
$$\mathbf{velocity}(j+n)\ \mathbf{duydy}(j));$$

2.4. calculate the timestep as

$$t \leftarrow \frac{\delta}{\max(||\mathbf{R}||)};$$

where the $||\cdot||$ operator returns a vector representing the Euclidian norm of each row of $\mathbf{R}$; a value of $\delta = 10^{-3}$ was used in this implementation;
2.5. calculate the proposed displacement:
for $j = 1 : n$

$$\mathbf{proposed\_displacement}(j,1) \leftarrow \mathbf{grid\_points}(j) \curvearrowright$$
$$-\mathbf{displacement}_i(j) - t\,\mathbf{R}(j,1),$$
$$\mathbf{proposed\_displacement}(j,2) \leftarrow \mathbf{grid\_points}(j+n) \curvearrowright$$
$$-\mathbf{displacement}_i(j+n) - t\,\mathbf{R}(j,2);$$

2.6. if the Jacobian of $m \cdot \mathbf{proposed\_displacement}$ is less than 0.5 at any point;
- then `% need to regrid`

  2.6.1. increment the regrid counter for the ith shape:

$$regrid\_counter_i = regrid\_counter_i + 1;$$

  2.6.2. store the current transformation into a matrix, whose rows contain all saved re-parameterisations:

$$\textbf{Saved\_Reparameterisation}_i(regrid\_counter_i, \cdot) \leftarrow \curvearrowright \\ \textbf{grid\_points} - \textbf{displacement}_i;$$

2.6.3. resample the shape image according to the accumulated re-parameterisation:

2.6.3.1. let $\textbf{Cumulative\_Reparameterisation} \leftarrow \textbf{Grid\_Points}$;

for $k = 1 \ldots regrid\_counter_i$

– reshape the saved re-parameterisation into a $n \times 2$ matrix:
for $j = 1 \ldots n$

$$\textbf{Reshaped\_Reparameterisation}(j, 1) \leftarrow \curvearrowright \\ \textbf{Saved\_Reparameterisation}_i(k, j), \\ \textbf{Reshaped\_Reparameterisation}(j, 2) \leftarrow \curvearrowright \\ \textbf{Saved\_Reparameterisation}_i(k, j+n), \\ \textbf{Reshaped\_Reparameterisation}(j, 3) \leftarrow \curvearrowright \\ \textbf{Saved\_Reparameterisation}_i(k, j+2n)$$

– accumulate the re-parameterisations:
for $j = 1 \ldots n$

$$\textbf{Cumulative\_Reparameterisation}(j, \cdot) \leftarrow \curvearrowright \\ bilinear\_interpolation(\textbf{Grid\_Points}, \curvearrowright \\ \textbf{Reshaped\_Reparameterisation}, \curvearrowright \\ \textbf{Cumulative\_Reparameterisation}(j, \cdot));$$

2.6.3.2. resample the *original* shape image to create a new shape image, sampled according to $\textbf{Resized\_Reparameterisation}$:
for $j = 1 \ldots n$

$$\textbf{Shape\_Image}_i(j, \cdot) \leftarrow bilinear\_interpolation \curvearrowright \\ (\textbf{Grid\_Points}, \textbf{Original\_Shape\_Image}_i, \curvearrowright \\ \textbf{Resized\_Reparameterisation}(j, \cdot));$$

2.6.4. set all elements of the displacement vector of the ith training example to zero:

$$\textbf{displacement}_i \leftarrow (0, 0, \ldots 0)^T;$$

2.6.5. goto step 2.2;

- else % accept the displacement
for $j = 1 \ldots n$

$$\textbf{displacement}_i(j) \leftarrow \textbf{displacement}_i(j) + t \cdot \mathbf{R}_i(j, 1), \\ \textbf{displacement}_i(j+n) \leftarrow \textbf{displacement}_i(j+n) + t \cdot \mathbf{R}_i(j, 2);$$

3. update the ith shape vector:
for $j = 1 \ldots n$

$$\mathbf{sample\_point} \leftarrow \curvearrowright$$
$$\begin{pmatrix} \mathbf{grid\_points}(j) - \mathbf{displacement}_i(j), \\ \mathbf{grid\_points}(j+n) - \mathbf{displacement}_i(j+n) \end{pmatrix},$$
$$\mathbf{v} \leftarrow \mathit{bilinear\_interpolate}\left(\mathbf{Grid\_Points}, \mathbf{Shape\_Image}_i, \mathbf{sample\_point}\right),$$
$$\mathbf{shape\_vector}_i(j) \leftarrow \mathbf{v}(1),$$
$$\mathbf{shape\_vector}_i(j+n) \leftarrow \mathbf{v}(2),$$
$$\mathbf{shape\_vector}_i(j+2n) \leftarrow \mathbf{v}(3),$$

POST-PROCESSING
for $i = 1 \dots n_S$

1. save the final re-parameterisations:

$$\mathit{regrid\_counter}_i \leftarrow \mathit{regrid\_counter}_i + 1,$$
$$\mathbf{Saved\_Reparameterisation}_i(\mathit{regrid\_counter}_i, \cdot) \leftarrow \curvearrowright$$
$$\mathbf{grid\_points} - \mathbf{displacement}_i;$$

2. calculate the accumulated re-parameterisation:

   2.1. let $\mathbf{Cumulative\_Reparameterisation} \leftarrow \mathbf{Grid\_Points}$;
   for $k = 1 \dots \mathit{regrid\_counter}_i$
   - reshape the saved re-parameterisation into a $n \times 2$ matrix:
     for $j = 1 \dots n$

$$\mathbf{Reshaped\_Reparameterisation}(j, 1) \leftarrow \curvearrowright$$
$$\mathbf{Saved\_Reparameterisation}_i(k, j),$$
$$\mathbf{Reshaped\_Reparameterisation}(j, 2) \leftarrow \curvearrowright$$
$$\mathbf{Saved\_Reparameterisation}_i(k, j+n),$$
$$\mathbf{Reshaped\_Reparameterisation}(j, 3) \leftarrow \curvearrowright$$
$$\mathbf{Saved\_Reparameterisation}_i(k, j+2n);$$

   - accumulate the re-parameterisations:
     for $j = 1 \dots n$

$$\mathbf{Cumulative\_Reparameterisation}(j, \cdot) \leftarrow \curvearrowright$$
$$\mathit{bilinear\_interpolation}(\mathbf{Grid\_Points}, \curvearrowright$$
$$\mathbf{Reshaped\_Reparameterisation}, \curvearrowright$$
$$\mathbf{Cumulative\_Reparameterisation}(j, \cdot));$$

3. let $\mathbf{Reparameterisation}_i \leftarrow \mathbf{Cumulative\_Reparameterisation}_i$.

return $\{\mathbf{Reparameterisation}_i\}$

8.7 Example Optimisation Routines Using Iterative Updating of Shape Images

As we noted above, it is often necessary to employ shape images within an iterative approach to optimisation. In this section, examples are given of how this can be implemented for both open and closed surfaces.

8.7.1 Example 3: Open Surfaces Using Shape Images

The dataset of open surfaces we will use is the same set of $n_S = 10$ distal femurs that was used in the example implementation in Sect. 7.4.2. The actual training shapes are shown in Fig. 7.8.

Pseudocode for the algorithm is given below, which is the algorithm described diagramatically in Fig. 8.2.

Algorithm 8.4 : An Example of Optimising Correspondence on Open Surfaces Using Shape Images.

---

procedure $\{\mathbf{shape\_vector}_i\} \leftarrow \mathit{optimisation\_example\_3}$ $(\{\mathbf{Shape\_Points}_i\}, \{\mathbf{Triangulation}_i\})$

VARIABLES

- $\mathbf{Shape\_Points}_i$ is a $n_P \times 3$ matrix whose rows contain the coordinates of nodes that define the surface of the ith training example;
- $\mathbf{Triangulation}_i$ is a $n_t \times 3$ matrix that defines the connectivity of the nodes: each row defines a triangle by indexing the rows of $\mathbf{Shape\_Points}_i$ – see glossary for description of format;
- the function returns a set of $3n$-dimensional shape vectors $\{\mathbf{shape\_vector}_i\}$, formed by sampling each training set according to its optimal parameterisation.

DECLARATIONS

- $\mathbf{y} = \mathit{transfer\_point}(\mathbf{Points\_A}, \mathbf{Points\_B}, \mathbf{Triangulation}, \mathbf{x})$
transfers the point $\mathbf{x}$ from triangulation $(\mathbf{Points\_A}, \mathbf{Triangulation})$ to triangulation $(\mathbf{Points\_B}, \mathbf{Triangulation})$ to form a new point $\mathbf{y}$ – the pseudocode is given in Algorithm 7.3;
- $\mathbf{b} = \mathit{bilinear\_interpolation}(\mathbf{X}, \mathbf{Y}; \mathbf{a})$
the $n \times 2$ matrix $\mathbf{X}$, holds values of a (vector-valued) function, evaluated at points whose coordinates are stored in the $n \times 2$ matrix $\mathbf{Y}$; the function returns $\mathbf{b}$, the value of the function at point $\mathbf{a}$, which is estimated by bilinear interpolation;
- $\{\mathbf{Warp}_i\} = \mathit{optimise\_fluid}(\{\mathbf{Shape\_Image}_i\}, \mathbf{Grid\_Points})$
given a set of shape images $\{\mathbf{Shape\_Image}_i\}$, with pixels at positions $\mathbf{Grid\_Points}$, this function optimises correspondence by re-parameterisation using a fluid-based scheme; the function returns the set of warps $\{\mathbf{Warp}_i\}$, which correspond to the optimal set of re-parameterisations – pseudocode is given in Algorithm 8.3.

INITIALIZATION

1. let ref be the index of a reference shape, whose pose and parameterisation do not change during optimisation;
2. parameterise and align the surfaces as described in step 2 of Algorithm 7.4; we end up with a set of parameterisations, $\{\mathbf{Parameterisation}_i\}$, represented as a set of $n_P \times 2$ matrices, whose rows contain the coordinates of nodes in parameter space (the unit square), corresponding to the surface nodes stored in $\mathbf{Shape\_Points}_i$;
3. create a $m \times m$ grid over the unit square ($m^2 = n = 1600$ was used here) and store the coordinates in a $n \times 2$ matrix $\mathbf{Grid\_Points}$:

$$\mathbf{Grid\_Points} = \left((0,0)^T, \left(0, \frac{1}{m-1}\right)^T, \ldots, \left(\frac{1}{m-1}, 0\right)^T, \right.$$
$$\left. \left(\frac{1}{m-1}, \frac{1}{m-1}\right)^T, \ldots, (1,1)^T \right)^T;$$

4. create a $n \times 2$ matrix $\mathbf{Sample\_Points}_i$ that holds the sample points for each training example – the sample points are initially set to be regular grid points:
for $i = 1 \ldots n_S$

$$\mathbf{Sample\_Points}_i \leftarrow \mathbf{Grid\_Points}$$

OPTIMISATION

for $it = 1 \ldots n_{iterations}$ (we used $n_{iterations} = 1500$ here)

1. create a shape image for each example by sampling each shape according to the current parameterisation:
for $i = 1 \ldots n_S$, (if $it > 1$, then $i \neq ref$)

   for $j = 1 \ldots n$

$$\mathbf{Shape\_Image}_i(j,\cdot) \leftarrow transfer\_point(\mathbf{Parameterisation}_i, \curvearrowright$$
$$\mathbf{Shape\_Points}_i, \mathbf{Triangulation}_i, \mathbf{Sample\_Points}_i(j,\cdot)).$$

2. optimise the set of shape images by re-parameterisation using fluid regularization; the function returns a set of $n \times 2$ matrices that represent the optimal re-parameterisation of each shape;

$$\{\mathbf{Reparameterisation}_i\} \leftarrow fluid\_optimisation \curvearrowright$$
$$(\{\mathbf{Shape\_Image}_i\}, \mathbf{Grid\_Points})$$

3. re-parameterise the sample points of each shape using the optimal re-parameterisation:
for $i = 1 \ldots n_S$, $i \neq ref$

   for $j = 1 \ldots n$

$$\mathbf{Sample\_Points}_i(j,\cdot) \leftarrow bilinear\_interpolate(\mathbf{Grid\_Points}, \curvearrowright$$
$$\mathbf{Reparameterisation}_i, \mathbf{Sample\_Points}_i(j,\cdot))$$

POST-PROCESSING

1. use the optimal sample points to create a shape vector for each example:
   for $i = 1 \dots n_S$

   for $j = 1 \dots n$

   $\mathbf{v} \leftarrow transfer\_point(\mathbf{Parameterisation}_i,$ ↷
   $\mathbf{Shape\_Points}_i, \mathbf{Triangulation}_i, \mathbf{Sample\_Points}_i(j, \cdot)),$
   $\mathbf{shape\_vector}_i(j) \leftarrow \mathbf{v}(1),$
   $\mathbf{shape\_vector}_i(j+n) \leftarrow \mathbf{v}(2),$
   $\mathbf{shape\_vector}_i(j+2n) \leftarrow \mathbf{v}(3)..$

return $\{\mathbf{shape\_vector}_i\}$.

The correspondence and the model produced by the above algorithm for the femur dataset are almost indistinguishable from those produced by the earlier approach described in Sect. 7.4.2, which used a parametric representation of re-parameterisation (clamped-plate splines) and triangulated shape surfaces. However, the fluid-based algorithm described above converged ≈ 100 times faster.

8.7.2 Example 4: Optimisation of Closed Surfaces Using Shape Images

Our dataset of closed surfaces is a set of $n_S = 82$ surfaces, representing the human hippocampus. The data was obtained by manually segmenting the left hippocampus from magnetic resonance images (MRIs). Some examples from the training set are shown in Fig. 8.5.

We establish a correspondence across the training set by extending the algorithm described above to deal with closed surfaces. The approach is similar, but additional steps are required, as was illustrated in Fig. 8.4. In order to follow this approach, we need a procedure for mapping points from the plane onto the unit sphere – pseudocode for achieving this is given below (Algorithm 8.5), followed by the pseudocode for the main optimisation procedure (Algorithm 8.6).

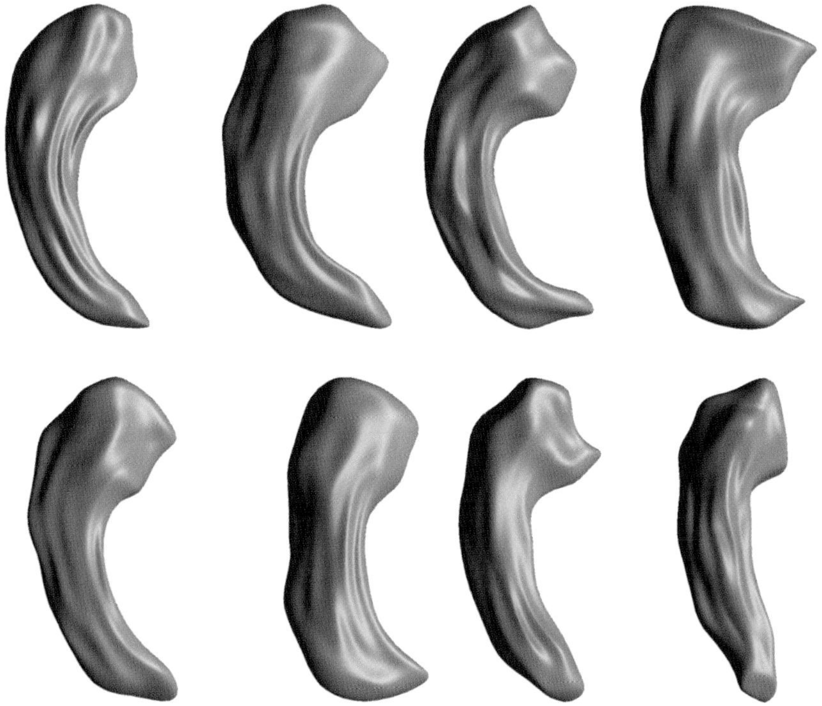

Fig. 8.5 Eight examples of the training set of 82 outlines of the left human hippocampus used in the example implementation described in Sect. 8.7.2.

Algorithm 8.5 : Transforming Points from the Plane onto a Sphere.

procedure **sphere_point** $\leftarrow$ *plane_to_sphere*(**plane_point**)

DESCRIPTION

Transfers a point from the plane onto the surface of a unit sphere via an octahedron.

VARIABLES

- **plane_point** is a two-dimensional vector, representing the coordinates of the point on the plane;
- **sphere_point** is a three-dimensional vector, representing the coordinates of the mapped point after projection onto the unit sphere

DECLARATIONS

- $\mathbf{y} = transfer\_point(\mathbf{Points\_A}, \mathbf{Points\_B}, \mathbf{Triangulation}, \mathbf{x})$
  transfers the point $\mathbf{x}$ from triangulation $(\mathbf{Points\_A}, \mathbf{Triangulation})$ to triangulation $(\mathbf{Points\_B}, \mathbf{Triangulation})$ to form a new point $\mathbf{y}$ – pseudocode is given in Algorithm 7.3.

INITIALIZATION[15]

1. construct a flattened triangulated octahedron, storing the coordinates in **Flattened_Octahedron_Points** , and the triangulation in a matrix, **Triangulation** (see Fig. 8.3);

$$\mathbf{Flattened\_Octahedron\_Points} \leftarrow \Big((0,1)^T, (0,\frac{1}{2})^T, (0,0)^T, \curvearrowright$$
$$(\frac{1}{2},1)^T, (\frac{1}{2},\frac{1}{2})^T, (\frac{1}{2},0)^T, (1,1)^T, (1,\frac{1}{2})^T, (1,0)^T \Big)^T;$$

$$\mathbf{Triangulation} \leftarrow \Big((5,8,6)^T, (5,8,4)^T, (5,2,4), (5,2,6)^T, \curvearrowright$$
$$(9,8,6)^T, (7,8,4)^T, (1,2,4)^T, (3,2,6)^T \Big)^T;$$

2. create an octahedron corresponding to the flattened version and store the nodes (with all nodes of the octahedron at unit length from the origin) in a matrix **Octahedron_Points**:

$$\mathbf{Octahedron\_Points} \leftarrow ((0,0,-1), (0,1,0), (0,0,-1), (1,0,0), \curvearrowright$$
$$(0,0,1), (-1,0,0), (0,0,-1), (0,-1,0), (0,0,-1));$$

PROJECTION

1. transfer the point from the flattened octahedron triangulation to the unflattened octahedron:

$$\mathbf{sphere\_point} \leftarrow transfer\_point(\mathbf{Flattened\_Octahedron\_Points}, \curvearrowright$$
$$\mathbf{Octahedron\_Points}, \mathbf{Triangulation}, \mathbf{plane\_point});$$

2. normalize the point to lie on the surface of the unit sphere:

$$\mathbf{sphere\_point} \leftarrow \frac{\mathbf{sphere\_point}}{||\mathbf{sphere\_point}||}.$$

return **sphere_point**;

---

[15] Note that this step can be computed once and saved

Algorithm 8.6 : An Example of Optimising Correspondence on Closed Surfaces Using Shape Images.

procedure $\{\mathbf{shape\_vector}_i\} \leftarrow \mathit{optimisation\_example\_4}$
$(\{\mathbf{Shape\_Points}_i\}, \{\mathbf{Triangulation}_i\})$

VARIABLES

- $\mathbf{Shape\_Points}_i$ is a $n_P \times 3$ matrix whose rows contain the coordinates of nodes that define the surface of the ith training example;
- $\mathbf{Triangulation}_i$ is a $n_t \times 3$ matrix that defines the connectivity of the nodes: each row defines a triangle by indexing the rows of $\mathbf{Shape\_Points}_i$
- the function returns a set of $3n$-dimensional shape vectors $\{\mathbf{shape\_vector}_i\}$, formed by sampling each training set according to its optimal parameterisation.

DECLARATIONS

- $\mathbf{sphere\_point} = \mathit{plane\_to\_sphere}(\mathbf{plane\_point})$
takes a point on the plane, whose coordinates are held in the two-dimensional vector $\mathbf{plane\_point}$, and projects it onto the sphere via the octahedron to create a point whose coordinates are held in the three-dimensional vector $\mathbf{sphere\_point}$. See Algorithm 8.5;
- $\mathbf{y} = \mathit{transfer\_point}(\mathbf{Points\_A}, \mathbf{Points\_B}, \mathbf{Triangulation}, \mathbf{x})$
transfers the point $\mathbf{x}$ from triangulation $(\mathbf{Points\_A}, \mathbf{Triangulation})$ to triangulation $(\mathbf{Points\_B}, \mathbf{Triangulation})$ to form a new point $\mathbf{y}$ – pseudocode is given in Algorithm 7.3;
- $\{\mathbf{Warp}_i\} = \mathit{optimise\_fluid}(\{\mathbf{Shape\_Image}_i\}, \mathbf{Grid\_Points})$
given a set of shape images $\{\mathbf{Shape\_Image}_i\}$, with pixels at positions $\mathbf{Grid\_Points}$, this function optimises correspondence by re-parameterisation using a fluid-based scheme; the function returns the set of warps $\{\mathbf{Warp}_i\}$, which correspond to the optimal set of re-parameterisations – pseudocode is given in Algorithm 8.3.

INITIALIZATION

1. let ref be the index of a reference shape, whose pose and parameterisation do not change during optimisation;
2. parameterise and align the surfaces as described in step 2 of Algorithm 7.4 – note, however, that we are dealing here with closed surfaces, hence there is no need to extract and optimise the boundary nodes; we end up with a set of parameterisations, $\{\mathbf{Parameterisation}_i\}$, represented as a set of $n_P \times 3$ matrices, whose rows contain the coordinates of nodes in parameter space (the surface of the unit sphere), corresponding to the surface nodes stored in $\mathbf{Shape\_Points}_i$;
3. create a $m \times m$ grid over the unit square ($m^2 = n = 1600$ was used here) and store the coordinates in a $n \times 2$ matrix $\mathbf{Grid\_Points}$:

$$\mathbf{Grid\_Points} = \left((0,0)^T, \left(0, \frac{1}{m-1}\right)^T, \ldots, \left(\frac{1}{m-1}, 0\right)^T, \right.$$
$$\left. \left(\frac{1}{m-1}, \frac{1}{m-1}\right)^T, \ldots, (1,1)^T\right)^T;$$

4. project the gridpoints onto the sphere via an octahedron:

$$\mathbf{Sphere\_Grid\_Points} \leftarrow \mathit{plane\_to\_sphere}(\mathbf{Grid\_Points});$$

5. create a $n_{tg} \times 3$ matrix $\mathbf{Grid\_Triangulation}$ that defines the connectivity of the spherical gridpoints: each row defines a triangle by indexing the rows of $\mathbf{Sphere\_Grid\_Points}$;
6. create two copies of $\mathbf{Sphere\_Grid\_Points}$: $\mathbf{Sphere\_Grid\_Points\_A}$ is an exact copy, whereas the signs of the z-coordinates in $\mathbf{Sphere\_Grid\_Points\_B}$ are reversed:

$$\mathbf{Sphere\_Grid\_Points\_A} \leftarrow \mathbf{Sphere\_Grid\_Points},$$
$$\mathbf{Sphere\_Grid\_Points\_B} \leftarrow \mathbf{Sphere\_Grid\_Points},$$
$$\mathbf{Sphere\_Grid\_Points\_B}(\cdot, 3) \leftarrow -\mathbf{Sphere\_Grid\_Points\_B}(\cdot, 3);$$

   what we end up with are two spherical projections of the gridpoints, with the (fixed) edges of the grid at a different position on the sphere;
7. create two $n \times 3$ matrices, $\mathbf{Sample\_Points\_A}_i$ and $\mathbf{Sample\_Points\_B}_i$ for each shape: their rows contain the coordinates of sample points; they are initialized to the spherical projection of the gridpoints:
   for $i = 1 \ldots n_S$

$$\mathbf{Sample\_Points\_A}_i \leftarrow \mathbf{Sphere\_Grid\_Points\_A},$$
$$\mathbf{Sample\_Points\_B}_i \leftarrow \mathbf{Sphere\_Grid\_Points\_B}.$$

Optimisation

for $it = 1 \ldots n_{iterations}$ (we used $n_{iterations} = 12500$ here)

1. for each training example: create a shape image by using the current sample points to sample the surface coordinates of the training example and store them in a $n \times 3$ matrix $\mathbf{Shape\_Image}_i$; note that if the iteration number, it, is an odd number, then sample points A are used otherwise sample points B are used:

   for $i = 1 \ldots n_S$, (if $it > 1$, then $i \neq ref$)
   if $((2\lceil it/2 \rceil - it) = 1)$
   then
   for $j = 1 \ldots n$

$$\mathbf{Shape\_Image}_i(j, \cdot) \leftarrow transfer\_point \curvearrowright$$
$$(\mathbf{Parameterisation}_i, \mathbf{Shape\_Points}_i, \curvearrowright$$
$$\mathbf{Triangulation}_i, \mathbf{Sample\_Points\_A}_i(j, \cdot))$$

   else
   for $j = 1 \ldots n$

$$\mathbf{Shape\_Image}_i(j, \cdot) \leftarrow transfer\_point \curvearrowright$$
$$(\mathbf{Parameterisation}_i, \mathbf{Shape\_Points}_i, \curvearrowright$$
$$\mathbf{Triangulation}_i, \mathbf{Sample\_Points\_B}_i(j, \cdot))$$

2. optimise the set of shape images by re-parameterisation; the function returns a set of $n \times 2$ matrices, which represent the optimal re-parameterisation of each shape;

$$\{\mathbf{Warp}_i\} \leftarrow optimise\_fluid(\{\mathbf{Shape\_Image}_i\}, \mathbf{Grid\_Points})$$

3. project the re-parameterisation $\mathbf{Warp}_i$ of each shape onto the sphere:
   for $i = 1 \ldots n_S, i \neq ref$

   for $j = 1 \ldots n$

$$\mathbf{Reparameterisation}_i(j, \cdot) \leftarrow plane\_to\_sphere(\mathbf{Warp}_i(j, \cdot))$$

4. now, apply the (spherical) re-parameterisation to both sets of sample points:
   if $((2\lceil it/2 \rceil - it) = 1)$
   then

   for $i = 1 \ldots n_S, i \neq ref$
   4.1. update both sets of sample points:
   for $j = 1 \ldots n$

$$\mathbf{Sample\_Points\_A}_i(j, \cdot) \leftarrow transfer\_point \curvearrowright$$
$$(\mathbf{Sphere\_Grid\_Points\_A}, \mathbf{Reparameterisation}_i, \curvearrowright$$
$$\mathbf{Grid\_Triangulation}, \mathbf{Sample\_Points\_A}_i(j, \cdot))$$

$$\mathbf{Sample\_Points\_B}_i(j, \cdot) \leftarrow transfer\_point \curvearrowright$$
$$(\mathbf{Sphere\_Grid\_Points\_A}, \mathbf{Reparameterisation}_i, \curvearrowright$$
$$\mathbf{Grid\_Triangulation}, \mathbf{Sample\_Points\_B}_i(j, \cdot))$$

   else

   for $i = 1 \ldots n_S, i \neq ref$
   4.1. since sample points B has been used, we need to reverse the sign of the z-coordinates of each re-parameterisation point:

$$\mathbf{Reparameterisation}_i(\cdot, 3) \leftarrow -\mathbf{Reparameterisation}_i(\cdot, 3)$$

   4.2. update both sets of sample points:
   for $j = 1 \ldots n$

$$\mathbf{Sample\_Points\_A}_i(j, \cdot) \leftarrow transfer\_point \curvearrowright$$
$$(\mathbf{Sphere\_Grid\_Points\_B}, \mathbf{Reparameterisation}_i, \curvearrowright$$
$$\mathbf{Grid\_Triangulation}, \mathbf{Sample\_Points\_A}_i(j, \cdot))$$

$$\mathbf{Sample\_Points\_B}_i(j, \cdot) \leftarrow transfer\_point \curvearrowright$$
$$(\mathbf{Sphere\_Grid\_Points\_B}, \mathbf{Reparameterisation}_i, \curvearrowright$$
$$\mathbf{Grid\_Triangulation}, \mathbf{Sample\_Points\_B}_i(j, \cdot))$$

POST-PROCESSING

1. subdivide a unit icosahedron and project its coordinates onto the unit sphere (see Sect. 7.3.3) to produce a $n_u \times 3$ matrix **Uniform_Sample_Points**;
2. for each shape: re-parameterise the uniform sample points with either sample points A or B:

   for $i = 1 \ldots n_S$
   for $j = 1 \ldots n_u$

$$\mathbf{Reparameterised\_Sample\_Points}_i(j, \cdot) \leftarrow transfer\_point \curvearrowright$$
$$(\mathbf{Sphere\_Grid\_Points\_A}, \mathbf{Sample\_Points\_A}_i, \curvearrowright$$
$$\mathbf{Grid\_Triangulation}, \mathbf{Uniform\_Sample\_Points}(j, \cdot))$$

3. now use the reparameterised sample points to sample the original shapes and produce a shape vector for each example:

   for $i = 1 \ldots n_S$

for $j = 1 \ldots n$

$\mathbf{v} \leftarrow transfer\_point$ ↷

$(\mathbf{Parameterisation}_i, \mathbf{Shape\_Points}_i,$ ↷

$\mathbf{Triangulation}_i, \mathbf{Reparameterised\_Sample\_Points}_i(j, \cdot)),$

$\mathbf{shape\_vector}_i(j) \leftarrow \mathbf{v}(1),$

$\mathbf{shape\_vector}_i(j+n) \leftarrow \mathbf{v}(2),$

$\mathbf{shape\_vector}_i(j+2n) \leftarrow \mathbf{v}(3).$

return $\{\mathbf{shape\_vector}_i\}$.

The correspondence found for the hippocampi is shown in Fig. 8.6. It can be seen from the figure that correspondence has been established between similar regions on each surface, and appears plausible. The first mode of variation of the model built from this correspondence is shown in Fig. 8.7. It can be seen that this mode represents changes in the thickness and curvature of the 'tail'. Comparison with the training set examples shown in Fig. 8.5, indicates that this mode reflects the shape variation found in the training set.

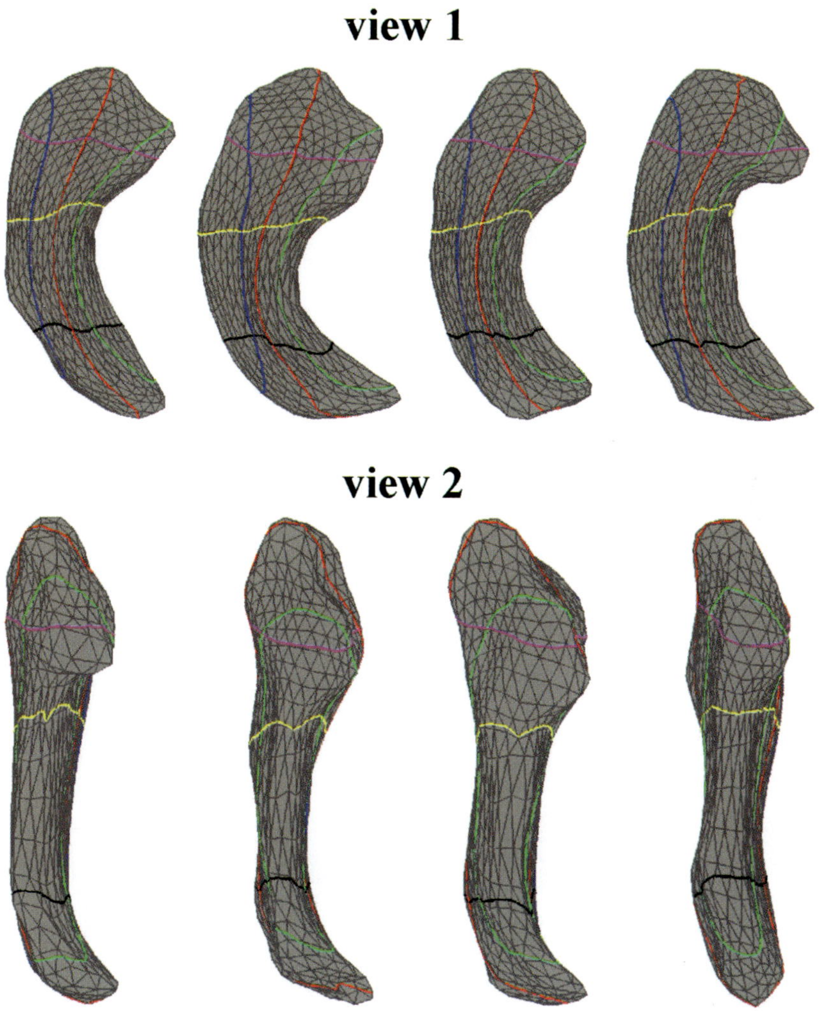

Fig. 8.6 The optimal correspondence found between the set of hippocampi. Four of the 82 training examples are shown here from two orthogonal viewpoints. Correspondence is denoted by the coloured lines. The figure shows that correspondence has been established between similar regions on each surface.

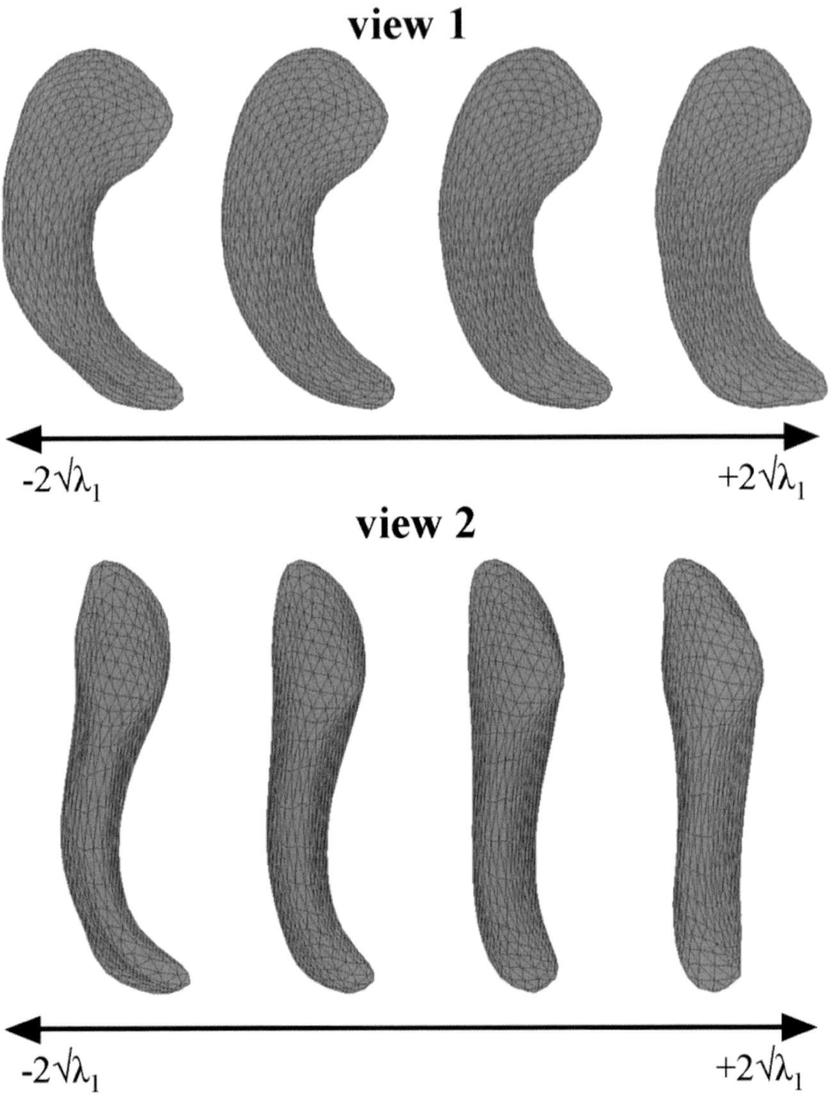

Fig. 8.7 The hippocampus model produced by Algorithm 8.6. The model is shown from two orthogonal viewpoints by varying the first mode of variation by ±[two standard deviations found over the training set]. The figure shows that the main mode of variation represents changes in the thickness and curvature of the 'tail'.

Chapter 9
Evaluation of Statistical Models

In this chapter, we consider the question of evaluating the statistical models of shape that we have built. However, it should be noted that this problem is not specific to the case of shape models, and the techniques we will present here can be applied to either statistical models of shapes, or statistical models of appearance (i.e., models of images).

If we consider the techniques of model-building that we presented in Chap. 2, the constructions presented there can be summarized as follows.

First, we construct a common *representation* of our training set of examples, so that each example can be mapped to a single point in our space of shapes or our space of images. Second, we then model this distribution of points by constructing a pdf on this space. There are hence two things which need to be evaluated, the quality of the *representation*, and the quality of the *model* based on that representation.

As regards the representation, the internal consistency is simple to evaluate, in that we can directly compare the actual examples in the training set with their representatives in shape space. But, there is also the issue of the *generalization ability* of the representation; can it also adequately represent example shapes or images which are not in the training set? The question of generalization ability also applies to the model, because even if the unseen example is adequately represented by some point in image or shape space, it is not necessarily the case that this point is included within the subspace spanned by the model. In this chapter, we first discuss the issue of evaluating the quality of the representation, then concentrate on the more challenging problem of evaluating the quality of the model.

Before we can evaluate the quality of the model, we have to decide what we mean by model quality. In general, this depends somewhat on what we intend to use the model for. For example, if we intend to use the model to analyse the training set itself, what is important is the quality of the model with respect to the training set; hence we have a self-referential definition of model quality. If we want to use the model to help us analyse *unseen* examples of shapes or images, then the issue is slightly more complicated, in that poor

R. Davies et al., *Statistical Models of Shape*,
DOI: 10.1007/978-1-84800-138-1_9, © Springer-Verlag London Limited 2008

performance in this task could imply either inadequacy of the training set itself, or an adequate training set, but an inadequate model built from this training set.

We consider first the use of ground truth data for evaluation, the advantages and some problems which can arise with this approach. We then show how it is possible to construct methods of evaluation which are independent of any ground truth.

9.1 Evaluation Using Ground Truth

We suppose that we have a training set, and also unseen examples which are not in the training set (what is usually referred to as a*test set*). In this case, we already have access to some ground truth data, in that we have the original shapes or images of the training set and test set.[1] This type of ground truth data is sufficient to establish the quality of the representation, as follows.

First, we build a model from our training set, and hence construct a shape space for our data (Chaps. 2 and (2.38)). For each training example, we then have our original shape vector $\mathbf{x} \in \mathbb{R}^{dn_P}$, and the corresponding parameter vector $\mathbf{b} \in \mathbb{R}^{n_m}$ which represents that shape, where:

$$\mathbf{x} \mapsto \mathbf{N}^T (\mathbf{x} - \bar{\mathbf{x}}) \doteq \mathbf{b}(\mathbf{x}), \;\; \mathbf{b} \mapsto \bar{\mathbf{x}} + \mathbf{N}\mathbf{b} \doteq \tilde{\mathbf{x}}(\mathbf{b}). \tag{9.1}$$

The mapping between shapes and shape space is determined by the matrix $\mathbf{N}$, which is the matrix of eigenvectors of our shape covariance matrix, where n_m is the number of eigenvectors retained, hence the number of modes of the representation. If we retain *all* the modes, then our reconstruction $\mathbf{x} \to \mathbf{b} \to \tilde{\mathbf{x}}$ will be exact by definition for all the examples in the training set. However in practice, we do not wish to retain all the modes, but only a sufficient number, since for a sufficiently large training set, the number of modes of variation is much less than the number of example shapes. It was shown previously (2.30) that the eigenvalues of the covariance matrix encode information about the variance in each of the modes, hence allowing a number of modes to be selected that retains a given proportion of the total variance across the training set. However, this does not necessarily mean that all of our training shapes are adequately represented. To check this, we must compare each original shape vector $\mathbf{x}$ with its reconstruction $\tilde{\mathbf{x}}$.

We can perform the same comparison process for test shapes. For this case, even if all the modes are retained, this does not guarantee exact recon-

[1] In cases where there is an insufficient number of examples to form both a test-set and a training set, a common approach is to use leave-one-out verification, where a model is built from all examples but one, then tested against the left-out example.

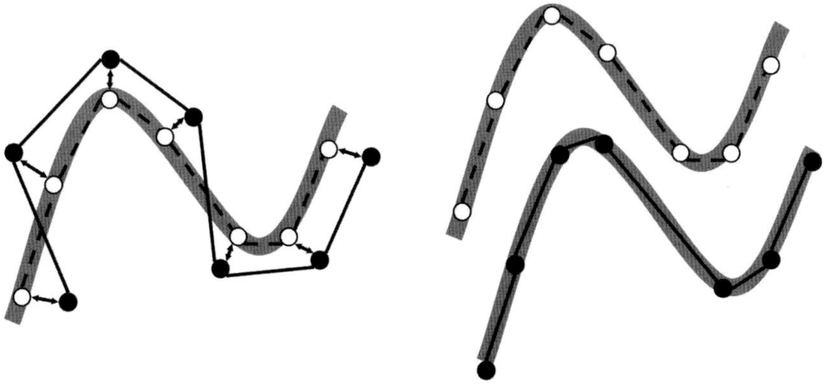

Fig. 9.1 Comparison of shapes and shape polygons. **Left:** An original shape with degree of tolerance indicated (thick grey line). The original polygonal representation is indicated by the white circles, whereas the reconstruction is given by the black circles. The point-to-point distances are indicated. **Right:** Original shape (**Top**) and reconstructed shape (**Bottom**), drawn separately for reasons of clarity. In this example, it can be seen that although the point-to-point distances are non-zero, the actual reconstructed shape lies within the same tolerance range as the original shape. In this case, a point-to-line distance would give a better measure of the validity of the shape reconstruction.

struction, since the test shape need not lie in the sub-space spanned by the training set.

When we compare two shapes $\mathbf{x}$ and $\tilde{\mathbf{x}}$, we are comparing the two polygons or the two triangulated meshes which these shape vectors describe, where the representation gives us a point-to-point correspondence between the vertices. The simplest quantitative measure for this comparison is the sum of the squares of the point-to-point distances (see Fig. 9.1), which is just the square of the Euclidean distance in shape space $\|\mathbf{x} - \tilde{\mathbf{x}}\|$. Note the we cannot use the statistical Mahalanobis distance (2.84) for this comparison, since by definition, this ignores separations in directions perpendicular to the model sub-space. It also scales separations in the model sub-space with respect to the variance of the training set in that direction. Hence it gives less weight to differences in directions which have higher variance. The Euclidean point-to-point distance treats equally separations perpendicular to the shape contour or surface, and separations tangential to the local shape. It can be argued that misplacing points by moving them along the tangent has a less deleterious effect on the validity of the represented shape than moving them perpendicular to the local shape (see Fig. 9.1). To allow for this, the perpendicular point-to-line or point-to-surface distance can be used instead for evaluating the quality of the representation. Note that using the point-to-point distance for *test shapes* also pre-supposes that we have the pointwise correspondence ground truth data for our test shapes, whereas the point-to-line distance does not require this correspondence data for unseen examples. When it comes to evaluating the model itself however, mis-correspondence does matter, so that we will use the Euclidean point-to-point distance in this case.

For appearance models, the quality of the representation can be evaluated by comparing the original shape and image patch, with the reconstructed shape and image patch, the comparison being performed in the frame of the original image.[2] We hence obtain two distances in this case, the Euclidean point-to-point or the point-to-line distance between the two shapes, and the Euclidean distance between the two image patches in image space. As for shape space, the Euclidean distance in image space is formed by considering the pixel-by-pixel differences between the two patches. For appearance models where the entirety of the image is modelled, rather than just a patch within a shape contour or surface, we will use only the Euclidean distance in image space.

Now let us look at the methods available for obtaining the required ground truth data. For the case of test shapes given above, we noted that use of the point-to-point distance assumed we had ground truth data about the pointwise correspondence for our test shapes. The automatic methods that exist for extracting shape information from images that also assign a correspondence to that shape, typically rely on using a shape or appearance model itself within the context of an active shape or appearance model (Sects. 2.4.1 and 2.4.2). Hence we cannot use such methods when it comes to evaluating the model itself.

Other automatic methods [169] can provide good segmentations and extraction of continuous shape outlines or shape surfaces, but without the ground truth correspondence information. Hence for ground truth data from such a source, we can use the point-to-line distance to verify the shape representation, but it does not provide the ground truth correspondence data that is required for detailed evaluation of the model.

Manual or semi-automatic annotation can provide reliable ground truth information in terms of shape outlines for shapes extracted from images, but only for certain classes of images and shapes. But there can still be significant variability between different annotators, just as there can be significant variation between different automatic methods. This can be incorporated into the evaluation of the representation, as is indicated in Fig. 9.1, where the shape contour is represented as a wide line, indicating that the ground truth data has a certain degree of uncertainty.

In summary, we see that the ground truth data required to assess the quality of the shape representation is obviously available, obtained either automatically or using human annotation, since it is the same type of data that is required to construct the model in the first place. Assessing the quality of the shape representation is equivalent to checking that our choice of the model sub-space within shape space is appropriate. We now move on to assessing the quality of the model that is built within this model sub-space.

[2] Note that we do not compare the *shape-free* image patches, since warping to the shape-free frame can hide errors in the representation.

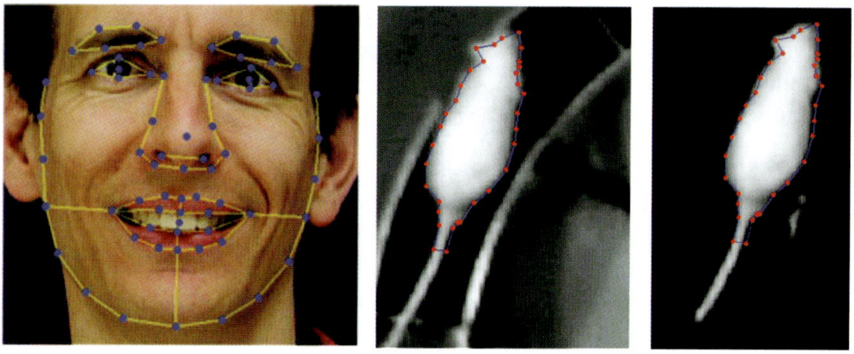

Fig. 9.2 Examples of manual annotation for shapes from images. **Left:** Annotation of a face with a set of points. In this case, the relevant corresponding features are easy to find, and the points can be placed quite accurately. **Centre & Right:** A much harder example, images of an experimental rat [187], viewed from above in a low-resolution video sequence. The method of annotation is semi-automatic, in that the fixed background is removed, the shape then being annotated manually. Given enough experience, features can be labeled fairly reliably, although the variability is greater than for the face example.

For certain classes of shapes or images, human annotation can also provide us with *limited* ground truth information about correspondence. Indeed, when SSMs were first introduced [38], hand-annotated data was all that was used, both to define the shape, and to define the correspondence. The importance of correct point placement and the existence of reproducible and consistent points of correspondence was stressed. Consider the example shapes from images shown in Fig. 9.2. For the face, there are landmark points that can be reliably located and annotated, such as the corners of the eyes, the nostrils, and the corners of the mouth. Other points can then be reliably placed partway between such points, such as the points around the eyebrows, or those along the outline of the face. But the ground truth correspondence obtainable by such manual annotation methods is limited. And even for the fairly simple annotation shown in the figure, the process is time-consuming, and can become prohibitive when hundreds of examples require annotation. For other types of two-dimensional image data, reproducible landmarks are much harder to find. And for three-dimensional data such as that provided in MR image volumes, the placement of landmark points becomes very difficult if not impossible. Other types of ground truth data are available for such three-dimensional data, such as the manual labelling of structures, the manual annotation of their surfaces, or the dense voxel-by-voxel labelling of tissue classes. Although this type of ground truth data can be useful for other types of evaluation, it is still not the dense point-to-point correspondence ground truth data that we would ideally like for model evaluation.

These problems with obtaining suitable ground truth data lead us to consider the question of whether evaluation of the model is possible in the absence of ground truth data.

9.2 Evaluation in the Absence of Ground Truth

We will suppose that all we have access to is our training set of examples, and the probabilistic model that we have built from this training set. It will be taken as given that the quality of the representation has been verified, according to the methods described above.

In the absence of ground truth data, it could be argued that we can generate artificial ground truth data (if that is not a misnomer!), by transforming the shape examples we do have. For example, for shapes which are curves in two-dimensional space, we can apply a homeomorphism to $\mathbb{R}^2$, which transforms the curve into some new curve, and carries the correspondence along with it. We hence obtain artificial unseen examples, with a correspondence, whose deformations are independent of the model and not constrained to lie within the model sub-space. If suitable homeomorphisms of the bulk space could be chosen, the deformations could be thought of as realistic deformations of the physical objects to which the shapes correspond. However, the problem is that we usually do not have any guarantee that the deformations and artificial examples that we generate are at all realistic, nor is it necessarily the case that the transformed correspondence is the correct one. Similar considerations hold for the case of artificially deformed images for appearance models. We hence reject this sort of artificial ground truth data as unsuitable for detailed model evaluation.

We are then left with just the training set data points in shape space $\{\mathbf{b}^{(i)} : i = 1, \ldots n_S\}$, and the corresponding model pdf $p(\mathbf{b})$. Note that for a fixed class of model, we fit the model to the data (i.e., determine the parameters of the model) by maximising the likelihood (see Theorem 2.2):

$$\sum_{i=1}^{n_S} \ln p(\mathbf{b}^{(i)}). \tag{9.2}$$

However we cannot use the likelihood itself to perform a comparison of model quality *between* classes of models. This is because the likelihood only depends on the values of the pdf at the data points $\{\mathbf{b}^{(i)}\}$. We can hence imagine many different pdfs that yield the same likelihood for a given data set. We can hence deduce that a successful measure of model quality will need to be sensitive to the entirety of the pdf $p(\mathbf{b})$, not just the values at the data points.

In order to construct such measures, let us consider a qualitative description of what distinguishes good statistical models from poorer models:

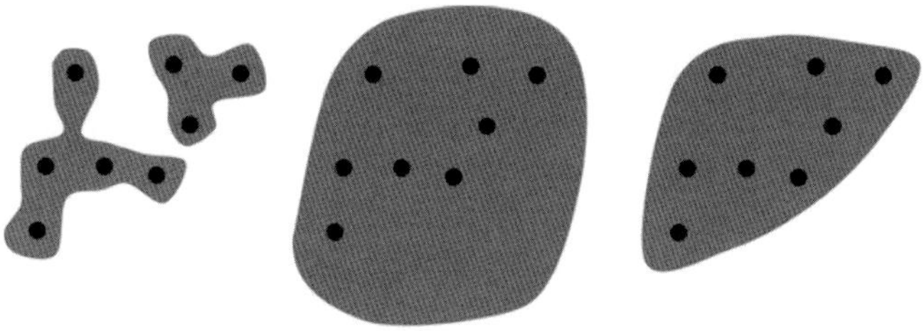

Fig. 9.3 Training set (black points) and a given probability isosurface for various model pdfs (grey fill). **Left:** A specific model, but not a general model. **Centre:** A general model, but not a specific one. **Right:** A model that achives a balance between generalization ability and specificity.

- **Compactness:** For fixed training data, a compact model pdf should describe the distribution of the data using the smallest possible number of parameters. This principle has already been applied when we used PCA (see Theorem 2.1), where we discarded principal directions of low variance. This means the dimensionality of the model sub-space has been reduced, hence the number of parameters required when, say, a multivariate Gaussian model (Sect. 2.2.1) is constructed within this sub-space.
- **Specificity:** This is the requirement that the model can only represent valid instances of the class(es) of objects presented in the training set, hence that the model be *specific* with respect to this training set.
- **Generalization Ability:** The model should be able to *generalize* from the examples given in the training set, hence describe any valid example of the class of object, not just those instances seen in the training set.

In Fig. 9.3, we give a simple diagrammatic illustration of models with varying generalization ability and specificity, for a fixed set of data points. As noted above, compactness is usually handled within the context of dimensional reduction. We hence concentrate on developing *quantitative* measures for the concepts of *specificity* and *generalization* ability.

9.2.1 Specificity and Generalization: Quantitative Measures

If we refer to Fig. 9.3, we see that a *specific* pdf is one that is concentrated around the data points, whereas one with good *generalization* ability is spread out between and around the data points.

In [52], a leave-one-out measure of generalization ability was used. Suppose $T = \{\mathbf{x}_i : i = 1, \dots n_S\}$ is our complete training set. If we build a model from all examples but the jth, we obtain a model pdf $p^{(j)}(\mathbf{b})$. We can then reconstruct the jth example $\mathbf{x}_j$ using this model, to obtain an approximate reconstruction $\tilde{\mathbf{x}}_j$. If we perform this leave-one-out process for each example $\mathbf{x}_j$ in turn, and for varying numbers n_m of modes in the model, we obtain the leave-one-out generalization measure:

$$\hat{G}(n_m) = \frac{1}{n_S} \sum_{j=1}^{n_S} \|\mathbf{x}_j - \tilde{\mathbf{x}}_j\|, \tag{9.3}$$

which is just the mean Euclidean distance between the original training example $\mathbf{x}_j$, and that example reconstructed using the leave-j-out model with n_m modes. Obviously the better the generalization ability, the smaller the difference between the original left-out shape and its reconstructions. However, this leave-one-out generalization measure still only probes the model pdfs at the data points.

Let us turn now to specificity. A model pdf is specific if it is concentrated around the data points. One way to measure this is by sampling from the model pdf $p(\mathbf{b})$ as follows.

Let $Y = \{\mathbf{y}_A : A = 1, \ldots M\}$ be a set of shapes sampled from the model pdf $p(\mathbf{b}; n_m)$, where n_m is the number of modes included in the model. The sample set Y hence has the same distribution as the model in the limit $M \to \infty$. The way these points do or do not cluster in the vicinity of the data is then quantified by the specificity measure:

$$\boxed{\hat{S}(n_m) \doteq \frac{1}{M} \sum_{A=1}^{M} \min_i \|\mathbf{y}_A - \mathbf{x}_i\|.} \tag{9.4}$$

This is just the distance from each sample point $\mathbf{y}_A$ to the nearest element of the training set. Hence the more specific the model, the smaller the value of the measure. Small values of S do not mean, however, that the pdf covers *all* the training set.

We can build an analogous measure for generalization:

$$\boxed{\hat{G}(n_m) \doteq \frac{1}{n_S} \sum_{i=1}^{n_S} \min_A \|\mathbf{y}_A - \mathbf{x}_i\|.} \tag{9.5}$$

This just swaps the rôles of the sample and training set, and only has small values when *all* of the training set are covered by the model pdf. We hence see that this is a measure of generalization ability, and unlike the leave-one-out generalization measure mentioned above, this measure is sensitive to the entirety of the model pdf.

It could be argued that these measures are somewhat deficient, in that a global minimum of both the specificity and generalization can be obtained for the case where the model pdf $p(\mathbf{b})$ is just the empirical distribution of the data, where:

$$p(\mathbf{b}) = \frac{1}{n_S} \sum_{i=1}^{n_S} \delta(\mathbf{b} - \mathbf{b}^{(i)}). \tag{9.6}$$

This pdf can be viewed as the zero-width limit of a Gaussian mixture model. For a multivariate Gaussian, the number of model parameters is just the number of PCA modes n_m. Hence the analogous quantity for the empirical pdf above is the number of Gaussians, which ranges from 1 to n_S. If we now consider the generalization for this empirical pdf as a function of the number of Gaussian components, we see that the minimum is only achieved for the maximum number of Gaussians, and for less than this the measure has a higher value, related to the nearest-neighbour distance for the training set. The specificity for the empirical pdf also only achieves the minimum value if all of the training points are considered in the evaluation.

We can now see the significance of the generalization and specificity explicitly depending on the number of modes/parameters of the model. We can deduce that a minimum value of the measures is not necessarily significant, but should only be considered significant if this behaviour persists as the number of modes/parameters is varied.

These measures of specificity (9.4) and generalization (9.5) have been used extensively in the literature to enable comparison between shape models built using different methods of establishing correspondence. For example, Davies et al. [52] compared shape models of the hand built using manual annotation, and models built using the MDL approach, but with two different optimisation schemes. They also compared shape models of hippocampi built using the SPHARM method [76] with those built using MDL. The important point about these comparisons is that they clearly show that specificity and generalization do show statistically significant differences between these difference approaches, and that these differences persist as the number of modes n_m is varied. Specificity and generalization have been employed to compare different methods of groupwise non-rigid registration for images, via assessment of the appearance models built from the registered images [153]. In particular, these measures were validated for the case of image registration by considering their behaviour as the registration was perturbed about the point of registration. It was shown that the specificity and generalization showed a monotonic relationship with the degree of mis-registration, and were correlated with a ground truth based measure of mis-registration, based on an evaluation of the overlap of a set of dense, voxel-by-voxel tissue labels [46].

In summary, the use of these two measures has been validated for the assessment of both shape and appearance models, and they have been shown to be both sensitive and discriminative in practice. However, the derivation given above is rather ad hoc, and it is not clear, for example, how these measures behave in the limit of a large training set n_S, or the limit of a large sample set M. Placing the meaning of these measures on a firm theoretical foundation is the subject of the next section.

9.3 Specificity and Generalization as Graph-Based Estimators

In this section, we consider the behaviour of the generalized specificity and generalization measures in the limit of both large sample sets, and large training sets.[3] This enables the expected scaling behaviour of these measures to be made explicit. It also enables us to derive an integral form for these measures in this limit, which establishes a link between these graph-based measures and other integral-based measures such as the cross-entropy and Kullback-Leibler divergence [103].

We start with a training set of examples $T^{(n_S)} = \{\mathbf{x}_i : i = 1, \dots n_S\}$. For shapes represented by n_P points in $\mathbb{R}^d$, the original shape vector $\mathbf{x}_i$ lies in $\mathbb{R}^{dn_P}$. Dimensional reduction via PCA maps this shape vector to a shape parameter vector $\mathbf{b}^{(i)} \in \mathbb{R}^{n_m}$ (2.34), where n_m is the number of PCA directions retained. For the purposes of this derivation, we will gloss over this dimensional reduction step, and suppose that our training set consists of points $\mathbf{x}_i$ in $\mathbb{R}^n$. A model built from this training set is then given by a pdf $p(\mathbf{z})$, $\mathbf{z} \in \mathbb{R}^n$. A sample set of M points generated from this model pdf is then given by $Y^{(M)} = \{\mathbf{y}_A \in \mathbb{R}^n : A = 1, \dots M\}$.

The ad hoc definitions of specificity (9.4) and generalization (9.5) given above can now be written in the form:

$$\hat{S}_1(T^{(n_S)}; Y^{(M)}) \doteq \frac{1}{M} \sum_{A=1}^{M} \min_i \|\mathbf{y}_A - \mathbf{x}_i\|,$$

$$\hat{G}_1(T^{(n_S)}; Y^{(M)}) \doteq \frac{1}{n_S} \sum_{i=1}^{n_S} \min_A \|\mathbf{y}_A - \mathbf{x}_i\|,$$

where $\|\cdot\|$ is the usual Euclidean vector norm in $\mathbb{R}^n$. This form makes explicit the dependance of these measures on the training set $T^{(n_S)}$ and the sample set $Y^{(M)}$, and in particular, the dependance on the *sizes* of the training set n_S and sample set M. It can now be seen that these measures are based on graphs which connect one point set (the training set or the sample set) to the other point set (the sample set or training set respectively).

An obvious generalization is to take the sum of *powers* of the distances, so that we obtain the γ-specificity and γ-generalization:

---

[3] The derivations given in this section are based on those presented by the current authors in [186].

$$\hat{S}_{\gamma}(T^{(n_S)};Y^{(M)}) \doteq \frac{1}{M}\sum_{A=1}^{M}\min_{i}(\|\mathbf{y}_A-\mathbf{x}_i\|)^{\gamma}, \tag{9.7}$$

$$\hat{G}_{\gamma}(T^{(n_S)};Y^{(M)}) \doteq \frac{1}{n_S}\sum_{i=1}^{n_S}\min_{A}(\|\mathbf{y}_A-\mathbf{x}_i\|)^{\gamma}. \tag{9.8}$$

Because of the symmetrical form of our definitions, and the symmetry of the two graph constructions, we have:

$$\hat{G}_{\gamma}(T^{(n_S)};Y^{(M)}) \equiv \hat{S}_{\gamma}(Y^{(M)};T^{(n_S)}).$$

We wish to consider the behaviour of these measures in the limit of large training sets $n_S \to \infty$ and large sample sets $M \to \infty$. It is hence sufficient to just consider the γ-specificity (9.7).

We first take the limit $M \to \infty$. It can be seen that this is just a Monte Carlo [123] estimator of the integral γ-specificity:

$$\lim_{M\to\infty}\hat{S}_{\gamma}(T^{(n_S)};Y^{(M)}) \doteq S_{\gamma}(T^{(n_S)};p) = \int_{\mathbb{R}^n} p(\mathbf{z})\min_{i}\left(\|\mathbf{z}-\mathbf{x}_i\|\right)^{\gamma} d\mathbf{z}. \tag{9.9}$$

We now wish to consider how $S_{\gamma}(T^{(n_S)};p)$ behaves in the limit of a large training set $n_S \to \infty$. We suppose that in this limit, the training set $T^{(n_S)}$ has a well-defined distribution $g(\mathbf{z})$, which is the distribution of the hypothetical process which generated our training data. We then have the following Theorem:

Theorem 9.1. Large-Numbers Limit of Specificity.
With the definitions given above:

$$\boxed{\lim_{n_S\to\infty}\left[(n_S)^{\frac{\gamma}{n}} S_{\gamma}(T^{(n_S)};p)\right] = \beta_{n,\gamma}\int_{\mathbb{R}^n} p(\mathbf{z})g^{-\frac{\gamma}{n}}(\mathbf{z})d\mathbf{z},} \tag{9.10}$$

where $\beta_{n,\gamma}$ are numerical coefficients which depend only on n and γ.

Proof. Consider first the integral (9.9):

$$S_{\gamma}(T^{(n_S)};p) \doteq \int_{\mathbb{R}^n} p(\mathbf{z})\min_{i}\left(\|\mathbf{z}-\mathbf{x}_i\|\right)^{\gamma} d\mathbf{z}.$$

If we consider a particular point $\mathbf{x}_i$ of the training set, we see that the contribution to the integral from this point is the integral over all points $\mathbf{z}$ which are closer to this point than to any other point of the training set. This set of points is the Voronoi Cell [190] Ω_i belonging to $\mathbf{x}_i$, where:

$$\Omega_i(T^{(n_S)}) \doteq \{\mathbf{z} \in \mathbb{R}^n : \|\mathbf{z} - \mathbf{x}_i\| \leq \|\mathbf{z} - \mathbf{x}_j\| \ \forall\ j = 1, \ldots n_S\}. \tag{9.11}$$

The integral can hence be decomposed in the form:

$$S_\gamma(T^{(n_S)}; p) = \sum_{i=1}^{n_S} \int_{\Omega_i(T^{(n_S)})} p(\mathbf{z}) \left(\|\mathbf{z} - \mathbf{x}_i\|\right)^\gamma d\mathbf{z}. \tag{9.12}$$

If we consider now the limit $n_S \to \infty$, then in some infinitesimal volume $\Delta\mathbf{z}$ about the point $\mathbf{x}_i$, other points of the training set will be distributed with a density $n_S g(\mathbf{x}_i)$. And provided that $\Delta\mathbf{z}$ is small enough, and n_S large enough, the process generating $n_S g(\mathbf{z})\Delta\mathbf{z}$ points within the volume $\Delta\mathbf{z}$ becomes indistinguishable from points generated by an infinite uniform *Poisson process.*[4] *We will denote the relevant uniform Poisson process by:*

$$\mathcal{P}_i = \mathcal{P}(\mathbf{x}_i, n_S g(\mathbf{x}_i)),$$

which is a process of intensity $n_S g(\mathbf{x}_i)$ about the point $\mathbf{x}_i$. This means that we can replace the Voronoi cell $\Omega_i(T^{(n_S)})$ by the Voronoi cell $\Omega_i(\mathcal{P}_i)$, which is the cell about $\mathbf{x}_i$ for the uniform Poisson process $\mathcal{P}_i$. If n_S is large enough, we can also neglect the variation of $p(\mathbf{z})$ across this cell, and finally we obtain:

$$\sum_{i=1}^{n_S} \int_{\Omega_i(T^{(n_S)})} p(\mathbf{z}) \left(\|\mathbf{z} - \mathbf{x}_i\|\right)^\gamma d\mathbf{z} \underset{n_S \to \infty}{\longrightarrow} \sum_{i=1}^{n_S} p(\mathbf{x}_i) \int_{\Omega_i(\mathcal{P}_i)} \left(\|\mathbf{z} - \mathbf{x}_i\|\right)^\gamma d\mathbf{z}.$$

We then make the approximation of replacing this integral across the Voronoi cell by its expectation value over the Poisson process:

$$\mathbb{E}\left[\int_{\Omega_i(\mathcal{P}_i)} \left(\|\mathbf{z} - \mathbf{x}_i\|\right)^\gamma d\mathbf{z}\right] = (l_i)^{\gamma+n}\, \mathbb{E}\left[\int_{\Omega_0(\mathcal{P}_1)} \left(\|\mathbf{z}\|\right)^\gamma d\mathbf{z}\right], \tag{9.13}$$

where we have used simple scaling to replace the Poisson process $\mathcal{P}_i$ of intensity $n_S g(\mathbf{x}_i)$ by a Poisson process of unit *intensity $\mathcal{P}_1$, and translational invariance of uniform Poisson processes to measure distances from the origin rather than from $\mathbf{x}_i$. The scaling factor is given in terms of $l_i \doteq (n_S g(\mathbf{x}_i))^{-\frac{1}{n}}$, where $(l_i)^n$ is the average volume of the Voronoi cell $\Omega_i(T^{(n_S)})$ about $\mathbf{x}_i$. We then define:*

$$\beta_{n,\gamma} \doteq \mathbb{E}\left[\int_{\Omega_0(\mathcal{P}_1)} \left(\|\mathbf{z}\|\right)^\gamma d\mathbf{z}\right], \tag{9.14}$$

which are the required set of numerical coefficients, depending only on the dimensionality of the space n and the power γ to which distances are raised.

[4] This is just the *objective method* approach of Steele [171], where the limiting behaviour of local functionals on finite point sets is described in terms of related functionals defined on infinite Poisson processes.

Putting all this together then gives the final result:

$$\lim_{n_S \to \infty} S_\gamma(T^{(n_S)}; p) = \beta_{n,\gamma} \sum_{i=1}^{n_S} p(\mathbf{x}_i) \, (l_i)^{\gamma+n} \, ,$$

$$= \frac{\beta_{n,\gamma}}{(n_S)^{\frac{\gamma}{n}}} \sum_{i=1}^{n_S} p(\mathbf{x}_i) g^{-\frac{\gamma}{n}}(\mathbf{x}_i) \, (l_i)^n \underset{n_S \to \infty}{\longrightarrow} \frac{\beta_{n,\gamma}}{(n_S)^{\frac{\gamma}{n}}} \int_{\mathbb{R}^n} p(\mathbf{z}) g^{-\frac{\gamma}{n}}(\mathbf{z}) d\mathbf{z}.$$

$$\textit{That is:} \lim_{n_s \to \infty} \left[(n_S)^{\frac{\gamma}{n}} S_\gamma(T^{(n_S)}; p) \right] = \beta_{n,\gamma} \int_{\mathbb{R}^n} p(\mathbf{z}) g^{-\frac{\gamma}{n}}(\mathbf{z}) d\mathbf{z}. \tag{9.15}$$

□

To summarize, we have shown that our ad hoc definition of the graph-based specificity (9.7) between a sample set generated by the model pdf $p(\mathbf{z})$ and the training set can be related to an integral definition of the specificity (9.9) of the training set with respect to the pdf $p(\mathbf{z})$. By then considering the limit of an infinitely large training set generated by some process with density $g(\mathbf{z})$, we have shown that the integral specificity can be related to an integral involving these two pdfs, the generating pdf $g(\mathbf{z})$ and the model pdf $p(\mathbf{z})$.

In order to relate this result to other quantities, we define the following divergences[5]:

$$D_\gamma(p, g) \doteq \frac{n}{\gamma} \log \int_{\mathbb{R}^n} p(\mathbf{z}) g^{-\frac{\gamma}{n}}(\mathbf{z}) d\mathbf{z}, \tag{9.16}$$

$$\widetilde{D}_\gamma(p, g) \doteq D_\gamma(p, g) - D_\gamma(p, p). \tag{9.17}$$

$$\lim_{\gamma \to 0} D_\gamma(p, g) = - \int_{\mathbb{R}^n} p(\mathbf{z}) \log g(\mathbf{z}) d\mathbf{z} = H(p, g), \tag{9.18}$$

$$\lim_{\gamma \to 0} \widetilde{D}_\gamma(p, g) = D_{KL}(p, g), \tag{9.19}$$

where $H(p, g)$ is the standard cross-entropy of two distributions, and $D_{KL}(p, g)$ the Kullback-Leibler (KL) divergence:

$$D_{KL}(p, g) = H(p, g) - H(p, p) = \int_{\mathbb{R}^n} p(\mathbf{z}) \log \frac{p(\mathbf{z})}{g(\mathbf{z})} d\mathbf{z}. \tag{9.20}$$

We hence see that our specificity can be used to obtain an estimator for cross-entropy:

[5] The term *divergence* relates to the separation of two distributions. It cannot be called a metric or a distance since in general divergences are not symmetric, that is, $D(p, g) \neq D(g, p)$.

$$H(p,g) = \lim_{\gamma\to 0} D_\gamma(p,g) \approx \lim_{\gamma\to 0}\left(\frac{n}{\gamma}\left[\log\left((n_S)^{\frac{\gamma}{n}} S_\gamma(T^{(n_S)};p)\right) - \log \beta_{n,\gamma}\right]\right). \tag{9.21}$$

Given the symmetry of the definitions of specificity and generalization, we see that corresponding limit of the generalization measure is just the other cross-entropy $H(g,p)$.

The divergence $D_\gamma(p,g)$ (9.16) is of a slightly different form to the Rényi α-divergence [141]:

$$D_\gamma(p,g) = \frac{n}{\gamma}\log\int\limits_{\mathbb{R}^n} p(\mathbf{z})g^{-\frac{\gamma}{n}}(\mathbf{z})d\mathbf{z}, \quad R_\alpha(p,g) = \frac{1}{\alpha-1}\log\int\limits_{\mathbb{R}^n} p^\alpha(\mathbf{z})g^{1-\alpha}(\mathbf{z})dz. \tag{9.22}$$

Hero et al. [91, 90] estimated $R_\alpha(p,g)$ using graph-based methods. However, they were interested in the slightly different question of the divergence of two point sets, rather than the divergence of a point set and a pdf. Both divergences tend towards the same expression in the limits $\gamma \to 0$ or $\alpha \to 1$.

Given the cross-entropy $H(p,g)$, we can then estimate the KL divergence if we also know $H(p,p)$ (which is just the Shannon entropy [160] – see Sect. 4.3.1). For certain classes of models, such as multivariate Gaussians, the KL divergence can be calculated in closed form.

An alternative is to use a second sample set generated from the model, and calculate the specificity between the two sample sets to estimate $H(p,p)$. In this case, we can see what our estimate of the divergence is doing – if we cannot tell the difference in terms of appropriately normalized graph lengths between two sample sets, or a sample set and a training set, then the model pdf that generated those sample sets is indeed our best estimate of the training set pdf $g(\mathbf{z})$.

The Kullback-Leibler divergence has the property that $D_{KL}(p,g) \geq 0$, with equality only when $p(\mathbf{z}) = g(\mathbf{z}) \;\forall\; \mathbf{z} \in \mathbb{R}^n$ [103]. However, it should be remembered that specificity and generalization are being used to assess the quality of models $p(\mathbf{z})$ which have been fitted to the data, not to fit the models themselves. If we did try to use the KL divergence to fit the models to the data over all possible models, we would encounter the same problem with the empirical distribution (9.6) that we discussed previously. In practice, we avoid this problem by considering only a limited class of models (multivariate Gaussians, say), and fit these models to the data in the usual way. When we vary the correspondence across the training set, this shifts the positions of the training points slightly, which we then fit to a slightly amended version of the same class of models. If we neglect the variation in $H(p,p)$ for these slightly different models, the specificity as an estimator for $H(p,g)$ (hence $D_{KL}(p,g)$), under the fitted-model restrictions, does provide a good measure of the quality of the degree of fit of the model and the training set. In effect, our assumption that the true training set distribution $g(\mathbf{z})$ is smooth and continuous is included by limiting ourselves to classes of models which display the same features. The empirical distribution can also be excluded on the grounds that we want models which vary continuously with the data, and in particular, that vary continuously if single points are either included in or excluded from the training set.

The derivation in Theorem 9.1 makes the assumption that the Voronoi cells of the training set are closed, and that their size scales appropriately with the size of the training set. This will not be true for points on the periphery of the training set, which have open Voronoi cells which will not have the size and shape as predicted in the proof of the Theorem. However, the contribution to the specificity from these points will be negligible provided that $p(\mathbf{z}) \to 0$ in this region, or equivalently, if the convex hull of the sample set $Y^{(M)}$ is not substantially larger than the convex hull of the training set $T^{(n_S)}$. The converse case holds for the limit of the generalization measure. Hence we see that we expect both graph-based estimators to be valid in the case where the model $p(\mathbf{z})$ is fitted to the data. This provides another reason why we cannot use the specificity and generalization to fit the model to the data, whilst they still will provide a good measure of the quality of fit near the point of fit itself.

However, we can only use these estimators if we know the values of $\beta_{n,\gamma}$ (9.14).

9.3.1 Evaluating the Coefficients $\beta_{n,\gamma}$

We start by recalling the definition of the coefficients (9.14):

$$\boxed{\beta_{n,\gamma} \doteq \mathbb{E}\left[\int_{\Omega_0(\mathcal{P}_1)} (\|\mathbf{z}\|)^{\gamma}\, d\mathbf{z}\right].} \tag{9.23}$$

There are two ways to look at the meaning of this expression. In the first case, we can view this integral as the limit of a sampling process, where the point $\mathbf{z}$ is generated randomly and uniformly. For each such $\mathbf{z}$, the distance is then measured to the nearest other point of the pre-existing uniform Poisson process $\mathcal{P}_1$. But this means that generating $\mathbf{z}$ just means adding a new point to $\mathcal{P}_1$. The distance $\|\mathbf{z}\|$ is then just the nearest-neighbour distance (*nearest-neighbour* since it is within the original Voronoi cell, hence all other points are further away by definition) for this amended Poisson process. Hence we can see $\beta_{n,\gamma}$ as the expectation value of the γ power of the nearest-neighbour distance for a uniform Poisson process.

The expectation value of the kth-nearest-neighbour distance has been calculated by Wade [191], and can be used to deduce the values for our

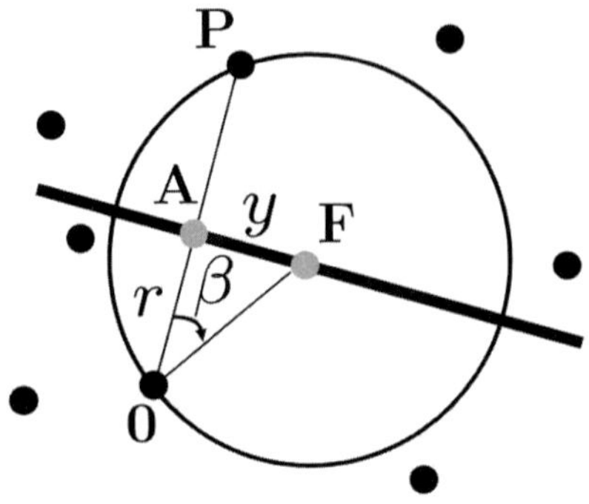

Fig. 9.4 Black dots: Seed points, central seed **0**, and neighbouring seed **P**. **Grey dots:** The mid-point **A**, possible face point **F**. **Thick black line:** Possible face points on $(n-1)$-dimensional hyperplane. **Medium black line:** Hypersphere through **0** and **P**, centre **F**.

coefficients $\beta_{n,\gamma}$. However, it is also possible to calculate the above integral explicitly, using the methods of Brakke [13], which yields our version of the proof of the following Theorem:

Theorem 9.2. Evaluating the Coefficients $\beta_{n,\gamma}$.
For a uniform Poisson process $\mathcal{P}_1$ of unit intensity in $\mathbb{R}^n$, the value for the following integral is:

$$\boxed{\beta_{n,\gamma} = \mathbb{E}\left[\int_{\Omega_{\mathbf{0}}(\mathcal{P}_1)} (\|\mathbf{z}\|)^{\gamma}\, d\mathbf{z}\right] = \frac{n}{(n+\gamma)} \cdot \Gamma\left(2 + \frac{\gamma}{n}\right) \cdot (v_n)^{-\frac{\gamma}{n}},} \tag{9.24}$$

where $\Gamma(\cdot)$ is the usual Gamma function, and v_n is the volume of the unit ball in $\mathbb{R}^n$, $v_n = \dfrac{2\pi^{\frac{n}{2}}}{n\Gamma\left(\frac{n}{2}\right)}$.

Proof. *The naïve way to calculate this expectation value would be to define the positions of a set of points of the Poisson process (seeds) about a seed at the origin* **0**. *For this particular set of seeds, we would compute the Voronoi cell about* **0**, *then attempt to perform the integration over this cell. At the final stage, we would then perform the statistical averaging over the seed positions. However, a simpler approach is to follow the methods of Brakke [13], which effectively combines these stages.*

Consider the situation shown in Fig. 9.4. We have the seed **0** *that we take as our origin. We have a neighbouring seed* **P** *whose position is held fixed for the moment. A point* **F** *lying on the plane midway between the seeds* **0** *and* **P** *is a point on a face of the Voronoi cell of* **0** *provided that no seed points lie in the hypersphere (medium black line) centred on* **F** *that passes through* *the pair of seeds. Let* $\mathbf{y}$ *denote the position of* $\mathbf{F}$ *relative to* $\mathbf{A}$*,* $y = |\mathbf{y}|$ *and* r *is the distance* $\mathbf{0A}$*.* β *is then defined as the angle between the lines* $\mathbf{0A}$ *and* $\mathbf{0F}$*.*

The required void is then a hypersphere centred on $\mathbf{F}$ *of radius* $\mathbf{OF}$*, where* $\mathbf{0F} = \sqrt{r^2 + y^2}$*. It hence has a volume of* $v_n(r^2+y^2)^{\frac{n}{2}}$*, where* v_n *is the volume enclosed by the unit hypersphere in* $\mathbb{R}^n$*. The probability that a void of this size exists within the Poisson process is the void probability:*

$$\exp(-v_n(r^2 + y^2)^{\frac{n}{2}}).$$

Let $d\mathbf{y}$ *be the infinitesimal element of hyperarea (*$(n-1)$*-dimensional volume lying wholly in the hyperplane) about* $\mathbf{F}$*. When joined to* $\mathbf{0}$*, this defines a hypercone, which forms part of the volume of the Voronoi cell. Since this cell is a convex polytope, integrating over all points* $\mathbf{F}$ *sweeps out the whole volume of the cell once and only once. We can hence calculate the contribution of this cone to the integral as follows:*

$$\int_0^{(r^2+y^2)^{\frac{1}{2}}} \cos\beta \cdot R^\gamma \cdot \left(\frac{R}{(y^2+r^2)^{\frac{1}{2}}}\right)^{n-1} dR d\mathbf{y} = \frac{r}{n+\gamma}\left(r^2+y^2\right)^{\frac{\gamma}{2}} d\mathbf{y}. \quad (9.25)$$

Here the $\cos\beta = r/(r^2 + y^2)^{\frac{1}{2}}$ *term projects* $d\mathbf{y}$ *perpendicular to* $\mathbf{0F}$*, and the integration variable* R *takes us along the cone from the tip at* $\mathbf{0}$ *to the base centred at* $\mathbf{F}$*.*

We then integrate over the angular part of $d\mathbf{y}$*, which takes* $\mathbf{F}$ *over the surface of the* $(n-1)$*-dimensional hypersphere centred at* $\mathbf{A}$ *of radius* y *to give:*

$$\frac{r}{n+\gamma} \cdot s_{n-1}\, y^{n-2} \cdot \left(r^2+y^2\right)^{\frac{\gamma}{2}} dy, \quad \text{where:} \quad s_p \doteq \frac{2\pi^{\frac{p}{2}}}{\Gamma(\frac{p}{2})}, \quad (9.26)$$

s_p *being the surface area of the unit hypersphere in* p *dimensions, and where* dy *is now just the scalar measure. We now include the void probability and integrate over the position of the neighbouring seed* $\mathbf{P}$*:*

$$\int \frac{s_{n-1}}{n+\gamma} \cdot r y^{n-2} \cdot \left(r^2+y^2\right)^{\frac{\gamma}{2}} \exp(-v_n(r^2+y^2)^{\frac{n}{2}})\, dy\, d\mathbf{P}. \quad (9.27)$$

Note that the measure $d\mathbf{P}$ *associated with the position of seed* $\mathbf{P}$ *is not a scalar. However, we can integrate over the angular part, taking* $\mathbf{P}$ *over the surface of a* n*-dimensional hypersphere of radius* $2r$ *centred on* $\mathbf{0}$*:*

$$\int_{\text{angular part}} d\mathbf{P} = s_n \cdot 2^n r^{n-1} dr. \quad (9.28)$$

The remaining variables to be integrated over are just the distances y and r, which give the complete expression:

$$\beta_{n,\gamma} = \frac{s_n s_{n-1} 2^n}{n+\gamma} \int_0^\infty dr \int_0^\infty dy\, r^n y^{n-2} \cdot (r^2+y^2)^{\frac{\gamma}{2}} \exp(-v_n (r^2+y^2)^{\frac{n}{2}}). \quad (9.29)$$

We use the standard change of variables:

$$r = x\cos\theta, \;\; y = x\sin\theta,$$

which gives:

$$\beta_{n,\gamma} = \frac{s_n s_{n-1} 2^n}{n+\gamma} \left(\int_0^{\frac{\pi}{2}} d\theta \cos^n\theta \sin^{n-2}\theta \right) \left(\int_0^\infty x^{\gamma+2n-1} \exp(-v_n x^n) dx \right). \quad (9.30)$$

The integral over θ ([79] 3.621 & 8.384) can be expressed in terms of the beta function, hence in terms of the Gamma function, so that:

$$\int_0^{\frac{\pi}{2}} d\theta \cos^n\theta \sin^{n-2}\theta = \frac{1}{2} \frac{\Gamma\left(\frac{n-1}{2}\right)\Gamma\left(\frac{n+1}{2}\right)}{\Gamma(n)}.$$

Using another change of variable, the integral over x can also be computed ([79] 3.381). Putting these results together, and with the aid of a little manipulation and use of the Legendre duplication formula,[6] yields the final result:

$$\beta_{n,\gamma} = \frac{n}{(n+\gamma)} \cdot \Gamma\left(2+\frac{\gamma}{n}\right) \cdot (v_n)^{-\frac{\gamma}{n}}. \quad (9.31)$$

As a final check on the normalization, we note that $\beta_{n,0}$ is just the expectation value of the volume of the Voronoi cell, which is correctly given as 1 by the above expression. □

Now we have an expression for the coefficients $\beta_{n,\gamma}$ in closed form, we have all that is required to use the specificity and generalization as graph-based estimators.

Consider the γ-specificity as an estimator of the divergence defined in (9.16):

---

[6] $\Gamma(z)\Gamma(z+\frac{1}{2}) = 2^{1-2z}\sqrt{\pi}\,\Gamma(2z)$.

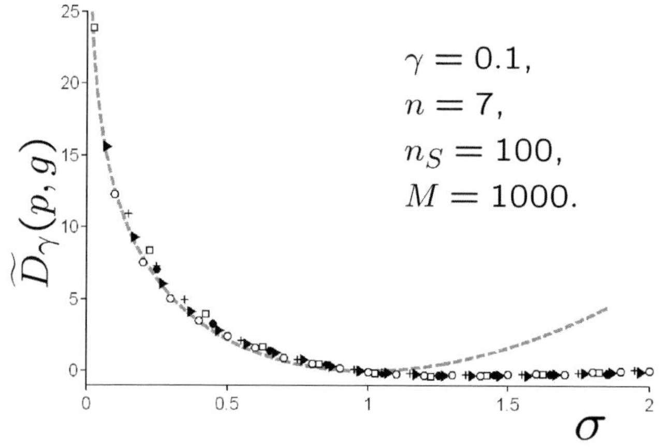

Fig. 9.5 $\widetilde{D}_\gamma(p, g)$ for $p(\mathbf{z})$ and $g(\mathbf{z})$ spherically symmetric Gaussians in $\mathbb{R}^7$. The training sets of size $n_S = 100$ are generated from a unit width Gaussian, whereas the model pdf $p(\mathbf{z})$ has a variable width σ. The divergence is estimated by generating sample-sets of size $M = 1000$. Different symbols indicate different instantiations of the training set. The dotted line shows the exact theoretical result. It can be seen that the estimate fits the prediction for values $\sigma \leq 1$, as was also predicted.

$$D_\gamma(p, g) \approx \left(\frac{n}{\gamma} \left[\log \left((n_S)^{\frac{\gamma}{n}} \hat{S}_\gamma (T^{(n_S)}; Y^{(M)}) \right) - \log \beta_{n,\gamma} \right] \right), \tag{9.32}$$

$$\widetilde{D}_\gamma(p, g) \doteq D_\gamma(p, g) - D_\gamma(p, p). \tag{9.33}$$

We will take $p(\mathbf{z})$ and $g(\mathbf{z})$ to be spherically symmetric Gaussian distributions in $\mathbb{R}^n$. The divergences $D_\gamma(p, g)$, $\widetilde{D}_\gamma(p, g)$ can then be calculated in closed form. If we take the model distribution to be of width σ, and the data generating distribution $g(\mathbf{z})$ to be of unit width, then:

$$\widetilde{D}_\gamma(p, g) = \frac{n^2}{\gamma} \left[\frac{1}{2} \ln \left(\frac{n - \gamma}{n - \gamma\sigma^2} \right) - \frac{\gamma}{n} \ln \sigma \right].$$

Artificial training sets with $n_S = 100$ and $n = 7$ were generated by sampling from $g(\mathbf{z})$. Sample sets of size $M = 1000$ were generated from $p(\mathbf{z})$ with values of σ up to 2. The comparison between the exact value of $\widetilde{D}_\gamma(p, g)$ and the value estimated from the γ-specificity for $\gamma = 0.1$ are shown in Fig. 9.5, plotted as a function of σ. For each estimate, several sample sets were generated, and the mean value of the γ-specificity used, with the error bars on the estimates being smaller than the size of the symbols. Note that we use the exact result for $D_\gamma(p, p)$, which means that the *estimated* value of $\widetilde{D}_\gamma(p, g)$ need not necessarily be positive.

It can be seen that the estimated values are in good agreement with the theoretical prediction for $\sigma \leq 1$. This result is also in accord with our comment on the proof of Theorem 9.1, that the derivation is valid provided that

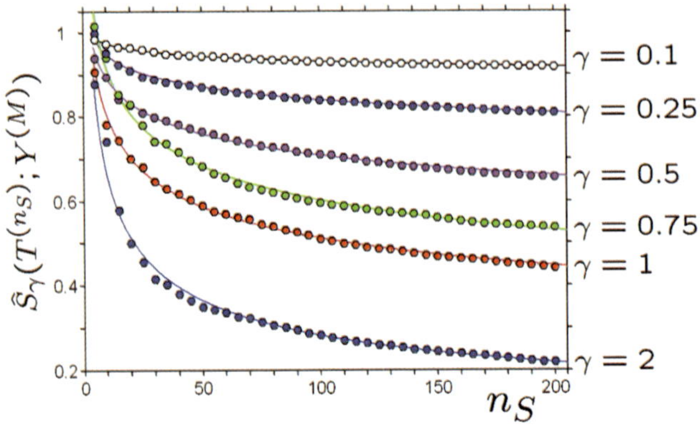

Fig. 9.6 $\hat{S}_\gamma(p, g)$ for $p(\mathbf{z})$ and $g(\mathbf{z})$ spherically symmetric Gaussians in $\mathbb{R}^5$. The training sets of size $n_S \leq 200$ are generated from Gaussians of width 0.5, whereas the model pdf $p(\mathbf{z})$ is of width 0.45. Sample sets are of size $M = 1000$. The plot shows how the measured specificity varies as the size of the training set is increased, for various values of γ. The lines are curves fitted to the data.

the convex hull of the sample set lies mostly within the convex hull of the training set. It confirms that the estimate cannot be used to fit the model, but that the estimate is reliable at the point of fit ($\sigma = 1$) as required.

Our analysis also makes a prediction as regards the way the specificity scales with n_S. From (9.15), we have:

$$\hat{S}_\gamma(T^{(n_S)}, Y^{(M)}) \propto (n_S)^{-\frac{\gamma}{n}} .$$

To test this prediction, we generated artificial training sets by sampling from spherically-symmetric Gaussians of width 0.5 in $n = 5$, for training set sizes up to $n_S = 200$. The sample sets of fixed size $M = 1000$ are generated from spherically symmetric Gaussians of width 0.45, so that we are close to fit but not exactly at the point of fit. The specificity is then evaluated for various values of γ. Figure 9.6 shows how the measured specificity varies as a function of the size of the training set. Each set of points was fitted to

Table 9.1 Table comparing the predicted scaling exponent for specificity with the predicted value, as a function of γ. The final column gives the percentage difference between values.

γ	Predicted power $\frac{\gamma}{n}$	Fitted power	Degree of Fit
0.10	0.02	0.0196	2.0%
0.25	0.05	0.0532	6.0%
0.50	0.10	0.1002	0.2%
0.75	0.15	0.1819	21.0%
1.00	0.20	0.1970	2.0%
2.00	0.40	0.3718	7.0%

a curve of the form given above. In Table 9.1, we compare the fitted value of the power in the scaling law with the actual value of $\frac{\gamma}{n}$. It can be seen that the specificity scales with the size of the training set as predicted, and that this scaling is valid even for small sizes of training set. Furthermore, the measured exponent is close to the theoretical prediction.

9.3.2 Generalized Specificity

The results given so far consider just the case of first-nearest-neighbours between the training and sample sets. We have considered already the extension of taking a power of the distance. It is obviously possible to consider graph constructions beyond the first-nearest-neighbour, such as the jth-nearest-neighbour, or the full set of all k-nearest-neighbours. The analysis follows through as before, except that the partitioning of $\mathbb{R}^n$ into higher-order voronoi cells is rather more complicated. The final results follows as before, except that the coefficients $\beta_{n,\gamma}$ are replaced by the more general coefficients as given by Wade ([191], page 6, Theorem 2.1).

Other extensions are possible, such as using a general, translation invariant distance other than the Euclidean distance. Interested readers should consult the detailed example given in [186].

For the case of using specificity to evaluate image registration, Schestowitz et al. [153] used not only the Euclidean distance, but also used the shuffle distance between images. The use of these distances was validated by considering random warps applied to registered sets of images, and showing that the measured specificity and generalization both showed a monotonic relationship with respect to degree of mis-registration, as did a ground truth measure based on label overlap.

If $\mathrm{I}(\mathbf{x})$ and $\mathrm{J}(\mathbf{x})$ are two image functions, and if $\mathbf{r}(\mathbf{x})$ represents the points in a region centred on the point $\mathbf{x}$, then the Euclidean distance between I and J is:

$$d(\mathrm{I},\mathrm{J}) = \sum_{\mathbf{x}} |\mathrm{I}(\mathbf{x}) - \mathrm{J}(\mathbf{x})|,$$

whereas the shuffle distance is:

$$d(\mathrm{I},\mathrm{J}) = \sum_{\mathbf{x}} \min_{\mathbf{y}\in\mathbf{r}(\mathbf{x})} |\mathrm{I}(\mathbf{x}) - \mathrm{J}(\mathbf{y})|.$$

The region $\mathbf{r}(\mathbf{x})$ is typically some circular or spherical region of fixed radius about $\mathbf{x}$. The advantage of the shuffle distance is that it tends to have a smoother behaviour with respect to mis-registration than the Euclidean distance. The form given here is not symmetric with respect to $\mathrm{I} \leftrightarrow \mathrm{J}$, but it is obviously trivial to construct a symmetric version. This symmetric shuffle distance then obviously scales linearly as we scale the image values, but it is not translation invariant. Nor is it a simple monotonic function of the Euclidean distance. Hence the above analysis cannot be applied to this instance. Nevertheless, it may be that locally on the space of actual images, the shuffle distance may be approximately monotonic for small displacements, and approximately translation invariant for small displacements lying in the space of images. If this were the case, then we would have an approximate relation of the same general form as that given for Euclidean distances. However, at the moment, this is still just a conjecture for the specificity of images, and obviously highly data-dependant.

9.4 Specificity and Generalization in Practice

In practice, what tends to be used for model evaluation is just the specificity (9.4) and generalization (9.5):

$$\hat{S}(n_m) \doteq \frac{1}{M}\sum_{A=1}^{M}\min_i \|\mathbf{y}_A - \mathbf{x}_i\|, \quad \hat{G}(n_m) \doteq \frac{1}{n_S}\sum_{i=1}^{n_S}\min_A \|\mathbf{y}_A - \mathbf{x}_i\|. \quad (9.34)$$

Both of these measures give an estimate of the mean nearest-neighbour distance between the two point sets. Each measurement that contributes to the mean is statistically independent, hence the error on these measures is just given by the standard error:

$$\sigma_S = \frac{\text{std}\left\{\min_i \|\mathbf{y}_A - \mathbf{x}_i\|\right\}}{\sqrt{M}},$$

where std is just the estimated standard deviation across the population. A similar expression holds for the error on the generalization.

Given the measurement, and the error on that measurement, we can now develop a measure of the *sensitivity* of the measure itself. Suppose $f(d)$ represents a measure, where d is some parameter representing the degree of mis-correspondence. The measure here could be specificity or generalization, but it could also be some ground truth based measure. For example, for the case of image registration mentioned above, Schestowitz et al. used specificity and generalization based on Euclidean and shuffle distances, plus a ground truth measure based on the overlap of manually-annotated dense tissue labels.

The sensitivity of a measure is then defined as [153]:

$$\frac{1}{\overline{\sigma_f}}\left(\frac{f(d) - f(0)}{d}\right),$$

where $\overline{\sigma_f}$ is the mean error on f over the range of d considered. The meaning of this is that it gives some indication of the degree of mis-correspondence that can be reliably detected as being different from zero. Schestowitz et al. found that specificity measured using the shuffle distance was more sensitive than sensitivity measured using Euclidean distance. They also found that generalization was the least-sensitive of the measures they considered. More importantly, they also found that the specificity was *more* sensitive than any of the ground truth measures based on the overlap of dense tissue labels. This is not totally surprising, given that model specificity uses the full image intensity information, whereas ground truth labelling by tissue type is a very impoverished representation of image structure. Nevertheless, this result indicates that in some cases, not only is reliable model evaluation possible without ground truth, but that even where ground truth information is available, the ground truth free measures may be superior.

We now give a simple concrete example for shapes of how specificity varies as the correspondence varies from the ground truth correspondence. The shapes we will use are based on the so-called box-bump (see Fig. 9.7), consisting of a rectangle with a semi-circular bump on the top edge, the bump being able to move along the top edge. These shapes were chosen since the ground truth correspondence can be reliably defined, in that ground truth correspondence gives one and only one mode of variation, whereas deviation from this generates spurious modes. Yet despite this simplicity, finding this ground truth correspondence using automatic methods such as MDL is still a challenging task [51].

Ground truth landmarks exist at the corners of the rectangle, and the corners where the bump meets the top edge. A continuous ground truth correspondence is then defined by interpolating between these landmarks, as illustrated in Fig. 9.7 by the colouring. A training set is constructed by varying the bump position between the left and right extreme positions. For the ground truth correspondence, this then gives only one mode of variation. Note that the distribution of shapes is not Gaussian, due to the cut-off provided by the limits on the position of the bump.

For a training set of such box-bumps, the initial parameterisation is defining using an arc-length parameterisation on the mean shape $\bar{\mathbf{S}}$, with the parameters then scaled to lie between 0 and 2π. This means that the shape covariance matrix (2.124) can be written in the form:

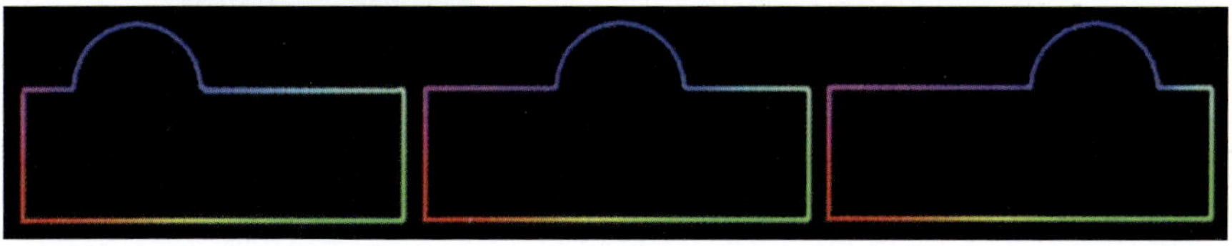

Fig. 9.7 Example box-bumps. The colours denote the ground truth correspondence.

$$\widetilde{D}_{ij} \doteq \frac{1}{2\pi} \int_0^{2\pi} (\mathbf{S}_i(u) - \bar{\mathbf{S}}(u)) \cdot (\mathbf{S}_j(u) - \bar{\mathbf{S}}(u)) du.$$

This is calculated numerically by sampling points densely and equidistantly on the mean shape, and then re-sampling each training shape at the equivalent positions. A multivariate Gaussian model using *all* the modes of variation is then built from this covariance matrix in the usual way (Sect. 2.2.1). Note however that the shapes are *not* Procrustes aligned (Sect. 2.1.1), since this would introduce extra, spurious modes of shape variation.

For the experiments, training sets of $n_S = 100$ continuous shapes were generated by placing the bumps in random positions between the two extremes. The ground truth correspondence was then perturbed using a re-parameterisation of the form:

$$\phi_i(u) = u + (\Delta u)_i,$$

where $(\Delta u)_i$ was chosen at random for each shape apart from the first shape, whose parameterisation was left unchanged. The degree of mis-correlation across the training set was quantified using the mean of the perturbations $\{|(\Delta u)_i| : i = 1, \ldots n_S\}$.

Hence a succession of perturbed training sets was created from each ground truth training set. For each such perturbed training set, the mean shape was calculated, and the shapes resampled using arc-length parameterisation

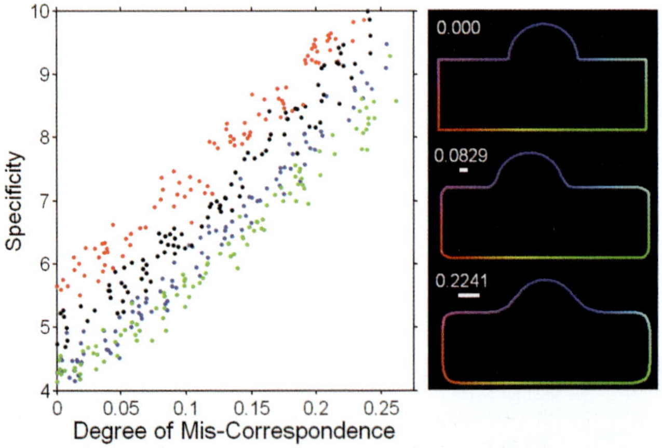

Fig. 9.8 Left: Specificity plotted against degree of mis-correspondence for box-bumps. The different colours indicate different ground truth training sets. See the text for a detailed explanation of the scale on the x-axis. **Right:** The mean shape for various degrees of mis-correspondence. The exact degree of mis-correspondence is indicated by the number, using the same measure as that in the plot, and the length of the white bar indicates the amount of movement on the actual shape. As the mis-correspondence increases, the corners of the mean shape and the corners of the bump becomes smoothed.

on this mean as before. This enabled the integral covariance matrix to be calculated and a multivariate Gaussian model to be built, as detailed above. The specificity of each perturbed model was then measured by generated sample sets of size $M = 300$ from the model. The results for the specificity are shown in the graph in Fig. 9.8. Four different ground truth training sets were generated, and the 100 perturbed results from each such training set are shown by the colours. Note that for reasons of clarity, the standard error on each measurement is not plotted.

It can be seen that there is significant scatter within each training set, which is not surprising given the stochastic nature of the perturbation process. There are also significant differences between the different ground truth training sets, again not surprising given the stochastic nature of their generation. However, the results do show a clear trend, where the measured specificity increases monotonically with the degree of mis-correspondence. As detailed above, mis-correspondence is defined with respect to arc-length parameterisation on the mean, which is then scaled so that parameter values lie between 0 and 2π. Hence a degree of mis-correspondence of 0.1 corresponds to displacement by $\frac{0.1}{2\pi} \approx 1.6\%$ of the total length of the shape.

9.5 Discussion

In this chapter, we have discussed the problems with obtaining ground truth correspondence data as would be required for detailed evaluation of statistical models. We considering the properties that a good statistical model should possess (*compactness, specificity, and generalization ability*), and showed how it was possible to build ad hoc quantitative measures based on the latter two concepts. Detailed consideration of these graph-based measures then revealed that these measures could be considered as graph-based estimators of the more familiar concepts of cross-entropy and Kullback-Leibler divergence. This places our formerly ad hoc measures on a sound theoretical footing, and explains why they are sensible measures for evaluating statistical models.

Finally, we gave a concrete example of how specificity degraded as the correspondence across a training set was perturbed from its ground truth values.

Throughout this chapter, it has been assumed that ground truth correspondence has a definite meaning, and that where it is possible, such ground truth correspondence is the correspondence that would be assigned to shapes by a skilled human annotator. However, it can be argued that the concept of ground truth is only relevant in the case where our ultimate aim is to reproduce such correspondence by an automatic method.

An alternative view is that automatic correspondence is instead something that is inferred from the data, and that the purpose of correspondence is to produce the most meaningful description of the entire training set. Hence

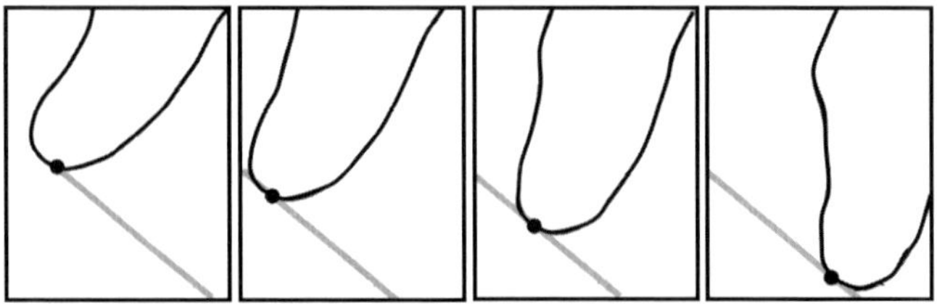

Fig. 9.9 Automatic MDL correspondence for hand outlines – detail of the end of the thumb. The automatic method allows points to slide around at the tip of the thumb, so as to linearize the variation corresponding to rotation of the thumb.

what we mean by the best correspondence can (and should) change as the size of the training set changes. And indeed, the correspondence that we find in the limit of infinite size need not converge to the correspondence that would be assigned by a human. In this view, human ground truth correspondence is seen as a consequence of the particular way that the human visual system extracts meaningful shapes from images or scenes, the way the human brain classifies and categorizes the variation of such sets of shapes, and the way humans include physical knowledge of the behaviour of the actual objects imaged into the process. But the algorithms that the human brain has evolved for such purposes are not the only way to process such information. And in the context of statistical models, comparison with human algorithms need not be the best way to assess our automatic results.

Davies et al. [51] included an interesting example which illustrates the difference between human and automatic approaches to shape correspondence. They considered the case of the outline of the hand, and compared MDL correspondence with manual annotation. It was shown that the automatic correspondence leads to models which are more compact and with better generalization ability (specificity was not evaluated in that particular paper) than the case of manual annotation.

The reasons for this difference becomes clear if we consider how people annotate such shapes. Given their knowledge of the physical object to which the shapes correspond, people tend to place landmarks at the tips of the fingers and thumb. Particularly for the thumb, there is considerable rotational movement as the hand moves, which leads to non-linear degrees of freedom when Cartesian coordinates are used in the shape representation (as noted in Sect. 2.1). Automatic methods attempt to linearize this rotational degree of freedom, by allowing the landmarks at the tips of the thumb and fingers to move slightly, depending on the exact pose of the digit (see Fig.9.9). This leads to quantitatively better multivariate-Gaussian models, although the correspondence is not one that we would assign based on physical considerations. This hence illustrates the difference between ground truth correspondence assigned by humans, based on the particulars of human understanding of shape variation, or experience of the physical properties of the actual objects imaged, versus automatic approaches where the main consideration is producing a compact and efficient statistical model of the variation. Hence comparison of a model with human ground truth is not necessarily the best method of evaluation, and the ground truth free methods described in this chapter provide a principled method of evaluation free of human biases.

Appendix A
Thin-Plate and Clamped-Plate Splines

We give here a brief description of the construction of thin-plate and clamped-plate splines, since particularly for the case of clamped-plate splines, a listing of the full set of Green's functions [9] can be hard to find in the more recent literature.

The basic thin or clamped-plate spline is simple to understand. We take the idealized physical case of a thin, flexible sheet of material, which is bent, but constrained to pass through a certain set of points (the spline control points). The actual shape of the plate, hence the *spline interpolant* of the control points, is then calculated by minimising a *bending energy* term, with respect to the constraint that the sheet passes exactly through the control points.

The thin-plate spline and the clamped-plate spline differ in terms of the boundary conditions applied to the plate. Specifically, the thin-plate spline corresponds to a plate of infinite extent with no particular boundary conditions, whereas the clamped-plate corresponds to a plate whose edges are clamped, as the name suggests.

The solution to this problem and generalizations of it, is obtained by using the Green's function technique.

A.1 Curvature and Bending Energy

First, we need to define the bending-energy term. Consider the simple case of a bent strip of material, which is constrained to lie in a single plane. The shape of the strip can then be described by a function $f(x)$. We will suppose that the deformation of the strip is not too extreme, so that the function $f(x)$ is everywhere a single-valued function of $x \in \mathbb{R}$.

At a point on the strip, we have the radius of curvature R, and its inverse, the curvature κ, where:

$$\kappa \doteq \frac{1}{R} = \frac{f_{xx}}{(1+f_x^2)^{\frac{3}{2}}}, \tag{A.1}$$

with $f_x \doteq \mathrm{d}_x f(x)$, $f_{xx} \doteq (\mathrm{d}_x)^2 f(x) = \mathrm{d}_x^2 f(x)$ and so on. If we make the approximation that the gradient is everywhere small, so that $|f_x| \ll 1$, we then have to leading order:

$$\kappa \approx f_{xx}. \tag{A.2}$$

We take as our expression for the bending energy density of the strip at a point, the *square* of the local curvature κ. A straight strip then corresponds to zero energy, and the more bent the strip, the higher the local curvatures, hence the higher the total energy.

The case of a two-dimensional plate can be handled in a similar fashion. Within a Monge patch, the shape of the plate is given by:

$$(x, y, f(x,y)) \in \mathbb{R}^3, \tag{A.3}$$

in Cartesian coordinates. The mean of the principal curvatures at a point of the plate is then given by:

$$H = \frac{(1+f_y^2)f_{xx} - 2f_x f_y f_{xy} + (1+f_x^2)f_{yy}}{2(1+f_x^2+f_y^2)^{\frac{3}{2}}}, \tag{A.4}$$

where $f_x \doteq \partial_x f(x,y)$ and so on. As before, the energy density is given by the square of the curvature, which when integrated over the whole plate gives the Willmore energy [195]. If we make the same vanishing-gradient approximation as above ($f_x \to 0$, $f_y \to 0$), we find:

$$H \approx \frac{1}{2\sqrt{2}} \left(f_{xx} + f_{yy}\right). \tag{A.5}$$

As an approximate form for the mean curvature, this is linear in the second-derivatives, and has the property that it is invariant to rotations in the (x, y) plane. This suggests that an appropriate form for the approximate Willmore energy density E should be quadratic in the second-derivatives thus:

$$E \doteq f_{xx}^2 + Af_{xy}(f_{xx}+f_{yy}) + Bf_{xy}^2 + Cf_{xx}f_{yy} + f_{yy}^2. \tag{A.6}$$

If we also impose invariance under rotations in the plane, we then obtain the general form for the rotationally-invariant energy density[1]:

$$E \doteq f_{xx}^2 + (2-\alpha)f_{xy}^2 + \alpha f_{xx}f_{yy} + f_{yy}^2, \tag{A.7}$$

where α is some number. Squaring the approximate expression for the mean curvature (A.5) gives us the above expression for the case $\alpha = 2$, whereas

[1] If $\mathbf{F} = \begin{pmatrix} f_{xx} & f_{xy} \\ f_{yx} & f_{yy} \end{pmatrix}$, the matrix of second derivatives, then a general quadratic rotationally-invariant scalar formed from F is of the form: $(\mathrm{Tr}F)^2 + (\alpha - 2)\det F$.

Bookstein [10] uses an energy term given by the case $\alpha = 0$. However as we shall see below, any value of α leads to the same variational problem and the same Green's functions.

A.2 Variational Formulation

We now have an expression for the bending energy of a strip or plate. Let us now consider the original constrained optimisation problem that defines our splines. Suppose we constrain our plate to pass through the set of points $\{(x_i, y_i, f_i) \in \mathbb{R}^3\}$. We then have to find the minimum of the expression:

$$\mathcal{L} = \sum_i l_i \left(f(x_i, y_i) - f_i\right) + \iint\limits_{\mathbb{R}^2} \left(f_{xx}^2 + (2-\alpha) f_{xy}^2 + \alpha f_{xx} f_{yy} + f_{yy}^2\right) dx dy, \tag{A.8}$$

where the $\{l_i\}$ are Lagrange multipliers. Setting the derivatives $\dfrac{\partial \mathcal{L}}{\partial l_i}$ to zero just imposes the corresponding constraint at (x_i, y_i). By the use of the Dirac δ-function, we can rewrite this as:

$$\begin{aligned} \mathcal{L} = & \iint\limits_{\mathbb{R}^2} \left(f_{xx}^2 + (2-\alpha) f_{xy}^2 + \alpha f_{xx} f_{yy} + f_{yy}^2\right) dx dy \\ & + \sum_i \iint\limits_{\mathbb{R}^2} l_i \left(f(x, y) - f_i\right) \delta(x - x_i, y - y_i) dx dy. \end{aligned} \tag{A.9}$$

Just as in finite-dimensional analytic minimisation, we require the derivative of $\mathcal{L}$ with respect to $f(x, y)$. To calculate this *functional* derivative, we consider the following infinitesimal variation:

$$f(x, y) \to f(x, y) + \epsilon(x, y), \ \text{ where: } |\epsilon(x, y)| \ll 1 \ \forall \ (x, y) \in \mathbb{R}^2, \tag{A.10}$$

and $\epsilon(x, y)$ obeys the boundary conditions that it vanishes either at the boundary of some region (the clamped-plate case), or at infinity (the thin-plate case). If we consider a term such as f_{xx}^2, we have:

$$f_{xx} \to f_{xx} + \epsilon_{xx} \ \Rightarrow \ (f_{xx})^2 \to f_{xx}^2 + 2 f_{xx} \epsilon_{xx} + O\left(\epsilon^2\right). \tag{A.11}$$

The important point to note is that since these terms occur inside an integral, we can integrate by parts and pass derivatives across. Since the variation $\epsilon(x, y)$ vanishes at the boundaries/infinity, we do not acquire any extra boundary terms. This means that the entire variation $\Delta\mathcal{L}$ can be written in the integral form:

$$\Delta\mathcal{L} = \iint\limits_{\mathbb{R}^2} \frac{\delta\mathcal{L}}{\delta f(x,y)}\,\epsilon(x,y)dxdy. \tag{A.12}$$

At the optimum, we require that $\Delta\mathcal{L}$ vanishes to leading order *for all* possible variations $\epsilon(x,y)$. We hence obtain the required solution in the form:

$$\frac{\delta\mathcal{L}}{\delta f(x,y)} = 0 \;\forall\; (x,y). \tag{A.13}$$

This is just equating the *functional derivative* of $\mathcal{L}$ to zero, where the functional derivative is calculated as described above.

For the specific case of the flexible plate (A.8), we find:

$$\frac{\delta\mathcal{L}}{\delta f(x,y)} = 2\left(\partial_x^4 + 2\partial_x^2\partial_y^2 + \partial_y^4\right) f(x,y) + \sum_i l_i\delta(x-x_i, y-y_i).$$

$$\therefore\quad \frac{\delta\mathcal{L}}{\delta f(x,y)} = 0 \;\Rightarrow\; (\partial_x^2 + \partial_y^2)^2 f(x,y) \propto \sum_i l_i\delta(x-x_i, y-y_i). \tag{A.14}$$

Note that this result is independent of the value of α chosen in (A.7), hence the Bookstein form and the other forms discussed above give equivalent results.

A.3 Green's Functions

The above equation (A.14) is of the general form:

$$\mathfrak{L}f(\mathbf{x}) = g(\mathbf{x}),\;\; \mathbf{x} \in \mathbb{R}^d, \tag{A.15}$$

where $\mathfrak{L}$ is some scalar differential operator, $g(\mathbf{x})$ is some known, fixed function, and $f(\mathbf{x})$ is the solution we are seeking, subject to appropriate boundary conditions. Equations of this form occur frequently in physics. For example, in electrostatics, we have Poisson's equation, with $\mathfrak{L} = \nabla^2$, $g(\mathbf{x})$ proportional to the electric charge density, and $f(\mathbf{x}) = \phi(\mathbf{x})$, the electrostatic potential. Related examples occur in the fields of quantum mechanics and quantum field theory [130].

Such equations are solved using the technique of Green's functions [82, 170]. For a differential operator $\mathfrak{L}$, the associated Green's function G is defined as:

$$\mathfrak{L}G(\mathbf{x},\mathbf{y}) = \delta(\mathbf{x}-\mathbf{y}), \tag{A.16}$$

where $G(\mathbf{x},\mathbf{y})$ obeys the same boundary conditions as the initial problem. The solution to (A.15) can then be written in the form:

$$f(\mathbf{x}) = f_0(\mathbf{x}) + \int\limits_{\mathbb{R}^d} G(\mathbf{x}, \mathbf{y}) g(\mathbf{y}) d\mathbf{y}, \text{ where } \mathfrak{L} f_0(\mathbf{x}) = 0. \tag{A.17}$$

Hence to calculate the solution, we need to compute the Green's function. We consider now the Green's functions for the thin and clamped-plate problems we defined previously, and for generalizations of that problem.

A.3.1 Green's Functions for the Thin-Plate Spline

For the thin-plate spline (as introduced by Duchon [62], and developed by Meinguet [120, 119, 121]), the plate is free to bend out to infinity. The differential operator in the example given above is $\mathfrak{L} = (\nabla^2)^2 \equiv (\triangle)^2$ (the *bi-harmonic* thin-plate spline). But this can obviously be generalized to the case:

$$\mathfrak{L} = (-\triangle)^m, \text{ derivatives wrt } \mathbf{x} \in \mathbb{R}^d, \tag{A.18}$$

which is the *polyharmonic* thin-plate spline in d-dimensions. The thin-plate spline as popularized by Bookstein [11] for shape analysis uses the radially-symmetric biharmonic Green's function in $\mathbb{R}^d = \mathbb{R}^2$, where:

$$G(\mathbf{x}, \mathbf{y}) = G(\|\mathbf{x} - \mathbf{y}\|), \ G(r) = -r^2 \ln r^2, \ r \geq 0. \tag{A.19}$$

From (A.17), the biharmonic thin-plate spline interpolant (see also [10], page 570) is then of the form:

$$f(x, y) = f(\mathbf{r}) = \sum_i a_i \|\mathbf{r} - \mathbf{r}_i\|^2 \ln \|\mathbf{r} - \mathbf{r}_i\| + Ax + By + C, \tag{A.20}$$

where $\{\mathbf{r}_i \in \mathbb{R}^2\}$ represents the positions $\{(x_i, y_i)\}$ of the spline control points, and the coefficients $\{a_i\}$ are adjusted so that the spline agrees when compared to the values $\{f_i\}$ at these points. The last three terms correspond to a flat plate, and represent the affine part of the deformation.

The problem with this form of interpolant, popular as it is, is that although the non-affine part is *asymptotically* flat,[2] it has no finite limit, and hence doesn't do anything as simple as approach a flat plate at infinity (unless the particular coefficients obey a certain special relation, as in the example given in [10]).

As a result, another version of such splines was introduced [114] into image and shape analysis, with the property that the spline interpolant was strictly bounded in its effects. This spline was based on solutions to the general *clamped-plate* problem, as will be considered next.

---

[2] Asymptotic flatness means that the curvature tends to zero at infinity, which is not the same as the gradient approaching some finite limit at infinity.

A.3.2 Green's Functions for the Clamped-Plate Spline

The clamped-plate problem is similar to the free plate considered above in terms of the bending-energy and so on – the significant difference is that the plate is now constrained to be motionless on the boundary of some region (e.g., a square, a rectangle, a circle, or more complicated shapes such as the limaçon [47]), hence the term *clamped-plate*. We will just consider the case of a circle, or in general, the unit ball in higher dimensions.

As regards the choice of differential operators, this is as above, with:

$$\mathfrak{L} = (-\triangle)^m. \tag{A.21}$$

But the general point $\mathbf{x}$, rather than being anywhere in $\mathbb{R}^d$, is now restricted to the interior of the unit ball $\|\mathbf{x}\| < 1$, in $\mathbb{R}^d$. This means that the Green's functions and the associated splines now have the property that they vanish on the boundary of the unit ball $\|\mathbf{x}\| = 1$ in $\mathbb{R}^d$.

The general solution for the Green's function in this case was given by Boggio [9] in 1905. If $\mathbf{x}$ and $\mathbf{y}$ are the positions of two general points within the unit ball in $\mathbb{R}^d$, then the Green's function can be written as follows:

$$\text{Define: } [XY] \doteq \left\| \mathbf{x}\|\mathbf{y}\| - \frac{\mathbf{y}}{\|\mathbf{y}\|} \right\| \equiv \sqrt{\|\mathbf{x}\|^2\|\mathbf{y}\|^2 - 2\mathbf{x}\cdot\mathbf{y} + 1}, \tag{A.22}$$

$$\& \ A(\mathbf{x},\mathbf{y}) \doteq \frac{[XY]}{\|\mathbf{x}-\mathbf{y}\|}, \tag{A.23}$$

$$\text{Then: } \boxed{G_{m,n}(\mathbf{x},\mathbf{y}) = k_{m,n}\|\mathbf{x}-\mathbf{y}\|^{2m-n} \int_1^{A(\mathbf{x},\mathbf{y})} \frac{(v^2-1)^{m-1}}{v^{n-1}} dv,} \tag{A.24}$$

where $\{k_{m,n}\}$ are numerical coefficients. These Green's functions not only vanish on the boundary of the unit ball, but so do all their derivatives up to order $m-1$, so they not only vanish, but vanish in a smooth fashion.

Let us consider specifically the biharmonic case (that is, $m = 2$), in two dimensions ($n = 2$). The *radially symmetric*, centred Green's function is then given by:

$$G_{2,2}(\mathbf{x},\mathbf{0}) = G_{2,2}(r) \propto r^2 \int_1^{\frac{1}{r}} \left(v - \frac{1}{v}\right) dv = \frac{1}{2}\left(1 - r^2 + r^2 \ln r^2\right), \tag{A.25}$$

and this is the function that was used in the main text (see Sect. 5.1.3, (5.18)).

The interested reader should consult [114, 112, 182, 113, 125, 147] for further applications of clamped-plate splines within the fields of shape analysis and image registration.

Appendix B
Differentiating the Objective Function

In this Appendix, we consider a few more properties of the covariance matrix of the data. In particular, we give a complete derivation of the computation of the derivatives of the covariance matrix, the eigenvectors, and eigenvalues with respect to variation of the input training shapes. The computation is given for both finite and infinite dimensional shape representations. We show how these results that we obtain from the point of view of PCA correspond to the results obtained by Ericsson and Åström [65], and by Hladůvka and Bühler [94], from the point of view of singular value decomposition (SVD).

We also compute the gradient of the Mahalanobis distance with respect to variation of the input training shapes.

B.1 Finite-Dimensional Shape Representations

We start by considering finite-dimensional shape representations. As in Sect. 2.1.3, we define our aligned training shapes as a set of vectors in $\mathbb{R}^{dn_P}$, $\{\mathbf{x}_i : i = 1, \dots n_S\}$, where:

$$\mathbf{x}_i = \{x_{i\mu} : \mu = 1, \dots d \times n_P\}.$$

The mean shape $\bar{\mathbf{x}}$ is defined as:

$$\bar{\mathbf{x}} \doteq \frac{1}{n_S} \sum_{i=1}^{n_S} \mathbf{x}_i. \tag{B.1}$$

The covariance matrix $\mathbf{D}$ (2.16) of our training set is then given by the $dn_P \times dn_P$ matrix with components:

$$D_{\mu\nu} \doteq \sum_{i=1}^{n_S} (\mathbf{x}_i - \bar{\mathbf{x}})_\mu (\mathbf{x}_i - \bar{\mathbf{x}})_\nu. \tag{B.2}$$

We define the set of orthonormal eigenvectors $\{\mathbf{n}^{(a)} : a = 1, \ldots n_m\}$ of $\mathbf{D}$ (2.17) thus:[1]

$$\mathbf{D}\mathbf{n}^{(a)} = \lambda_a \mathbf{n}^{(a)}, \; D_{\mu\nu} n_\nu^{(a)} = \lambda_a n_\mu^{(a)}, \; \mathbf{n}^{(a)} \cdot \mathbf{n}^{(b)} = \delta_{ab}.$$

Note that here we specifically consider only the *complete* set of eigenvectors with *non-zero* eigenvalues, $\lambda_a \neq 0$. There are at most $n_S - 1$ such eigenvectors for n_S data points, so that $n_m \leq n_S - 1$.

This complete set of eigenvectors hence spans the sub-space which contains the data, so that any training shape can be expanded *exactly* in terms of this eigenvector basis:

$$\mathbf{x}_i \doteq \bar{\mathbf{x}} + \sum_{a=1}^{n_m} b_a^{(i)} \mathbf{n}^{(a)}, \tag{B.3}$$

$$\mathbf{b}^{(i)} \doteq \{b_a^{(i)} : a = 1, \ldots n_m\}, \; b_a^{(i)} \doteq (\mathbf{x}_i - \bar{\mathbf{x}}) \cdot \mathbf{n}^{(a)}, \tag{B.4}$$

where $\mathbf{b}^{(i)}$ is the vector of the full set of shape parameters for the ith shape (2.34).

B.1.1 The Pseudo-Inverse

The important point to note about $\mathbf{D}$ is that in general it is not invertible, nor is its determinant non-zero. This is a consequence of the fact that the data sub-space $\mathbb{R}^{n_m}$ is a *sub*-space of $\mathbb{R}^{dn_P}$. There hence exist eigenvectors of $\mathbf{D}$ with zero eigenvalues, which are perpendicular to the data sub-space. Nevertheless, it is still possible to define a type of inverse, as follows.

Substituting from (B.3) into (B.2), the covariance matrix can be written in the form:

$$D_{\mu\nu} = \sum_{i=1}^{n_S} \sum_{a,c=1}^{n_m} b_a^{(i)} b_c^{(i)} n_\mu^{(a)} n_\nu^{(c)},$$

$$\Rightarrow D_{\mu\nu} n_\nu^{(b)} = \sum_{i=1}^{n_S} \sum_{a=1}^{n_m} b_a^{(i)} b_b^{(i)} n_\mu^{(a)} = \lambda_b n_\mu^{(b)}.$$

$$\therefore \sum_{i=1}^{n_S} b_a^{(i)} b_b^{(i)} = \lambda_b \delta_{ab}, \Rightarrow \boxed{D_{\mu\nu} = \sum_{a=1}^{n_m} \lambda_a n_\mu^{(a)} n_\nu^{(a)}.} \tag{B.5}$$

Consider the matrix $\mathbf{M}$, where:

[1] As in the main text, we use the summation convention that repeated indices $D_{\mu\nu} n_\nu^{(a)}$ are summed over unless otherwise stated, whilst bracketed indices $\lambda_a n_\mu^{(a)}$ are only summed over if explicitly stated.

$$M_{\mu\nu} \doteq \sum_{a=1}^{n_m} \frac{1}{\lambda_a} n_\mu^{(a)} n_\nu^{(a)}. \tag{B.6}$$

Then:

$$(MD)_{\mu\nu} = M_{\mu\alpha} D_{\alpha\nu} = \sum_{a,b=1}^{n_m} \frac{1}{\lambda_a} n_\mu^{(a)} n_\alpha^{(a)} \lambda_b n_\alpha^{(b)} n_\nu^{(b)} = \sum_{a=1}^{n_m} n_\mu^{(a)} n_\nu^{(a)}. \tag{B.7}$$

We hence see that acting on any vector $\mathbf{v} \in \mathbb{R}^{dn_P}$:

$$\mathbf{MDv} = \sum_{a=1}^{n_m} \mathbf{n}^{(a)} \left(\mathbf{v} \cdot \mathbf{n}^{(a)} \right). \tag{B.8}$$

The matrix **MD** hence projects $\mathbf{v}$ into the data sub-space. It therefore acts as if it were the *identity* for vectors lying wholly in the data sub-space, whilst annihilating vectors which are perpendicular to that sub-space.

The matrix **M** is the Moore-Penrose pseudo-inverse [129, 133] of **D**. It appears when we calculate the Mahalanobis distance [110]. From (2.84), the square of the Mahalanobis distance from a general point $\mathbf{x}$ in the data sub-space to the mean shape can be written in the form:

$$\begin{aligned} l^2(\mathbf{x}) \doteq n_S \sum_{a=1}^{n_m} \frac{1}{\lambda_a} \left((\mathbf{x} - \bar{\mathbf{x}}) \cdot \mathbf{n}^{(a)} \right)^2 &= n_S \sum_{a=1}^{n_m} \frac{1}{\lambda_a} (\mathbf{x} - \bar{\mathbf{x}})_\mu n_\mu^{(a)} n_\nu^{(a)} (\mathbf{x} - \bar{\mathbf{x}})_\nu, \\ &= n_S (\mathbf{x} - \bar{\mathbf{x}})_\mu M_{\mu\nu} (\mathbf{x} - \bar{\mathbf{x}})_\nu. \end{aligned} \tag{B.9}$$

Note that the factor of n_S appears since we did not include a factor of $\frac{1}{n_S}$ in our definition of the covariance matrix (B.2). For the ith data point, from (B.4), we have:

$$l^2(\mathbf{x}_i) = n_S \sum_{a=1}^{n_m} \frac{1}{\lambda_a} b_a^{(i)} b_a^{(i)}, \tag{B.10}$$

which is a result we will use later.

B.1.2 Varying the Shape

Now we will consider the effect on the covariance matrix and its eigenvalues and eigenvectors of an infinitesimal variation of just the ith shape. Specifically, we consider:

$$\mathbf{x}_i \mapsto \mathbf{x}_i + \Delta \mathbf{x}_i.$$

Note that the mean shape (B.1), hence the position of our origin, also changes:

$$\bar{\mathbf{x}} \mapsto \bar{\mathbf{x}} + \frac{1}{n_S}\Delta\mathbf{x}_i,$$

so that:

$$\mathbf{x}_j - \bar{\mathbf{x}} \mapsto \mathbf{x}_j - \bar{\mathbf{x}} - \frac{1}{n_S}\Delta\mathbf{x}_i,\ j \neq i,\ \&\ \mathbf{x}_i - \bar{\mathbf{x}} \mapsto \mathbf{x}_i - \bar{\mathbf{x}} + \Delta\mathbf{x}_i - \frac{1}{n_S}\Delta\mathbf{x}_i. \quad \text{(B.11)}$$

Substituting into (B.2), we find, after some algebra, that to leading order:

$$D_{\mu\nu} \mapsto D_{\mu\nu} + (\mathbf{x}_i - \bar{\mathbf{x}})_\mu(\Delta\mathbf{x}_i)_\nu + (\mathbf{x}_i - \bar{\mathbf{x}})_\nu(\Delta\mathbf{x}_i)_\mu,$$

$$\boxed{\Delta D_{\mu\nu} \doteq (\mathbf{x}_i - \bar{\mathbf{x}})_\mu(\Delta\mathbf{x}_i)_\nu + (\mathbf{x}_i - \bar{\mathbf{x}})_\nu(\Delta\mathbf{x}_i)_\mu.} \quad \text{(B.12)}$$

Since the covariance matrix changes,[2] so do the eigenvectors and eigenvalues:

$$\lambda_a \mapsto \lambda_a + \Delta\lambda_a, \quad \mathbf{n}^{(a)} \mapsto \mathbf{n}^{(a)} + \Delta\mathbf{n}^{(a)}.$$

The orthonormality constraint on the eigenvectors still has to hold, hence:

$$\left(\mathbf{n}^{(a)} + \Delta\mathbf{n}^{(a)}\right) \cdot \left(\mathbf{n}^{(b)} + \Delta\mathbf{n}^{(b)}\right) = \delta_{ab}$$

$$\Rightarrow \mathbf{n}^{(a)} \cdot \Delta\mathbf{n}^{(b)} = -\mathbf{n}^{(b)} \cdot \Delta\mathbf{n}^{(a)}\ \&\ \mathbf{n}^{(a)} \cdot \Delta\mathbf{n}^{(a)} = 0. \quad \text{(B.13)}$$

Now let us consider the modified eigenvector equation:

$$(D_{\mu\nu} + \Delta D_{\mu\nu})(\mathbf{n}^{(a)} + \Delta\mathbf{n}^{(a)})_\nu = (\lambda_a + \Delta\lambda_a)(\mathbf{n}^{(a)} + \Delta\mathbf{n}^{(a)})_\mu$$

$$\Rightarrow \boxed{D_{\mu\nu}\Delta n_\nu^{(a)} + \Delta D_{\mu\nu} n_\nu^{(a)} = \lambda_a \Delta n_\mu^{(a)} + \Delta\lambda_a n_\mu^{(a)}.} \quad \text{(B.14)}$$

If we take the dot product of the above equation with $\mathbf{n}^{(a)}$, we then obtain an expression for the variation of the eigenvalues:

$$\boxed{\Delta\lambda_a = n_\mu^{(a)} \Delta D_{\mu\nu} n_\nu^{(a)}.} \quad \text{(B.15)}$$

If we instead take the dot product with $\mathbf{n}^{(b)},\ b \neq a$, we obtain a relation for the variation of the eigenvectors:

$$\boxed{\mathbf{n}^{(b)} \cdot \Delta\mathbf{n}^{(a)} = \frac{n_\mu^{(b)} \Delta D_{\mu\nu} n_\nu^{(a)}}{\lambda_a - \lambda_b}.} \quad \text{(B.16)}$$

Note that we have assumed that none of the eigenvalues are degenerate, since as well as leading to problems with the above expression, it would also lead to such pairs of eigenvectors being undetermined up to a mutual rotation.

[2] In [101], Kotcheff and Taylor assumed in their proof that the covariance matrix could be held constant during the differentiation with respect to shape variation, which is obviously incorrect.

We also wish to consider the variation of the parameter vector $\mathbf{b}^{(i)}$ for the ith shape. From (B.4):

$$b_a^{(i)} + \Delta b_a^{(i)} = \left(\mathbf{x}_i - \bar{\mathbf{x}} + \left(1 - \frac{1}{n_S}\right)\Delta\mathbf{x}_i\right) \cdot \left(\mathbf{n}^{(a)} + \Delta\mathbf{n}^{(a)}\right)$$

$$\Rightarrow \boxed{\Delta b_a^{(i)} = \sum_{b \neq a} b_b^{(i)}(\mathbf{n}^{(b)} \cdot \Delta\mathbf{n}^{(a)}) + \left(1 - \frac{1}{n_S}\right)\Delta\mathbf{x}_i \cdot \mathbf{n}^{(a)}.} \tag{B.17}$$

Finally, we take the dot product of (B.14) with a unit vector $\mathbf{m}$, where $\mathbf{m}$ is perpendicular to the data sub-space, hence perpendicular to all the eigenvectors $\{\mathbf{n}^{(a)}\}$. This gives:

$$m_\mu \Delta D_{\mu\nu} n_\nu^{(a)} = \lambda_a(\mathbf{m} \cdot \Delta\mathbf{n}^{(a)}). \tag{B.18}$$

Now let us suppose that the shape $\mathbf{x}_i$ is varied in a direction that is also perpendicular to the data sub-space, so that $\Delta\mathbf{x}_i = \epsilon\mathbf{m},\ \epsilon \ll 1$. Substituting from (B.12) into (B.15) and into (B.16), we find that:

$$\Delta\lambda_a = 0 \ \&\ \mathbf{n}^{(b)} \cdot \Delta\mathbf{n}^{(a)} = 0.$$

That is, for this particular variation, the eigenvalues do not vary, and the eigenvectors only change in a direction perpendicular to the data sub-space. This is what we might have expected intuitively, in that moving a single point out of the original data sub-space causes the sub-space to tip slightly. Furthermore, we see from (B.17), that $\Delta b_a^{(i)} = 0\ \forall a$.

We can hence conclude that the gradient of either the Mahalanobis distance (B.10) or any function of the eigenvalues $\mathcal{L}(\{\lambda_a\})$ lies wholly within the data sub-space. We will now proceed to calculate these gradients.

Let us consider the variation:

$$\Delta\mathbf{x}_i = \epsilon\mathbf{n}^{(d)},\ \epsilon \ll 1.$$

From (B.12) and (B.15) we find:

$$\Delta\lambda_a = 0\ \forall\ a \neq d,\ \Delta\lambda_d = 2\epsilon\mathbf{n}^{(d)} \cdot (\mathbf{x}_i - \bar{\mathbf{x}}) = 2\epsilon b_d^{(i)}. \tag{B.19}$$

The variation of the eigenvectors is computed from (B.16) to give:

$$\Delta\mathbf{n}^{(d)} = \sum_{a \neq d} \frac{\epsilon b_a^{(i)}}{\lambda_d - \lambda_a}\mathbf{n}^{(a)},\ \Delta\mathbf{n}^{(a)} = -\frac{\epsilon b_a^{(i)}}{\lambda_d - \lambda_a}\mathbf{n}^{(d)},\ a \neq d.$$

The variation of the parameter vectors is computed from (B.17):

$$\Delta b_a^{(i)} = -\frac{\epsilon b_a^{(i)} b_d^{(i)}}{\lambda_d - \lambda_a}, \quad a \neq d$$

$$\Delta b_d^{(i)} = \sum_{a \neq d} \frac{\epsilon b_a^{(i)} b_a^{(i)}}{\lambda_d - \lambda_a} + \epsilon \left(1 - \frac{1}{n_S}\right).$$

To compute the variation of $l^2(\mathbf{x}_i)$ (B.10), we need to consider the variation of terms such as:

$$\frac{b_c^{(i)} b_c^{(i)}}{\lambda_c},$$

remembering that both $b_c^{(i)}$ and λ_c vary. Using the results above, we find after some algebra that:

$$\Delta l^2(\mathbf{x}_i) = -\left(\frac{2\epsilon n_S b_d^{(i)}}{\lambda_d}\right)\left[\frac{l^2(\mathbf{x}_i)}{n_S} + \frac{1}{n_S} - 1\right],$$

Or:
$$\boxed{\Delta l^2(\mathbf{x}_i) = -n_S \left(\frac{\Delta \lambda_d}{\lambda_d}\right)\left[\frac{l^2(\mathbf{x}_i)}{n_S} + \frac{1}{n_S} - 1\right].} \quad \text{(B.20)}$$

The important point to note about this result is that only the first bracketed term depends on the exact direction of the variation $\Delta\mathbf{x}_i = \epsilon\mathbf{n}^{(d)}$, whereas the second bracketed term depends on the point chosen.

At first sight, this expression can seem rather puzzling – we might naïvely expect that the gradient of the Mahalanobis distance should be of constant sign, directed towards the mean. Yet the above expression can change sign. This reflects the fact that there are several contributions to the variation of the Mahalanobis distance – the movement of the point is only one, and there are other contributions from the changes to the eigenvectors and eigenvalues.

As a check on this result, let us consider the totally trivial case of $n_S = 2$ points. Without loss of generality, their initial positions can be written as $-x$ and x. We then perturb one point, so that $x \mapsto x + \Delta x$. Without any calculation, we know that the Mahalanobis distance is unaffected by this change, since it just corresponds to a scaling and translation of the dataset. The covariance matrix is just the scalar $\mathbf{D} = 2x^2$, with a single eigenvalue $\lambda = 2x^2$, with eigenvector $\mathbf{n} = 1$. It then follows that $l^2(\mathbf{x}_i) = 1$, which is independent of x as stated previously. Substituting into (B.20):

$$\left[\frac{l^2(\mathbf{x}_i)}{n_S} + \frac{1}{n_S} - 1\right] = \left[\frac{1}{2} + \frac{1}{2} - 1\right] = 0 \;\Rightarrow\; \Delta l^2(\mathbf{x}_i) = 0,$$

which is as we predicted. This completes our check.

Let us now return to the full expression (B.20), and consider the limit $n_S \mapsto \infty$, since in this limit, the effects of the movement of one point on the distribution of the data will vanish. Remember that from Theorem 2.2:

$$\sigma_a^2 = \frac{1}{n_S}\lambda_a,$$

where σ_a is the width parameter of the distribution.We hence conclude that with our present definition of the covariance matrix, the eigenvalues scale as n_S. So, taking the large n_S limit, we find:

$$\Delta l^2(\mathbf{x}_i) \approx \left(\frac{2\epsilon b_d^{(i)}}{\sigma_d^2}\right),$$

which is the change in the square of the Mahalanobis distance for moving a *test point* (as opposed to a data point) from $\mathbf{x}_i$ to $\mathbf{x}_i + \epsilon\mathbf{n}^{(d)}$. It has the sign we would expect, since moving a point away from the origin ($\epsilon > 0$) increases the Mahalanobis distance. We can also see that the other $O\left(\frac{1}{n_S}\right)$ terms in (B.20) are of the correct relative sign, since the mean moves to follow the point a little, hence decreases the Mahalanobis distance slightly.

Note that this result can also be written in the form:

$$\begin{aligned}\Delta l^2(\mathbf{x}_i) \approx \frac{2\epsilon n_s b_d^{(i)}}{\lambda_d} &= 2n_s(\mathbf{x}_i - \bar{\mathbf{x}})_\mu n_\mu^{(d)} \frac{1}{\lambda_d} n_\nu^{(d)} (\Delta\mathbf{x}_i)_\nu \\ &= 2n_s(\mathbf{x}_i - \bar{\mathbf{x}})_\mu M_{\mu\nu}(\Delta\mathbf{x}_i)_\nu,\end{aligned}$$

where $\mathbf{M}$ is the pseudo-inverse (B.6) of $\mathbf{D}$. This expression (apart from a factor of n_S as regards the definition of Mahalanobis distance) is that given by Kotcheff and Taylor, although they mistakenly used the full inverse rather than the pseudo-inverse. Note also that we *cannot* use this result for comparison with the gradient of objective functions based on the covariance matrix. This is because to obtain this result, we have to explicitly assume that the covariance *does not vary*, which of course makes the variation of any such objective functions zero to the same order. We hence have to use the full result given above for any such comparison.

From (B.20), the full result for a *general* variation $\Delta\mathbf{x}_i$ can then be derived, which hence yields the required derivative:

$$\Delta l^2(\mathbf{x}_i) = (2n_s(\mathbf{x}_i - \bar{\mathbf{x}})_\mu M_{\mu\nu}(\Delta\mathbf{x}_i)_\nu)\left[1 - \frac{1}{n_S}\left(1 + l^2(\mathbf{x}_i)\right)\right]$$

$$\therefore \quad \boxed{\frac{\delta l^2(\mathbf{x}_i)}{\delta\mathbf{x}_i} = 2n_S\left[1 - \frac{1}{n_S}\left(1 + l^2(\mathbf{x}_i)\right)\right]\mathbf{M}(\mathbf{x}_i - \bar{\mathbf{x}}).} \tag{B.21}$$

Let us now consider the variation of the determinant of the covariance matrix $\mathbf{D}$, or, more correctly, the variation of:

$$\mathcal{L} = \left(\prod_{a=1}^{n_m}\lambda_a\right).$$

If we consider just the specific variation $\Delta \mathbf{x}_i = \epsilon \mathbf{n}^{(d)}$, then using the above results, we find:

$$\boxed{\Delta \mathcal{L} = \frac{\Delta \lambda_d}{\lambda_d} \left(\prod_{a=1}^{n_m} \lambda_a \right).} \tag{B.22}$$

Comparing this with (B.20), we can hence conclude that the gradient of the determinant of the covariance matrix is in the *same* direction as the gradient of the square of the Mahalanobis distance.[3]

We now move on to consider how our analysis relates to the work of others who considered the problem of obtaining the gradient of the objective function.

B.1.3 From PCA to Singular Value Decomposition

In [65], Ericsson and Åström tackled the problem of differentiating objective functions based on the eigenvalues of the covariance matrix for finite-dimensional shape representations by using a singular-value decomposition (SVD) of the data matrix $\{(\mathbf{x}_i - \bar{\mathbf{x}})_\mu : i = 1, \ldots n_S, \mu = 1, dn_P\}$. We will proceed to show how the analysis given above is equivalent to their treatment, by demonstrating the equivalence of singular value decomposition of the data matrix versus principal component analysis of the covariance matrix of the data.

Following Ericsson and Åström, we will define a $n_S \times dn_P$ data matrix $\mathbf{A}$, where:

$$A_{i\mu} \doteq (\mathbf{x}_i - \bar{\mathbf{x}})_\mu.$$

From the definition of the parameter vector (B.4), the data matrix can be written as:

$$A_{i\mu} = (\mathbf{x}_i - \bar{\mathbf{x}})_\mu = \sum_{a=1}^{n_m} b_a^{(i)} n_\mu^{(a)} = \sum_{a=1}^{n_m} \frac{b_a^{(i)}}{\sqrt{\lambda_a}} \sqrt{\lambda_a} n_\mu^{(a)}. \tag{B.23}$$

The set of eigenvectors $\{\mathbf{n}^{(a)}\}$ are a set of n_m orthonormal vectors in $\mathbb{R}^{dn_P}$ that span the data sub-space. We can hence complete this set, forming an orthonormal basis for $\mathbb{R}^{dn_P}$. The first n_m such vectors then have non-zero eigenvalues, whereas the remaining basis vectors are orthogonal to the sub-space, hence have zero eigenvalues.[4] From these vectors we then construct the matrix $\mathbf{V}$ of size $dn_P \times dn_P$, where:

---

[3] In [101], Kotcheff and Taylor stated this result, but their proof, as we have shown, was incorrect.

[4] If $dn_P < n_S$, we can consider just artificially increasing the dimensionality of the data space $\mathbb{R}^{dn_P}$ so that the above analysis is still valid.

$$V_{\mu a} \doteq n_{\mu}^{(a)}.$$
$$\mathbf{n}^{(a)} \cdot \mathbf{n}^{(b)} = \delta_{ab} \;\Rightarrow\; \mathbf{V}^T\mathbf{V} = \mathbf{V}\mathbf{V}^T = \mathbb{I}.$$

That is, $\mathbf{V}$ is an orthogonal matrix. We next define a diagonal matrix $\mathbf{S}$ of size $n_S \times dn_P$, where the only non-zero elements are formed from the non-zero eigenvalues:

$$s_a \doteq S_{aa} = \sqrt{\lambda_a}, \;\; a = 1, \ldots n_m.$$

Finally, consider the vectors $\mathbf{u}^{(a)}$ in $\mathbb{R}^{n_S}$ where:

$$u_i^{(a)} \doteq \frac{b_a^{(i)}}{\sqrt{\lambda_a}}, \;\; a = 1, \ldots n_m.$$

From (B.5), we see that these n_m vectors are a set of orthonormal vectors in $\mathbb{R}^{n_S}$. We can hence complete the set, giving n_S orthonormal vectors, the first n_m of these as above. These vectors are then collected into the matrix $\mathbf{U}$ where:

$$U_{ia} = u_i^{(a)}, \;\; i, a = 1, \ldots n_S,$$

which is then also an orthogonal matrix.

We can then rewrite $\mathbf{A}$ (B.23) in terms of these matrices to give:

$$\mathbf{A} = \mathbf{U}\mathbf{S}\mathbf{V}^T.$$

This is the singular value decomposition of $\mathbf{A}$. The diagonal matrix $\mathbf{S}$ is the diagonal matrix of *singular values* $\{s_a : a = 1, \ldots n_S\}$, which are just the square roots of the eigenvalues.

We now return to the variation of the eigenvalues. From (B.19), it can be seen that for a general variation $\Delta\mathbf{x}_i$:

$$\Delta\lambda_d = 2(\Delta\mathbf{x}_i \cdot \mathbf{n}^{(d)})b_d^{(i)} \;\Rightarrow\; \boxed{\frac{\delta\lambda_d}{\delta\mathbf{x}_i} = 2b_d^{(i)}\mathbf{n}^{(d)}} \;\Rightarrow\; \frac{\partial\lambda_d}{\partial x_{i\mu}} = 2s_d u_{id} v_{\mu d}, \quad \text{(B.24)}$$

which is in agreement with the result given by Ericsson and Åström.

B.2 Infinite Dimensional Shape Representations

The analysis in the previous section used the $dn_P \times dn_P$ covariance matrix $D_{\mu\nu}$ (2.16), which is only defined for finite-dimensional shape representations. If we are using infinite-dimensional shape representations, we instead have to use the $n_S \times n_S$ covariance matrix $\widetilde{D}_{jk}$ (2.124):

$$\widetilde{D}_{jk} \doteq \frac{1}{A}\int (\mathbf{S}_j(\mathbf{x}) - \bar{\mathbf{S}}(\mathbf{x})) \cdot (\mathbf{S}_k(\mathbf{x}) - \bar{\mathbf{S}}(\mathbf{x}))dA(\mathbf{x}),$$

where $dA(\mathbf{x})$ is the area measure on the mean shape $\bar{\mathbf{S}}$. The eigenproblem is then given as:

$$\widetilde{\mathbf{D}}\widetilde{\mathbf{n}}^{(a)} = \lambda_a \widetilde{\mathbf{n}}^{(a)}.$$

We have shown previously in Theorem 2.6 that there is a correspondence between the eigenproblems of $\widetilde{\mathbf{D}}$ and $\mathbf{D}$, and in Theorem 2.7 we showed how to write the eigenproblem as an eigenfunction problem for the case of infinite-dimensional shape representations.

Hence for the infinite-dimensional case, we work with the eigenproblem:

$$\widetilde{\mathbf{D}}\widetilde{\mathbf{n}}^{(a)} = \lambda_a \widetilde{\mathbf{n}}^{(a)}, \;\; \widetilde{\mathbf{n}}^{(a)} \cdot \widetilde{\mathbf{n}}^{(b)} = A\lambda_a \delta_{ab}.$$

In the finite-dimensional case, we were able to calculate $\dfrac{\partial \lambda_a}{\partial x_{i\mu}}$ directly. In the infinite-dimensional case, we have the corresponding functional derivative:

$$\frac{\delta \lambda_a}{\delta \mathbf{S}_i(\mathbf{x})} = \sum_{j,k=1}^{n_S} \frac{\partial \lambda_a}{\partial \widetilde{D}_{jk}} \frac{\delta \widetilde{D}_{jk}}{\delta \mathbf{S}_i(\mathbf{x})}.$$

We hence first need to calculate the partial derivative:

$$\frac{\partial \lambda_a}{\partial \widetilde{D}_{jk}}.$$

Let us consider a general variation $\widetilde{\mathbf{D}} \mapsto \widetilde{\mathbf{D}} + \Delta\widetilde{\mathbf{D}}$. The modified eigenproblem is then:

$$(\widetilde{\mathbf{D}} + \Delta\widetilde{\mathbf{D}})_{jk}(\widetilde{\mathbf{n}}^{(a)} + \Delta\widetilde{\mathbf{n}}^{(a)})_k = (\lambda_a + \Delta\lambda_a)(\widetilde{\mathbf{n}}^{(a)} + \Delta\widetilde{\mathbf{n}}^{(a)})_j.$$

Taking the dot product with $\widetilde{\mathbf{n}}^{(a)}$, we obtain:

$$\Delta\lambda_a = \frac{\widetilde{\mathbf{n}}^{(a)}\Delta\widetilde{\mathbf{D}}\widetilde{\mathbf{n}}^{(a)}}{\|\widetilde{\mathbf{n}}^{(a)}\|^2} \Rightarrow \boxed{\frac{\partial \lambda_a}{\partial \widetilde{D}_{jk}} = \frac{\widetilde{n}_j^{(a)}\widetilde{n}_k^{(a)}}{\|\widetilde{\mathbf{n}}^{(a)}\|^2}}. \tag{B.25}$$

Consider the orthonormal set of eigenvectors:

$$\mathbf{e}^{(a)} = \frac{\widetilde{\mathbf{n}}^{(a)}}{\|\widetilde{\mathbf{n}}^{(a)}\|}.$$

As in the finite-dimensional case, the covariance matrix $\widetilde{\mathbf{D}}$ can be decomposed in terms of the these eigenvectors:

$$\widetilde{D}_{jk} = \sum_{a=1}^{n_S} \lambda_a e_j^{(a)} e_k^{(a)}.$$

This is just the special case of singular value decomposition noted by Hladůvka and Bühler [94], and shows that the above result we have obtained via PCA is identical to their result:

$$\frac{\partial \lambda_a}{\partial \widetilde{D}_{jk}} = e_j^{(a)} e_k^{(a)}.$$

We then have the remaining part of the functional derivative:

$$\frac{\delta \widetilde{D}_{jk}}{\delta \mathbf{S}_i(\mathbf{x})}.$$

As in Theorem 2.7, we define the shape difference functions:

$$\widetilde{\mathbf{S}}_i(\mathbf{x}) \doteq \mathbf{S}_i(\mathbf{x}) - \bar{\mathbf{S}}(\mathbf{x}).$$

From the definition of $\widetilde{\mathbf{D}}$, we then find:

$$\frac{\delta \widetilde{D}_{jk}}{\delta \mathbf{S}_i(\mathbf{x})} = \frac{1}{A n_S} \left[(n_S \delta_{ij} - 1)\widetilde{\mathbf{S}}_k(\mathbf{x}) + (n_S \delta_{ik} - 1)\widetilde{\mathbf{S}}_j(\mathbf{x}) \right].$$

In order to put this all together, we define the vector-valued functions:

$$\mathbf{e}^{(a)}(\mathbf{x}) \doteq \frac{1}{\sqrt{\lambda_a}} \sum_{i=1}^{n_S} e_i^{(a)} \widetilde{\mathbf{S}}_i(\mathbf{x}).$$

We also define the infinite-dimensional analog of the dot product, the inner product for functions:

$$\left\langle \mathbf{e}^{(a)}, \mathbf{e}^{(b)} \right\rangle \doteq \frac{1}{A} \int \mathbf{e}^{(a)}(\mathbf{x}) \cdot \mathbf{e}^{(b)}(\mathbf{x}) dA(\mathbf{x}) = \delta_{ab},$$

so that these functions form an orthonormal set. We then have the identity:

$$e_i^{(a)} = \frac{1}{\sqrt{\lambda_a}} \left\langle \mathbf{e}^{(a)}, \widetilde{\mathbf{S}}_i \right\rangle.$$

Remember that $\sum_i \widetilde{\mathbf{S}}_i(\mathbf{x}) \equiv 0$, therefore we can deduce that:

$$\sum_{i=1}^{n_S} e_i^{(a)} = 0.$$

Putting together the results for $\dfrac{\partial \lambda_a}{\partial \widetilde{D}_{jk}}$ and $\dfrac{\delta \widetilde{D}_{jk}}{\delta \mathbf{S}_i(\mathbf{x})}$, we hence find that:

$$\boxed{\frac{\delta \lambda_a}{\delta \mathbf{S}_i(\mathbf{x})} = \frac{2}{A} \left\langle \mathbf{e}^{(a)}, \widetilde{\mathbf{S}}_i \right\rangle \mathbf{e}^{(a)}(\mathbf{x}).}$$

If we compare this with the result for the finite-dimensional case:

$$\frac{\delta\lambda_a}{\delta\mathbf{x}_i} = 2b_a^{(i)}\mathbf{n}^{(a)} = 2\left(\mathbf{n}^{(a)}\cdot(\mathbf{x}_i - \bar{\mathbf{x}})\right)\mathbf{n}^{(a)},$$

we see that it is a direct analog, if we identify the orthonormal vectors $\mathbf{n}^{(a)}$ with the orthonormal functions $\mathbf{e}^{(a)}(\mathbf{x})$.

Glossary

Summation Convention

In general, we adopt the Einstein summation convention, that twice-repeated indices are summed over. For example:

$$(\mathbf{x}_i - \bar{\mathbf{x}})_\mu (\mathbf{x}_j - \bar{\mathbf{x}})_\mu \equiv \sum_\mu (\mathbf{x}_i - \bar{\mathbf{x}})_\mu (\mathbf{x}_j - \bar{\mathbf{x}})_\mu.$$

But the convention is also taken to hold for the indices i, j, so that:

$$(\mathbf{x}_i - \bar{\mathbf{x}})_\mu (\mathbf{x}_i - \bar{\mathbf{x}})_\mu \equiv \sum_i \sum_\mu (\mathbf{x}_i - \bar{\mathbf{x}})_\mu (\mathbf{x}_i - \bar{\mathbf{x}})_\mu \equiv \sum_i (\mathbf{x}_i - \bar{\mathbf{x}}) \cdot (\mathbf{x}_i - \bar{\mathbf{x}}).$$

The major exception to this rule is that indices in brackets $\cdot^{(a)}$ are not summed over, so that:

$$\lambda_a \mathbf{n}^{(a)}$$

is a single specific vector $\mathbf{n}^{(a)}$ multiplied by a scalar λ_a, unless specifically stated that there is a sum. Triple indices, such as:

$$c_{aa} \mathbf{n}^{(a)}$$

are also not summed over unless explicitly stated.

Symbols and Abbreviations

$\mathbf{a}$	A general vector or vector field.
$\{a_\mu : \mu = 1, \dots d\}$	The components of a vector $\mathbf{a}$. In most cases, Cartesian components will be used.

$(a^\alpha, b^\alpha, c^\alpha)$	The barycentric, areal coordinates for some triangle $\mathbf{t}^\alpha$.
$\mathbf{A}$	A general matrix.
A_{ij}, $A_{\mu i}$	The elements of a matrix $\mathbf{A}$.
$\mathbf{b}^{(i)}$, $\mathbf{b}$	The parameter vector for the ith shape, a general shape parameter vector.
$b_k^n(x)$	The kth Bernstein polynomial of order n.
$\beta_{n,\gamma}$	The expectation value of the γ-power of the nearest-neighbour distance for a Poisson process in $\mathbb{R}^n$.
cdf	Cumulative Distribution Function.
$\mathcal{C}^n$	The property of being n-times differentiable.
${}^n\mathrm{C}_k$	The binomial coefficient, given by ${}^n\mathrm{C}_k \doteq \frac{n!}{(n-k)!k!}$.
$\delta(\mathbf{x})$	The Dirac δ-function.
δ_{ab}	The Kronecker delta, with $\delta_{ab} = 0$ if $a \neq b$, and 1 otherwise.
$\mathbf{D}$, $\widetilde{\mathbf{D}}$	Shape covariance matrices (see Table 2.2).
$\mathbf{D}(\mathbf{y}, \mathbf{x})$	Shape covariance function for infinite-dimensional shapes (see Table 2.2).
$D_\gamma(p, g)$, $\widetilde{D}_\gamma(p, g)$, $D_{KL}(p, g)$	The γ and the Kullback-Leibler divergences of two distributions p and g.
$\mathbb{E}[\cdot]$	The expectation value.
$\mathcal{F}$	A feature space.
$\mathbf{g}(\mathbf{x})$, $\mathbf{g}(\boldsymbol{\tau})$	For the case of fluid regularization, the induced Riemannian metric of the shape, in parameter-space coordinates $\mathbf{x}$, or tangent-space coordinates $\boldsymbol{\tau}$. The metric is matrix-valued, being an $n \times n$ symmetric matrix, where n is the dimensionality of the shape.
$G(\mathbf{x}, \mathbf{y})$	A Green's function.
$\hat{G}$, $\hat{G}_\gamma$, G_γ	The generalization and γ-generalization. See also $\hat{S}$.
$\Gamma(\cdot)$	The Gamma function.
$\Gamma_{\alpha\beta\mu}$, $\Gamma^\mu_{\alpha\beta}$	The Christoffel symbols of the first and second kinds respectively.
$\mathbb{I}$	The identity matrix.
$\mathcal{I}$	Input data space.
(i, j), (i, j, k)	Coordinates on a regular pixel or voxel grid. See also $\mathbf{X}$.
$\mathcal{K}, \mathcal{K}(\mathbf{b}, \mathbf{c})$	A Mercer kernel function.

$\mathbf{K}_{ij}$, $\widetilde{\mathbf{K}}_{ij}$	For KPCA and SVM methods, the $(ij)^{\text{th}}$ component of the kernel matrix of the non-centred ($\{\mathbf{\Phi}^{(i)}\}$) and centred ($\{\widetilde{\mathbf{\Phi}}^{(i)}\}$) mapped data.
KPCA	Kernel Principal Component Analysis.
λ_a	The ath eigenvalue in some set of eigenvalues.
λ	For fluids, the second viscosity coefficient. See also μ.
$\mathcal{L}$	An objective function.
L,$L^\dagger$, $\mathfrak{L}$	A general differential operator, the Lagrange dual operator, and a self-dual differential operator, for example $\mathfrak{L} = L^\dagger L$.
l, $\mathcal{L}$	For MDL, the codeword and total message length.
MDL	Minimum Description Length.
μ	For fluids, the shear viscosity. See also λ.
$\boldsymbol{\mu}$	For multivariate Gaussian distributions, the mean of the distribution. See also $\mathcal{N}$.
n_m	Number of modes of variation of a model.
$\mathbf{n}^{(a)}$, $\widetilde{\mathbf{n}}^{(a)}$	The ath eigenvector (see Table 2.2).
$\mathbf{n}^{(a)}(\mathbf{x})$	The ath vector-valued eigenfunction (see Table 2.2).
$\mathbf{N}$	A matrix formed by a collection of eigenvectors.
$\mathcal{N}(\mathbf{x}; \boldsymbol{\mu}, \mathbf{D})$	The functional form of a multivariate Gaussian distribution, with mean $\boldsymbol{\mu}$ and covariance matrix $\mathbf{D}$.
$\mathcal{N}(\boldsymbol{\mu}, \mathbf{D})$	A multivariate Gaussian distribution, as distinct from the functional form of such a distribution.
$\Omega_i(\mathcal{P})$, $\Omega_i(T^{(n_S)})$	The Voronoi cell about a point $\mathbf{x}_i$ for the Poisson process $\mathcal{P}$, or for the set of points $T^{(n_S)}$.
$\mathcal{P}(\cdot)$, $\mathcal{P}_1$	A Poisson process, a Poisson process of unit intensity.
p_α, $\mathbf{p}_\alpha$	For linear or recursive-linear re-parameterisation, the αth control node.
$p(\mathbf{x})$, p_α	A probability density function, the probability for some event α.
PCA	Principal Component Analysis.
pdf	Probability Density Function.
PDE	Partial Differential Equation.
PDM	Point Distribution Model, see also SSM.
$\mathbf{\Phi}$	For KPCA and SVM methods, the mapping between the input space $\mathcal{I}$ and the feature space $\mathcal{F}$.

$\boldsymbol{\Phi}^{(i)}$, $\boldsymbol{\Phi}(\mathbf{b})$, $\widetilde{\boldsymbol{\Phi}}^{(i)}$, $\widetilde{\boldsymbol{\Phi}}(\mathbf{b})$	For KPCA, the point that the ith data point $\mathbf{b}^{(i)}$ or a general point $\mathbf{b}$ maps to in feature space, and the non-centred and centred positions of this mapped point.
$\Phi_{\alpha}^{(i)}$, $\Phi_{\alpha}(\mathbf{b})$, $\widetilde{\Phi}_{\alpha}^{(i)}$, $\widetilde{\Phi}_{\alpha}(\mathbf{b})$	For KPCA, the non-centred and centred components of the ith data point $\mathbf{b}^{(i)}$ or a general point $\mathbf{b}$.
ϕ, ϕ_i, ϕ^{-1}	A re-parameterisation function, the ith such function, and its inverse. It acts on the parameter space $\mathfrak{X}$, so that $\mathbf{x} \in \mathfrak{X}$ maps to $\phi(\mathbf{x})$ under the re-parameterisation. Also, by extension, acts on parameterised shapes, and the correspondence between such shapes.
$\phi(\mathbf{x})$, $\phi(\mathbf{x};t)$	For the case of fluids, a (time-dependant) mapping, the track of a fluid particle over time, with $\phi(\mathbf{x};0) = \mathbf{x}$. See also $\mathbf{u}(\mathbf{x},t)$.
$\dot{\phi}(\mathbf{x};t)$	Temporal partial derivative, $\dot{\phi}(\mathbf{x};t) \doteq \left.\frac{\partial}{\partial t}\right\|_{\mathbf{x}} \phi(\mathbf{x};t)$.
$\mathbb{R}$, $\mathbb{R}^{+}$	The real numbers, and the positive real numbers.
$\mathbb{R}^{d}$	d-dimensional Euclidean space.
$\square^{\alpha\beta}$,$\square^{\alpha\gamma\beta\delta}$	A general rectangle, labelled by the identities of a pair of diagonal nodes, or by all four corner nodes.
S, S_i	An entire shape, the ith such shape, either in two or three dimensions.
$\mathbf{S}(\cdot)$	A shape function, so that $\mathbf{S}(\mathbf{x})$ represents a single point on the shape.
$\mathcal{S}$	A shape when considered as a manifold.
$\mathbf{S}$	A specific embedding of a shape manifold into Euclidean space.
$\overline{\mathbf{S}}$	The mean of a set of shapes.
$\mathbb{S}^{d}$	The sphere in d dimensions.
$\hat{S}$, $\hat{S}_{\gamma}$, S_{γ}	The specificity and γ-specificity. See also $\hat{G}$.
SSM	Statistical Shape Model (see also PDM).
SVD	Singular Value Decomposition (see Appendix B).
SVM	Support Vector Machines (see also KPCA).
t	Time.
$\mathbf{t}$, $\mathbf{t}^{\alpha}$	A triangle, or the αth such triangle in a triangulated mesh. See also $\mathbf{v}^{\alpha}$.

$\tau^{\alpha\beta}, \boldsymbol{\tau}^{\alpha\beta}, \boldsymbol{\tau}^{\alpha\beta\gamma}$	For recursive linear re-parameterisation, the fractional position in one or two dimensions of a daughter node relative to its parents. For one dimension, or two-dimensional rectangular tiling, two parents are sufficient, whereas all three are used for two-dimensional triangulation.
$\boldsymbol{\tau}$	For the specific case of fluid regularization, the position of a point in the tangent space to the shape, with tangent-space coordinates $\{\tau_\alpha\}$.
$T = \{\mathbf{x}_i\}$, $T^{(n_S)}$	The training set, consisting of the n_P points represented by shape vectors $\{\mathbf{x}_i : i = 1, \dots n_S\}$. See also $Y^{(M)}$.
T_P	The tangent plane/space to a manifold at a point P.
(θ, ψ)	In three dimensions, polar coordinates defined on the surface of the unit sphere, where $x \doteq \sin\theta\cos\psi$, $y \doteq \sin\theta\sin\psi$, and $z \doteq \cos\theta$. Coordinates lie within the ranges: $0 \leq \theta \leq \pi$ and $0 \leq \psi < 2\pi$.
$\mathbf{u}(\mathbf{x})$, $\mathbf{u}(\mathbf{x}, t)$	For a fluid particle with track $\phi(\mathbf{x}; t)$, the corresponding displacement field, the inverse mapping to ϕ. So that if $\mathbf{y} \stackrel{\phi}{\mapsto} \phi(\mathbf{y})$, then $\mathbf{x} \doteq \phi(\mathbf{y})$, $\mathbf{x} - \mathbf{u}(\mathbf{x}) \doteq \mathbf{y} = \phi^{-1}(\mathbf{x})$.
$\mathbf{U}(t)$	For the case of fluids, the concatenated values of the displacement field $\mathbf{u}(\mathbf{x}, t)$ taken at every point on the pixel-voxel grid $\mathbf{X}$, so that $\mathbf{U}(t) \doteq \{\mathbf{u}(\mathbf{x}, t)\}\ \forall\ \mathbf{x} \in \mathbf{X}$.
$\mathcal{U}$	A uniform distribution over some range of values, or a uniform distribution over a surface.
$\mathbf{v}^\alpha$, $\mathbf{v}^{\alpha a}$	The set of vertices of a triangle $\mathbf{t}^\alpha$, the ath such vertex. Typically, $\mathbf{v}^{\alpha a} \in \mathbb{R}^3$, so a single vertex is represented by a vector.
$\{\mathbf{v}^A\}$, $\mathbf{v}^A$, $\mathbf{v}_i^A$	The total set of vertices of a single triangulated mesh, the Ath such vertex. A triangulation then consists of assigned the vertices to triangles, so that the triangles form a complete, continuous surface. Where we consider multiple shapes, $\mathbf{v}_i^A$ refers to the Ath node of the triangulation of the ith shape.
$\mathbf{v}(\mathbf{x}, t)$	The Eulerian velocity field of a fluid, given by $\mathbf{v}(\mathbf{x}, t) \doteq \dot{\phi}(\phi^{-1}(\mathbf{x}); t)$.
$\mathbf{V}(t)$	Similarly to $\mathbf{U}(t)$, the collected values of $\mathbf{v}(\mathbf{x}, t)$ across a pixel/voxel grid.
(x, y), (x, y, z)	Cartesian coordinates of a point in two or three dimensions.
$\mathbf{x}^{(i)}$	The vector position of the ith shape point in a collection of points, with coordinates $\{x_\mu^{(i)}\}$.

$\mathbf{x}_j^{(i)}$	The vector position of the ith shape point of the jth shape.
$\mathbf{x}$	A general point in a space, such as $\mathbf{x} \in \mathbb{R}^2$ or $\mathbf{x} \in \mathbb{S}^2$.
	For the case of finite-dimensional shape representations, the shape vector consisting of a collection of shape points $\mathbf{x} = \{\mathbf{x}^{(i)} : i = 1, \dots n_P\}$.
	For the case of images, the position of a single pixel or voxel in the entire grid $\mathbf{x} \in \mathbf{X}$.
$\mathbf{x}_i$	For finite-dimensional shape representations, the ith shape vector in a set of shape vectors.
X	The parameter space for a set of parameterised shapes.
$\mathbf{X}$	The entire regular pixel or voxel grid of an image, the collection of all pixels/voxels in an image. See also (i, j).
$Y = \{\mathbf{y}_A\}$, $Y^{(M)}$	A sample set of examples $\{\mathbf{y}_A : A = 1, \dots M\}$ generated by a model, sharing the same distribution as the model pdf.
$\mathfrak{X}$, $\mathfrak{X}_i$	The mapping between a shape S and the parameter space X, or between a shape S_i and X. This mapping is usually one-to-one, so that we will often use the same notation to represent the mapping in either direction.

Miscellaneous Symbols

$\forall$	For all.
$\therefore$	Therefore.
$\exists$	There exists.
$\doteq$	Defined as.
$\propto$	Proportional to.
$\approx$	Approximately equal to.
$\equiv$	Equivalent to.
$\bullet$ or $\cdot$	The vector dot product.
$\sim$	Symbol indicating the dense, pointwise correspondence between two shapes, so that $\mathbf{S}_i(\mathbf{x}) \sim \mathbf{S}_j(\mathbf{x})$ means that the point $\mathbf{S}_i(\mathbf{x})$ on shape S_i corresponds to the point $\mathbf{S}_j(\mathbf{x})$ on shape S_j.

$\mapsto$	Mapping, such as that between a parameter space $\mathfrak{X} \ni \mathbf{x}$ and a shape S, $\mathfrak{X} \mapsto S$, $\mathbf{x} \mapsto \mathbf{S}(\mathbf{x}) \in S$. Also between points in the same parameter space, when considered as a re-parameterisation (see ϕ).

Derivatives

∂_x, d_x	In general, partial and total derivatives with respect to x, also written as $\dfrac{\partial}{\partial x}$ and $\dfrac{\mathrm{d}}{\mathrm{d}x}$.
∂_α, d_α, D_α	For the specific case of fluid regularization, ∂_α represents the αth partial derivative wrt the parameter space coordinates, whereas d_α represents the αth *partial* derivative wrt some other coordinate system (e.g., tangent-space coordinates). D_α is the covariant derivative.
$\dfrac{\delta}{\delta \mathbf{u}(\mathbf{x})}$	Functional derivative with respect to the field $\mathbf{u}(\mathbf{x})$.
∇	The vector derivative operator, $\nabla f(\mathbf{x}) = (\partial_x f, \partial_y f, \partial_z f)$ or $(\partial_x f, \partial_y f)$ as appropriate.
∇^2,$\triangle$	The Laplacian, so that $\nabla^2 \doteq \partial_x^2 + \partial_y^2 + \partial_z^2$ in $\mathbb{R}^3$.

Distances

$\lVert \cdot \rVert$	The usual Euclidean norm in $\mathbb{R}^d$.
$\lvert \cdot \rvert$	A general norm, or the modulus of a scalar.

Pseudocode Notation

Variable Data Types

a, *alpha*	Any variable not in boldface is a scalar.
a, **alpha**	Any variable in **boldface**, without a capitalized first letter is a vector.
$\mathbf{a}(i)$	The ith element of vector $\mathbf{a}$.
A, **Alpha**	Any variable in **boldface** with a capitalized first letter is a matrix.
$\mathbf{A}(i,j)$	The entry at the ith row and j^{th} column of $\mathbf{A}$.

$\mathbf{A}(i,\cdot)$	The ith row of $\mathbf{A}$ as a vector.
$\mathbf{A}(\cdot,j)$	The jth column of $\mathbf{A}$ as a vector.
$\{A_k\}$	A set of variables, indexed by k.
A_k	The kth element of the set $\{A_k\}$.
$\mathbf{A}_k(i,j)$	For a set of matrices, the element at the ith row and j^{th} column of the kth member of the set. Note that all members of the set are not necessarily the same size.

Operators and Symbols

$a \leftarrow b$	Assigns the value of variable b to variable a.
AND, NOT, OR	The standard logical operators.
$\lceil a \rceil$	The ceiling operator, where a is rounded to the next highest integer.
$\curvearrowright$	If the symbol $\curvearrowright$ appears at the end of a line, it means that a procedure call is split over that line and the one below.
%	Indicates a comment, so any text appearing after % is not part of the code.

Surface Representation

Triangulation	Surfaces are represented as triangulated meshes. A list of triangles is stored in this $n_t \times 3$ matrix.
Shape_Points	A $n_p \times d$ matrix, containing a list of point coordinates. Each row of **Triangulation** contains three integers that reference the row indices of **Shape_Points**.

The representation is easiest to illustrate with the following example, which shows how the coordinates of the three nodes of the kth triangle in **Triangulation** are accessed:

Shape_Points(**Triangulation**$(k,1),\cdot)$, % node 1

Shape_Points(**Triangulation**$(k,2),\cdot)$, % node 2

Shape_Points(**Triangulation**$(k,3),\cdot)$, % node 3

References

1. P.D. Allen, J. Graham, D.J.J. Farnell, E.J. Harrison, R. Jacobs, K. Karayianni, C. Lindh, P.F. van der Stelt, K. Horner, and H. Devlin. Detecting reduced bone mineral density from dental radiographs using statistical shape models. *IEEE Transactions on Information Technology in Biomedicine*, 11:601–610, 2007.
2. P. Alliez, M. Meyer, and M. Desbrun. Interactive geometry remeshing. In *SIGGRAPH*, volume 21, pages 347–354, 2002.
3. S. Angenent, S. Haker, A. Tannenbaum, and R. Kikinis. On the Laplace-Beltrami operator and brain surface flattening. *IEEE Transactions on Medical Imaging*, 18:700–711, 1999.
4. G.B. Arfken and H.J. Weber. *Mathematical Methods for Physicists*. Academic Press, 5th edition, 2001.
5. A. Baumberg and D. Hogg. Learning flexible models from image sequences. In 3nd *European Conference on Computer Vision*, volume 1, pages 299–308. Springer-Verlag, Berlin, 1994.
6. A. Baumberg and D. Hogg. An adaptive eigenshape model. In 6th *British Machine Vision Conference*, pages 87–96, BMVA Press, 1995.
7. S. Belongie, J. Malik, and J. Puzicha. Shape matching and object recognition using shape contexts. *IEEE Transactions on Pattern Analysis and Machine Intelligence*, 42(4):509–522, 2002.
8. P.J. Besl and N.D. McKay. A method for registration of 3D shapes. *IEEE Transactions on Pattern Analysis and Machine Intelligence*, 14(2):239–256, 1992.
9. T. Boggio. Sulle funzioni di Green d'ordine m. *Rendiconti - Circolo Matematico di Palermo*, 20:97–135, 1905.
10. F.L. Bookstein. Principal warps: Thin-plate splines and the decomposition of deformations. *IEEE Transactions on Pattern Analysis and Machine Intelligence*, 11(6):567–585, 1989.
11. F.L. Bookstein. *Morphometric Tools for Landmark Data*. Cambridge University Press, 1991.
12. F.L. Bookstein. Landmark methods for forms without landmarks: morphometrics of group differences in outline shape. *Medical Image Analysis*, 1(3):225–244, 1997.
13. K.A. Brakke. Random Voronoi tessellations in arbitrary dimension. `http://www.susqu.edu/brakke/aux/downloads/papers/arbitrary.pdf`, 2005.
14. C. Brechbühler, G. Gerig, and O. Kübler. Parameterisation of closed surfaces for 3-D shape description. *Computer Vision, Graphics and Image Processing*, 61:154–170, 1995.
15. L. Breiman, W. Meisel, and E. Purcell. Variable kernel estimates of multivariate densities. *Technometrics*, 19:135–144, 1977.

16. A.D. Brett, A. Hill, and C.J. Taylor. A method of automatic landmark generation for automated 3D PDM construction. *Image and Vision Computing*, 18:739–748, 2000.
17. A.D. Brett and C.J. Taylor. Construction of 3D shape models of femoral articular cartilage using harmonic maps. In *International Conference on Medical Image Computing and Computer Aided Intervention (MICCAI)*, pages 1205–1214, 2000.
18. C. Broit. Optimal registration of deformed images. Ph.D. Dissertation, Computer and Information Science Department, University of Pennsylvania, Philadelphia, PA., 1981.
19. R. Buckminster Fuller. US patent #2393676, 1946.
20. R. Buckminster Fuller. *Synergetics; Explorations in the Geometry of Thinking.* Macmillan, New York, 1975.
21. A. Caunce and C.J. Taylor. Using local geometry to build 3D sulcal models. In 16th *Conference on Information Processing in Medical Imaging*, pages 196–209. Springer, 1999.
22. G.E. Christensen, R.D. Rabbitt, and M.I. Miller. Deformable templates using large deformation kinematics. *IEEE Transactions on Image Processing*, 5:1435–1447, 1996.
23. U. Clarenz, M. Droske, S. Henn, M. Rumpf, and K. Witsch. *Computational Methods for Nonlinear Image Registration*, volume 10 of *Mathematics in Industry*, pages 81–102. Springer, 2006.
24. I. Cohen, N. Ayache, and P. Sulger. Tracking points on deformable objects using curvature information. In 2nd *European Conference on Computer Vision*, pages 458–466. Springer-Verlag, 1992.
25. T.F. Cootes, C. Beeston, G.J. Edwards, and C.J. Taylor. A unified framework for atlas matching using active appearance models. In 16th *Conference on Information Processing in Medical Imaging*, pages 322–333. Springer, 1999.
26. T.F. Cootes, G.J. Edwards, and C.J. Taylor. Active appearance models. In 5th *European Conference on Computer Vision*, volume 2, pages 484–498. Springer, Berlin, 1998.
27. T.F. Cootes, G.J. Edwards, and C.J. Taylor. Active appearance models. *IEEE Transactions on Pattern Analysis and Machine Intelligence*, 23(6):681–685, 2001.
28. T.F. Cootes, A. Hill, and C.J. Taylor. Medical image interpretation using active shape models: Recent advances. In 14th *Conference on Information Processing in Medical Imaging*, pages 371–372. Kluwer Academic Publishers, 1995.
29. T.F. Cootes, A. Hill, C.J. Taylor, and J. Haslam. The use of active shape models for locating structures in medical images. *Image and Vision Computing*, 12(6):276–285, July 1994.
30. T.F. Cootes, S. Marsland, C.J. Twining, K. Smith, and C.J. Taylor. Groupwise diffeomorphic non-rigid registration for automatic model building. In 8th *European Conference on Computer Vision*, volume 4, pages 316–327, 2004.
31. T.F. Cootes and C.J. Taylor. Active shape models. In 3rd *British Machine Vision Conference*, pages 266–275. Springer-Verlag, London, 1992.
32. T.F. Cootes and C.J. Taylor. Active shape models: A review of recent work. In *Current Issues in Statistical Shape Analysis*, pages 108–114. Leeds University Press, 1995.
33. T.F. Cootes and C.J. Taylor. Combining point distribution models with shape models based on finite-element analysis. *Image and Vision Computing*, 13(5):403–409, 1995.
34. T.F. Cootes and C.J. Taylor. Data driven refinement of active shape model search. In 7th *British Machine Vision Conference*, pages 383–392, BMVA Press, 1996.
35. T.F. Cootes and C.J. Taylor. A mixture model for representing shape variation. In 8th *British Machine Vision Conference*, pages 110–119, BMVA Press, 1997.
36. T.F. Cootes and C.J. Taylor. A mixture model for representing shape variation. *Image and Vision Computing*, 17(8):567–574, 1999.
37. T.F. Cootes and C.J. Taylor. Statistical models of appearance for computer vision. Technical report, Department of Imaging Science and Biomedical Engineering, University of Manchester, 2004.

38. T.F. Cootes, C.J. Taylor, D.H. Cooper, and J. Graham. Training models of shape from sets of examples. In 3rd *British Machine Vision Conference*, pages 9–18. Springer-Verlag, London, 1992.
39. T.F. Cootes, C.J. Taylor, D.H. Cooper, and J. Graham. Active shape models – their training and application. *Computer Vision and Image Understanding*, 61(1):38–59, Jan. 1995.
40. T.F. Cootes, C.J. Taylor, and A. Lanitis. Active shape models : Evaluation of a multi-resolution method for improving image search. In 5th *British Machine Vision Conference*, pages 327–336, BMVA Press, 1994.
41. T.F. Cootes, C.J. Twining, and C.J. Taylor. Diffeomorphic statistical shape models. In 15th *British Machine Vision Conference*, pages 447–456, BMVA Press, 2004.
42. N. Costen, T.F. Cootes, G.J. Edwards, and C.J. Taylor. Simultaneous extraction of functional face subspaces. In *IEEE Conference on Computer Vision and Pattern Recognition*, volume 1, pages 492–497, 1999.
43. N.P. Costen, M. Brown, and S. Akamatsu. Sparse models for gender classification. In *Sixth IEEE International Conference on Automatic Face and Gesture Recognition*, pages 201–206, 2004.
44. N.P. Costen, T.F. Cootes, G.J. Edwards, and C.J. Taylor. Automatic extraction of the face identity-subspace. *Image and Vision Computing*, 20(5–6):319–329, 2002.
45. H.S.M. Coxeter. *Introduction to Geometry*. Wiley classic library. Wiley, New York, 2nd edition, 1989.
46. W.R. Crum, O. Camara, D. Rueckert, K. Bhatia, M. Jenkinson, and D.L.G. Hill. Generalised overlap measures for assessment of pairwise and groupwise image registration and segmentation. In *International Conference on Medical Image Computing and Computer Aided Intervention (MICCAI)*, volume 3749, pages 99–106. Springer, 2005.
47. A. Dall'Acqua and G. Sweers. The clamped-plate equation for the limaçon. *Journal Annali di Matematica Pura ed Applicata*, 184(3):361–374, 2005.
48. R.H. Davies, T.F. Cootes, and C.J. Taylor. A minimum description length approach to statistical shape modelling. In 17th *Conference on Information Processing in Medical Imaging*, pages 50–63. Springer, 2001.
49. R.H. Davies, T.F. Cootes, C.J. Twining, and C.J. Taylor. An information theoretic approach to statistical shape modelling. In 12th *British Machine Vision Conference*, pages 1–10, BMVA Press, 2001.
50. R.H. Davies, T.F. Cootes, J.C. Waterton, and C.J. Taylor. An efficient method for constructing optimal statistical shape models. In *International Conference on Medical Image Computing and Computer Aided Intervention (MICCAI)*, pages 57–65, 2001.
51. R.H. Davies, C.J. Twining, P.D. Allen, T.F. Cootes, and C.J. Taylor. Building optimal 2D statistical shape models. *Image and Vision Computing*, 21(13):1171–1182, 2003.
52. R.H. Davies, C.J. Twining, P.D. Allen, T.F. Cootes, and C.J. Taylor. Shape discrimination in the hippocampus using an MDL model. In 18th *Conference on Information Processing in Medical Imaging*, pages 38–50. Springer, 2003.
53. R.H. Davies, C.J. Twining, T.F. Cootes, J.C. Waterton, and C.J. Taylor. A minimum description length approach to statistical shape modelling. *IEEE Transactions on Medical Imaging*, 21(5):525–537, 2002.
54. R.H. Davies, C.J. Twining, and C.J. Taylor. Consistent spherical paramaterisation for statistical shape modelling. In *IEEE Symposium on Biomedical Imaging*, pages 1388–1391, 2006.
55. R.H. Davies, C.J. Twining, and C.J. Taylor. Building optimal 3D statistical shape models. In *IEEE Transactions on Medical Imaging (under review)*, 2007.
56. R.H. Davies, C.J. Twining, T.G. Williams, and C.J. Taylor. Group-wise correspondence of surfaces using non-parametric regularisation and shape images. In *Proceedings of the* 4th *IEEE International Symposium on Biomedical Imaging (ISBI): From Nano to Macro*, pages 1208–1211, 2007.

57. C. de Boor. *A Practical Guide to Splines*, volume 21 of *Applied Mathematical Sciences*. Springer, revised edition, 2001.
58. E.D. Demaine and J. O'Rourke. *Geometric Folding Algorithms: Linkages, Origami, Polyhedra*. Cambridge University Press, 2007.
59. H. Devlin, P.D. Allen, J. Graham, R. Jacobs, K. Karayianni, C. Lindh, P.F. van der Stelt, E. Harrison, J.E. Adams, S. Pavitt, and K. Horner. Automated osteoporosis risk assessment by dentists: A new pathway to diagnosis. *Bone*, 40:835–842, 2007.
60. E.W. Dijkstra. A note on two problems in connexion with graphs. *Numerische Mathematik*, 1:269–271, 1959.
61. G. Donato and S. Belongie. Approximate thin plate spline mappings. In 7th *European Conference on Computer Vision*, pages 13–31, 2002.
62. J. Duchon. Interpolation des fonctions de deux variables suivant le principe de la flexion des plaques minces. *Revue Française d'Automatique, Informatique, Recherche Opérationelle (RAIRO) Analyse Numerique*, 10:5–12, 1976.
63. N. Duta, A. Jain, and M. Dubuisson-Jolly. Automatic construction of 2D shape models. *IEEE Transactions on Pattern Analysis and Machine Intelligence*, 23:433–445, 2001.
64. G.J. Edwards, C.J. Taylor, and T.F. Cootes. Learning to identify and track faces in image sequences. In *Proceedings of the* 3rd *International Workshop on Automatic Face and Gesture Recognition*, pages 260–265, Japan, 1998.
65. A. Ericsson and K. Åström. Minimizing the description length using steepest descent. In 14th *British Machine Vision Conference*, pages 93–102, BMVA Press, 2003.
66. M. Figueiredo, J. Leito, and A. Jain. Unsupervised contour representation and estimation using B-splines and a minimum description length criterion. *IEEE Transactions on Image Processing*, 9:1075–1087, 2000.
67. B. Fischer and J. Modersitzki. Fast inversion of matrices arising in image processing. *Numerical Algorithms*, 22:1–11, 1999.
68. B. Fischer and J. Modersitzki. Curvature based image registration. *Journal of Mathematical Imaging and Vision*, 18(1):81–85, 2003.
69. N.I. Fisher. *Statistical Analysis of Circular Data*. Cambridge University Press, 1993.
70. M. Floater and K. Horman. Surface parameterization: A tutorial and survey. In *Advances in Multiresolution for Geometric Modelling*, Mathematics and Visualization, pages 157–186. Springer, 2005.
71. M.S. Floater. Parametrization and smooth approximation of surface triangulations. *Foundations and Trends in Computer Graphics and Vision*, 14:231–250, 1997.
72. A.F. Frangi, D. Rueckert, J. Schnabel, and W. Niessen. Automatic construction of multiple-object three dimensional statistical shape models: Application to cardiac modelling. *IEEE Transactions on Medical Imaging*, 21(9):1151 – 1166, 2002.
73. T. Gatzke, C. Grimm, M. Garland, and S. Zelinka. Curvature maps for local shape comparison. In *International Conference on Shape Modelling and Applications*, pages 244–253, 2005.
74. Y. Gdalyahu and D. Weinshall. Flexible syntactic matching of curves and its application to automatic hierarchical classification of silhouettes. *IEEE Transactions on Pattern Analysis and Machine Intelligence*, 21(12):1312–1328, 1999.
75. J. Gee, M. Reivich, and R. Bajcsy. Elastically deforming 3D atlas to match anatomical brain images. *Journal of Computer Assisted Tomography*, 17(2):225–236, 1993.
76. G. Gerig, M. Styner, D. Jones, D. Weinberger, and J. Lieberman. Shape analysis of brain ventricles using SPHARM. In *Mathematical Methods in Biomedical Image Analysis*, pages 171–178, 2001.
77. P. Golland, W.E.L. Grimson, M.E. Shenton, and R. Kikinis. Small sample size learning for shape analysis of anatomical structures. In *International Conference on Medical Image Computing and Computer Aided Intervention (MICCAI)*, pages 72–82, 2000.
78. C. Goodall. Procrustes methods in the statistical analysis of shape. *Journal of the Royal Statistical Society B*, 53(2):285–339, 1991.

79. I.S. Gradshteyn and I.M. Ryzhik. *Table of Integrals, Series, and Products.* Academic Press Ltd, fifth edition, 1994.
80. K. Grauman and T. Darrell. Fast contour matching using approximate earth movers distance. In *IEEE Conference on Computer Vision and Pattern Recognition*, pages 220–227, 2004.
81. R.W. Gray. Exact transformation equations for Fuller's world map. *Cartographica: The International Journal for Geographic Information and Geovisualization*, 32(3):17–25, 1995.
82. G. Green. An essay on the application of mathematical analysis to the theories of electricity and magnetism. Privately published by the author, 1828.
83. P. Grünwald. Model selection based on minimum description length. *Journal of Mathematical Psychology*, 44:133–152, 2000.
84. X. Gu, S. Gortler, and H. Hoppe. Geometry images. In *Proceedings of SIGGRAPH*, pages 355–361, 2002.
85. X. Gu, Y. Wang, T.F. Chan, P.M. Thompson, and S.-T. Yau. Genus zero surface conformal mapping and its application to brain surface mapping. *IEEE Transactions on Medical Imaging*, 23(8):949–958, 2004.
86. J. Hadamard. Sur les problèmes aux dérivées partielles et leur signification physique. *Princeton University Bulletin*, 13:49–52, 1902.
87. T. Heap and D.C. Hogg. Automated pivot location for the cartesian-polar hybrid point distribution model. In 7th *British Machine Vision Conference*, pages 97–106, BMVA Press, 1996.
88. T. Heimann, I. Wolf, and H-P. Meinzer. Automatic generation of 3D statistical shape models with optimal landmark distributions. *Methods of Information in Medicine*, 46(3):275–281, 2007.
89. T. Heimann, I. Wolf, T.G. Williams, and H.P. Meinzer. 3D active shape models using gradient descent optimization of description length. In 19th *Conference on Information Processing in Medical Imaging*, pages 566–577. Springer, 2005.
90. A.O. Hero, B. Ma, O. Michel, and J.D. Gorman. Alpha-divergence for classification, indexing and retrieval. *Technical Rcport CSPL-328 Communications and Signal Processing Laboratory, The University of Michigan, 48109-2122,* `http://www.eecs.umich. edu/~hero/det_est.html`, 2001.
91. A.O. Hero and O. Michel. Estimation of Rényi information divergence via pruned minimal spanning trees. In *IEEE Workshop on Higher Order Statistics, Caesaria, Israel*, 1999.
92. A. Hill and C.J. Taylor. Automatic landmark generation for point distribution models. In 5th *British Machine Vision Conference*, pages 429–438, BMVA Press, 1994.
93. A. Hill, C.J. Taylor, and A.D. Brett. A framework for automatic landmark identification using a new method of non-rigid correspondence. *IEEE Transactions on Pattern Analysis and Machine Intelligence*, 22(3):241–251, 2000.
94. J. Hladůvka and K. Bühler. MDL spline models: Gradient and polynomial reparameterisations. In 17th *British Machine Vision Conference*, pages 869–878, BMVA Press, 2006.
95. P. Horkaew and G-Z. Yang. Construction of 3D dynamic statistical deformable models for complex topological shapes. In *International Conference on Medical Image Computing and Computer Aided Intervention (MICCAI)*, pages 217–224, 2004.
96. M.K. Hurdal, P.L. Bowers, K. Stephenson, D.W.L. Sumners, K. Rehm, K. Schaper, and D.A. Rottenberg. Quasi-conformal flat mapping the human cerebellum. In *International Conference on Medical Image Computing and Computer Aided Intervention (MICCAI)*, pages 279–286, 1999.
97. V. Jain and H. Zhang. Robust 2D shape correspondence using geodesic shape context. In *Proceedings of Pacific Graphics*, pages 121–124, 2005.
98. M. Kass, A. Witkin, and D. Terzopoulos. Active contour models. *International Journal of Computer Vision*, 1(4):321–331, 1987.

99. A. Kelemen, G. Szekely, and G. Gerig. Elastic model-based segmentation of 3D neurological data sets. *IEEE Transactions on Medical Imaging*, 18(10):828–839, 1999.
100. J.J. Koenderink. *Solid Shape*. MIT Press, 1990.
101. A.C.W. Kotcheff and C.J. Taylor. Automatic construction of eigenshape models by direct optimisation. *Medical Image Analysis*, 2(4):303–314, 1998.
102. W.J. Krzanowski and F.H.C. Marriott. *Multivariate Analysis, Part I, Distributions, Ordination and Inference*. Edward Arnold, London, 1994.
103. S. Kullback and R.A. Leibler. On information and sufficiency. *The Annals of Mathematical Statistics*, 22(1):79–86, 1951.
104. A. Lanitis, C.J. Taylor, and T.F. Cootes. Towards automatic simulation of aging effects on face images. *IEEE Transactions on Pattern Analysis and Machine Intelligence*, 24:442–454, 2002.
105. T.C.M. Lee. An introduction to coding theory and the two-part minimum description length principle. *International Statistical Review / Revue Internationale de Statistique*, 69(2):169–183, 2001.
106. M.E. Leventon, W.E.L. Grimson, and O. Faugeras. Statistical shape influence in geodesic active contours. In *IEEE Conference on Computer Vision and Pattern Recognition*, pages 316–323, 2000.
107. P. Lévy. L'addition des variables aléatoires définies sur une circonférence. *Bulletin de la Société Mathématique de France*, 67:1–41, 1939.
108. M. Li. Minimum Description Length Based 2D Shape Description. In *Fourth International Conference on Computer Vision*, pages 512–517, Berlin, Germany, 1993.
109. C. Lorenz and N. Krahnstover. Generation of point-based 3D statistical shape models for anatomical objects. *Computer Vision and Image Understanding*, 77:175–191, 2000.
110. P.C. Mahalanobis. On the generalised distance in statistics. *Proceedings of the National Institute of Sciences of India*, 2(1):49–55, 1936.
111. K.V. Mardia. *Statistics of Directional Data*. Academic Press, 1972.
112. S. Marsland and C.J. Twining. Constructing data-driven optimal representations for iterative pairwise non-rigid registration. In J.C. Gee, J.B. Antoine Maintz, and M.W. Vannier, editors, *Second International Workshop on Biomedical Image Registration (WBIR)*, pages 50–60. Springer, 2003.
113. S. Marsland and C.J. Twining. Constructing diffeomorphic representations for the groupwise analysis of non-rigid registrations of medical images. *IEEE Transactions on Medical Imaging*, 23(8):1006–1020, 2004.
114. S. Marsland, C.J. Twining, and C.J. Taylor. Groupwise non-rigid registration using polyharmonic clamped-plate splines. In *International Conference on Medical Image Computing and Computer Aided Intervention (MICCAI)*, pages 771–779, 2003.
115. S. Marsland, C.J. Twining, and C.J. Taylor. A minimum description length objective function for groupwise non-rigid image registration. *Image and Vision Computing*, 26(3):333–346, 2008.
116. P. McCullagh. Möbius transformation and cauchy parameter estimation. *The Annals of Statistics*, 24(2):787–808, 1996.
117. G.J. McLachlan and K.E. Basford. *Mixture Models: Inference and Applications to Clustering*. Dekker, New York, 1988.
118. D. Meier and E. Fisher. Parameter space warping: Shape-based correspondence between morphologically different objects. *IEEE Transactions on Medical Imaging*, 21:31–47, 2002.
119. J. Meinguet. *An intrinsic approach to multivariate spline interpolation at arbitrary points*, pages 163–190. Proceedings of the NATO Advanced Study Institute. Reidel, Dordrecht, The Netherlands, 1979. Ed. B. Sahney.
120. J. Meinguet. Multivariate interpolation at arbitrary points made simple. *Zeitschrift für Angewandte Mathematik und Physik (ZAMP)*, 30:292–304, 1979.

121. J. Meinguet. *Surface spline interpolation: Basic theory and computational aspects*, pages 127–142. Proceedings of the NATO Advanced Study Institute. Reidel, Dordrecht, The Netherlands, 1984. Eds. S.P. Singh, J.W.H. Burry, and B. Watson.
122. J. Mercer. Functions of positive and negative type and their connection with the theory of integral equations. *Philosophical Transactions of the Royal Society of London, A*, 209:415–446, 1909.
123. N. Metropolis and S. Ulam. The Monte Carlo method. *Journal of the American Statistical Association*, 44(247):335–341, 1949.
124. S. Mika, B. Schölkopf, A.J. Smola, K.-R. Müller, M. Scholz, and G. Rätsch. Kernel PCA and de-noising in feature spaces. In M. S. Kearns, S. A. Solla, and D. A. Cohn, editors, *Advances in Neural Information Processing Systems*, volume 11, pages 536–542. MIT Press, Cambridge, MA, 1999.
125. A. Mills. *Image Registration by the Geodesic Interpolating Spline*. PhD thesis, Department of Mathematics, University of Manchester, U.K., 2007.
126. C.W. Misner, K.S. Thorne, and J.A. Wheeler. *Gravitation*. W. H. Freeman, 1973.
127. J. Modersitzki. *Numerical methods for image registration*. Numerical Mathematics and Scientific Computation. Oxford University Press, 2004.
128. P. Moerland. *Mixture Models for Unsupervised and Supervised Learning*. PhD thesis, Swiss Federal Institute of Technology, Lausanne, 2000. Available as IDIAP Research Report IDIAP-RR 00-18.
129. E.H. Moore. Abstract #18, On the reciprocal of the general algebraic matrix. *Bulletin of the American Mathematical Society*, 26(9):394–395, 1920.
130. Y. Nambu. Structure of Green's functions in quantum field theory. *Physical Review*, 100(1):394–411, 1955.
131. T. Papadopoulo and M.I.A. Lourakis. Estimating the Jacobian of the singular value decomposition: Theory and applications. In 6th *European Conference on Computer Vision*, pages 554–570, 2000.
132. R. Paulsen and K. Hilger. Shape modelling using Markov random field restoration of point correspondences. In 18th *Conference on Information Processing in Medical Imaging*, pages 1–12. Springer, 2003.
133. R. Penrose. A generalized inverse for matrices. *Proceedings of the Cambridge Philosophical Society*, 51:406–413, 1955.
134. A.P. Pentland and S. Sclaroff. Closed-form solutions for physically based modelling and recognition. *IEEE Transactions on Pattern Analysis and Machine Intelligence*, 13(7):715–729, 1991.
135. A. Pitiot, H. Delingette, and P. Thompson. Learning shape correspondence for n-D curves. *International Journal of Computer Vision*, 71(1):71–88, 2007.
136. A. Pitiot, H. Delingette, A. Toga, and P. Thompson. Learning object correspondences with the observed transport shape measure. In 18th *Conference on Information Processing in Medical Imaging*, pages 25–37. Springer, 2003.
137. S.M. Pizer, P.T. Fletcher, S. Joshi, A. Thall, J.Z. Chen, Y. Fridman, D.S. Fritsch, A.G. Gash, J.M. Glotzer, M.R. Jiroutek, C. Lu, K.E. Muller, G. Tracton, P. Yushkevich, and E.L. Chaney. Deformable M-Reps for 3D medical image segmentation: Special Issue on Research at the University of North Carolina, Medical Image Display & Analysis Group (MIDAG). *International Journal of Computer Vision*, 55(2–3):85–106, 2003.
138. E. Praun and H. Hoppe. Spherical reparameterization and remeshing. In *Proceedings of SIGGRAPH*, pages 340–349, 2003.
139. W.H. Press, S.A. Teukolsky, W.T. Vetterling, and B.P. Flannery. *Numerical Recipes in C (2nd Edition)*. Cambridge University Press, 1992.
140. A. Rangagajan, H. Chui, and F.L. Bookstein. The softassign procrustes matching algorithm. In 15th *Conference on Information Processing in Medical Imaging*, pages 29–42. Springer, 1997.
141. A. Rényi. On measures of entropy and information. *Proceedings 4th Berkeley Symposium Math. Stat. and Prob.*, 1:547–561, 1961.

142. T. Richardson and S. Wang. Nonrigid shape correspondence using landmark sliding, insertion and deletion. In *International Conference on Medical Image Computing and Computer Aided Intervention (MICCAI)*, pages 435–442, 2005.
143. J.R. Rissanen. A universal prior for integers and estimation by minimum description length. *Annals of Statistics*, 11:416–431, 1983.
144. J.R. Rissanen. *Stochastic Complexity in Statistical Inquiry*. World Scientific, 1989.
145. M.G. Roberts, T.F. Cootes, and J.E. Adams. Vertebral morphometry: semi-automatic determination of detailed shape from DXA images using active appearance models. *Investigative Radiology*, 41:849–859, 2006.
146. M.G. Roberts, T.F. Cootes, E.M. Pacheco, and J.E. Adams. Quantitative vertebral fracture detection on DXA images using shape and appearance models. *Academic Radiology*, 14:1166–1178, 2007.
147. M. Rogers and J. Graham. Robust and accurate registration of 2-D electrophoresis gels using point-matching. *IEEE Transactions on Image Processing*, 16(3):624–635, 2007.
148. S. Romdhani, S. Gong, and A. Psarrou. A multi-view non-linear active shape model using kernel PCA. In 10th *British Machine Vision Conference*, volume 2, pages 483–492, BMVA Press, 1999.
149. D. Rueckert, P. Aljabar, R.A. Heckemann, J.V. Hajnal, and A. Hammers. Diffeomorphic registration using B-splines. In *International Conference on Medical Image Computing and Computer Aided Intervention (MICCAI)*, pages 702–709, 2006.
150. D. Rueckert, A.F. Frangi, and J.A. Schnabel. Automatic construction of 3D statistical deformation models of the brain using non-rigid registration. *IEEE Transactions on Medical Imaging*, 22(8):1014–1025, 2003.
151. D. Rueckert, L.I. Sonoda, C. Hayes, D.L.G. Hill, M.O. Leach, and D.J. Hawkes. Nonrigid registration using free-form deformations: Application to breast MR images. *IEEE Transactions on Medical Imaging*, 18(8):712–721, 1999.
152. S. Sain. Multivariate locally adaptive density estimation. Technical Report, Department of Statistical Science, Southern Methodist University, 1999.
153. R. Schestowitz, C.J. Twining, T.F. Cootes, V. Petrovic, C.J. Taylor, and W.R. Crum. Assessing the accuracy of non-rigid registration with and without ground truth. In *IEEE Symposium on Biomedical Imaging*, pages 836–839, 2006.
154. W. Schmidt. Statistische methoden beim gefügestudium kristalliner schiefer. *Sitzungsberichte der Kaiserliche Akademie der Wissenschaften in Wien, Mathematisch-Naturwissenschaftliche Klasse, Abtheilung 1, v. 126*, pages 515–538, 1917.
155. B. Schölkopf, S. Mika, C.J.C. Burges, P. Knirsch, K.-R. Müller, G. Rätsch, and A.J. Smola. Input space vs. feature space in kernel-based methods. *IEEE Transactions on Neural Networks*, 10(5):1000–1017, 1999.
156. B. Scholkopf, A. Smola, and K. Muller. Nonlinear component analysis as a kernel eigenvalue problem. *Neural Computation*, 10(5):1299–1319, 1998.
157. B. Schölkopf, A.J. Smola, and K.-R. Müller. Kernel principal component analysis. In B. Schölkopf, C.J.C. Burges, and A.J. Smola, editors, *Advances in Kernel Methods - Support Vector Learning*, pages 327–352. MIT Press, Cambridge, MA, 1999.
158. S. Sclaroff and A. Pentland. Modal matching for correspondence and recognition. *IEEE Transactions on Pattern Analysis and Machine Intelligence*, 17(6):545–561, 1995.
159. G.L. Scott and H.C. Longuet-Higgins. An algorithm for associating the features of two images. *Proceedings of the Royal Society of London*, 244:21–26, 1991.
160. C.E. Shannon. A mathematical theory of communication. *Bell Systems Technical Journal*, 27:379–423 and 623–656, 1948.
161. L.S. Shapiro and J.M. Brady. A modal approach to feature-based correspondence. In 2nd *British Machine Vision Conference*, pages 78–85. Springer-Verlag, 1991.

162. A. Sheffer, T. Praun, and K. Rose. Mesh parameterization methods and their applications. *Foundations and Trends in Computer Graphics and Vision*, 2(2):105–171, 2006.
163. D. Shi, S.R. Gunn, and R.I. Damper. Handwritten chinese radical recognition using nonlinear active shape models. *IEEE Transactions on Pattern Analysis and Machine Intelligence*, 25:277–280, 2003.
164. W. Sierpiński. Sur une courbe cantorienne dont tout point est un point de ramification. *Comptes Rendus Hebdomadaires des Séances de l'Académie des Sciences, Paris*, 160:302–305, 1915.
165. B.W. Silverman. *Density Estimation for Statistics and Data Analysis*. Chapman and Hall, London, 1986.
166. A.J. Smola and B. Schölkopf. Sparse greedy matrix approximation for machine learning. In P. Langely, editor, *Proceedings of the 17th International Conference on Machine Learning (ICML '00)*, pages 911–918. Morgan Kaufmann, San Francisco CA, 2000.
167. P.P. Smyth, C.J. Taylor, and J.E. Adams. Vertebral shape: automatic measurement with active shape models. *Radiology*, 211:571–578, 1999.
168. J.P. Snyder. Map projections–a working manual. *U. S. Geological Survey Professional Paper*, 1395:145–153, 1987.
169. M. Sonka, V. Hlavac, and R. Boyle. *Image Processing, Analysis, and Machine Vision*. Brooks/Cole, 1999.
170. I. Stakgold. *Green's functions and boundary value problems*. Pure and Applied Mathematics. Wiley, New York, 2nd edition, 1998.
171. J.M. Steele. *Probability Theory and Combinatorial Optimization*. SIAM, Philadelphia, 1997.
172. M. Styner, K. Rajamani, L. Nolte, G. Zsemlye, G. Szekely, C.J. Taylor, and R.H. Davies. Evaluation of 3D correspondence methods for model building. In 18th *Conference on Information Processing in Medical Imaging*, pages 63–75. Springer, 2003.
173. H.D. Tagare. Shape-based nonrigid correspondence with application to heart motion analysis. *IEEE Transactions on Medical Imaging*, 18:570–579, 1999.
174. J.M.F. Ten Berge. Orthogonal procrustes rotation for two or more matrices. *Psychometrika*, 42(2):267–276, 1977.
175. G.R. Terrell and D.W. Scott. Variable kernel density estimation. *The Annals of Statistics*, 20(3):1236–1265, 1992.
176. N.A. Thacker, P.A. Riocreux, and R.B. Yates. Assessing the completeness properties of pairwise geometric histograms. *Image and Vision Computing*, 13(5):423–429, 1995.
177. H. Thodberg. MDL shape and appearance models. In 18th *Conference on Information Processing in Medical Imaging*, pages 51–62. Springer, 2003.
178. H. Thodberg and H. Olafsdottir. Adding curvature to MDL shape models. In 14th *British Machine Vision Conference*, volume 2, pages 251–260, 2003.
179. M. Turk and A. Pentland. Eigenfaces for recognition. *Journal of Cognitive Neuroscience*, 3(1):71–86, 1991.
180. C.J. Twining, T.F. Cootes, S. Marsland, R. Schestowitz, V. Petrovic, and C.J. Taylor. A unified information-theoretic approach to groupwise non-rigid registration and model building. In G. Christensen and M. Sonka, editors, 19th *Conference on Information Processing in Medical Imaging*, volume 3565 of *Lecture Notes in Computer Science*, pages 1–14. Springer, 2005.
181. C.J. Twining, R.H. Davies, and C.J. Taylor. Non-parametric surface-based regularisation for building statistical shape models. In 20th *Conference on Information Processing in Medical Imaging*, pages 738–750. Springer, 2007.
182. C.J. Twining and S. Marsland. Constructing diffeomorphic representations of non-rigid registrations of medical images. In 18th *Conference on Information Processing in Medical Imaging*, pages 413–425. Springer, 2003.

183. C.J. Twining and C.J. Taylor. Kernel principal component analysis and the construction of non-linear active shape models. In 12th *British Machine Vision Conference*, pages 23–32, BMVA Press, 2001.
184. C.J. Twining and C.J. Taylor. Kernel principal component analysis and the construction of non-linear active shape models. In 12th *British Machine Vision Conference*, pages 23–32, BMVA Press, 2001.
185. C.J. Twining and C.J. Taylor. The use of kernel principal component analysis to model data distributions. *Pattern Recognition*, 36:217–227, 2003.
186. C.J. Twining and C.J. Taylor. Specificity as a graph-based estimator of cross-entropy and KL divergence. In 17th *British Machine Vision Conference*, volume 2, pages 459–468, BMVA Press, 2006.
187. C.J. Twining, C.J. Taylor, and P. Courtney. Robust tracking and posture description for laboratory rodents using active shape models. *Behavior Research Methods, Instruments, & Computers*, 33:381–391, 2001.
188. V. Vapnik and S. Mukherjee. Support vector method for multivariate density estimation. In S. A. Solla, T. K. Leen, and K.-R. Müller, editors, *Advances in Neural Information Processing Systems*, volume 12. MIT Press, Cambridge, MA, 2000.
189. R. von Mises. Über die 'ganzzahligkeit' der atomgewicht und verwandte fragen. *Physikalische Zeitschrift*, 19:490–500, 1918.
190. G. Voronoi. Nouvelles applications des paramètres continus à la théorie des formes quadratiques, premier et deuxième mémoires. *Journal für die Reine und Angewandte Mathematik*, 133 & 134 : 97–178 & 198–287 respectively, 1908.
191. A.R. Wade. Explicit laws of large numbers for random nearest-neighbour type graphs. *Advances in Applied Probability*, 39(2):326–342, 2007.
192. Y. Wang, B.S. Peterson, and L.H. Staib. Shape-based 3D surface correspondence using geodesics and local geometry. In *IEEE Conference on Computer Vision and Pattern Recognition*, pages 644–651, 2000.
193. E.W. Weisstein. Orthographic projection. *From MathWorld–A Wolfram Web Resource.* `http://mathworld.wolfram.com/OrthographicProjection.html`.
194. J. Weston, A. Gammerman, M. Stitson, V. Vapnik, V. Vovk, and C. Watkins. Support vector density estimation. In B. Schölkopf, C.J.C. Burges, and A.J. Smola, editors, *Advances in Kernel Methods - Support Vector Learning*, pages 293–306. MIT Press, Cambridge, MA, 1999.
195. T.J. Willmore. Note on embedded surfaces. *Analele Ştiinţifice ale Universităţii "Alexandru Ioan Cuza" din Iaşi. Serie Nouă. Matematică.*, 11B:493–496, 1965.
196. L. Zhu, S. Haker, and A. Tannenbaum. Area-preserving mappings for the visualization of medical structures. In *International Conference on Medical Image Computing and Computer Aided Intervention (MICCAI)*, pages 277–284, 2003.

Index

Printed in the United States of America